Charles William MacCord

Kinematics

A treatise on the modification of motion, as affected by the forms and modes of

connection of the moving parts of machines

Charles William MacCord

Kinematics
A treatise on the modification of motion, as affected by the forms and modes of connection of the moving parts of machines

ISBN/EAN: 9783337163891

Printed in Europe, USA, Canada, Australia, Japan

Cover: Foto ©berggeist007 / pixelio.de

More available books at **www.hansebooks.com**

KINEMATICS.

A TREATISE ON THE MODIFICATION OF MOTION, AS AFFECTED BY
THE FORMS AND MODES OF CONNECTION OF THE
MOVING PARTS OF MACHINES.

FULLY ILLUSTRATED

BY ACCURATE DIAGRAMS OF MECHANICAL MOVEMENTS,
AS PRACTICALLY CONSTRUCTED ;

FOR THE USE OF DRAUGHTSMEN, MACHINISTS, AND STUDENTS OF
MECHANICAL ENGINEERING.

BY

CHARLES WILLIAM MAC CORD, A.M., Sc.D.,

PROFESSOR OF MECHANICAL DRAWING IN THE STEVENS INSTITUTE OF TECHNOLOGY, HOBOKEN, N. J.
AUTHOR OF "LESSONS IN MECHANICAL DRAWING," "A PRACTICAL TREATISE ON THE
SLIDE VALVE AND ECCENTRIC," AND VARIOUS MONOGRAPHS ON MECHANISM.

NEW YORK:

JOHN WILEY & SONS.

1883.

Respectfully Inscribed

TO THE REV. JOHN T. DUFFIELD, D.D., LL.D.,

OF PRINCETON COLLEGE,

BY HIS FRIEND AND FORMER PUPIL.

PREFACE.

A word of explanation is due to the reader, in view of the fact that the following pages relate to but a small number of the vast array of devices included in the broad term Mechanism.

Having in the opening chapters considered the methods by which motion, and the modification of motion, may be represented and analyzed, and the basis upon which a proper classification of elementary mechanical combinations may be made, the next question was, what classes of such combinations should be first examined. So large a proportion of these consist of pieces rotating in contact about fixed axes, that they seemed to have a natural claim to precedence, which was duly recognized.

Attention was accordingly next directed to the discussion of the pitch surfaces, and in natural sequence to the forms of the teeth, of gear wheels of all kinds. Which having been done, it appeared proper to publish so much as was completed, because notwithstanding that many treatises upon these special topics have been written, there would seem to be room for another; the more particularly since even in sweeping out this part of the shop, scraps of new material and cuttings of old have been found, in quantity and of quality to be worth using.

The endeavor has been made to treat the theory of the subject in a practical manner, for the benefit of the practical man. That is to say, the demonstrations are made as far as possible directly dependent upon the diagrams; and the latter, in most cases reduced from work actually executed upon a large scale, are accompanied by explanations which it is hoped will enable any ordinarily expert draughtsman to "lay out" the movements with ease and accuracy.

In order to avoid interrupting the argument by subordinate discussions, as well as for more ready reference, an Appendix has been added, containing the methods of construction, and other graphic

CHAPTER XII.

PRACTICAL KINEMATICS.

CHAPTER I.

1. Mechanism is the science which treats of the designing and construction of machinery. Its objects are, to investigate those abstract principles which are involved in planning correctly, and to describe the practical operations involved in successful execution.

2. A Machine is very properly said, in a general way, to be an artificial work which serves to apply or to regulate moving power.

This definition will not answer our purpose, for which it is not sufficiently minute ; but it is correct so far as it goes, and close and clear enough for the ordinary employment of the word.

From its terms we infer that a machine applies or regulates extraneous power for some useful purpose. That is to say, the existence of the machine presupposes the existence of something to be done and of power to do it ; and it also implies the necessity of modifying, in some way or other, both the force and the motion caused by the force. No machine can move itself, nor can it create motive power ; this must be derived from external sources, such as the falling of a weight, the uncoiling of a spring, or the expansion of steam.

3. Motive power has different characteristics, according to the nature of the source. It may be constant, as in the case of a head of water kept at the same level by an unfailing stream ; it may vary according to regular laws, as when derived from expanding steam ; it may vary irregularly, like the strength of animals : or it may be wholly fitful and uncertain, like the wind.

But these characteristics, as well as the supply of power itself, are beyond our control. We cannot create power as we want it, but must take it whence we can get it. We cannot stipulate conditions, but must take the power as we find it, be thankful for it at that, and adapt it to our needs and purposes as best we can.

4. This is done by the use of machinery : and it is clear that in the construction of every machine, reference must be had to the characteristics of the motive power as well as to the nature of the work to be done. We may therefore amplify the definition above given, and say that—A Machine is an assemblage of moving parts, interposed between the power and the work, for the purpose of adapting the one to the other.

It is not, however, always necessary to trace back the source of power to its origin. For example, a line of shafting, whether itself driven by wind, water, or steam, may properly be considered as the "source of power" in reference to the various drills, lathes, planers, etc., driven by it.

Pure and Constructive Mechanism.

5. The operation of any machine depends upon two things, viz. : *definite force* and *determinate motion*. And in the process of designing, due consideration must be given to both these, so that each part may be adapted to bear the strains put upon it, as well as to move properly in relation to other parts.

But the nature of the movements does not depend upon the strength nor upon the absolute dimensions of the moving pieces, and may often be clearly illustrated by a model whose proportions are very unlike those of the actual working machine. Consequently the *force* and the *motion* may be considered separately ; and thus the science of Mechanism is divided into two branches, called respectively *Pure* and *Constructive.*

6. The selection of materials, and the proportioning of details with reference to strength and durability, are governed directly by considerations relating to the forces involved. Closely connected with these are other considerations relating to facility in manufacturing, convenience in repairing, and kindred features essential to practical excellence ; and the whole fall properly within the scope of **Constructive Mechanism.**

But we may examine the action of a machine by merely putting it in motion, without actually setting it at work ; and we can plan its movements without regard to the requisite strength of the parts.

The laws of motion may be discussed quite independently of any consideration of the force involved, and without reference to either the power or the work; and this constitutes the branch called **Pure Mechanism.**

7. Purposing now to confine our attention to this latter branch of the science, it is necessary, before entering upon its study, to define and explain the sense in which certain terms and phrases of frequent occurrence shall be used. And first of all, we perceive that we have not, as yet, assigned to the word *machine* a meaning which is precise and in accordance with the above limitation.

We must therefore modify the definition once more; and considering it with reference to its motion only,

A **Machine** *is a combination of parts so connected that when one moves according to a given law, the others must move according to certain other laws.*

8. **Motion** and **Rest** are essentially relative terms, within the limits of our knowledge. We can conceive a body to remain in a fixed position in space, but we cannot know that there is one which does. If there be any such body, it is in a state of *absolute rest.*

If two bodies, although both are moving in space, retain the same relative positions, each is said to be *at rest* with respect to the other : if they do not, either may be said to be *in motion* with respect to the other.

Path.—A point moving in space describes a line, called its *path;* which may be rectilinear or curvilinear. The motion of a body, or geometrical magnitude, may be defined by the paths of one or more of its points, selected at pleasure.

Direction.—In a given path, a point can move in either of two *directions* only, which may be defined in various ways, as up or down, to the right or left, with the clock or the reverse; direction, as well as motion, being entirely relative.

9. **Velocity,** however, is not essentially relative. Whatever the form of the path, the speed of the motion is estimated by comparing the distance, or *space,* through which a point or body moves, with the *time* occupied in doing it. And since both space and time are absolute magnitudes, the velocity itself is absolute.

Velocity is either *uniform,* equal spaces being traversed in equal times; *accelerated,* the spaces increasing, or *retarded,* the spaces decreasing, while the times remain equal. And the rate of acceleration or of retardation may itself be either uniform or fluctuating. But it is not necessary to consider all the complications which may arise in this way; for our purposes it will suffice to make one general distinction, viz.: that between motions with *uniform* velocities and motions with *variable* velocities.

10. In the case of uniform motion, the space varies directly with both the time and the velocity. Thus if one body move twice as long and twice as fast as another, it will clearly travel four times as far. This is otherwise expressed by saying that the space increases in the compound ratio of the time and the velocity; or, still better, in the form of an equation, thus,

$$\text{Space} = \text{Time} \times \text{Velocity.}$$

The space and the time are measured by comparing them with fixed standards, or units, and may therefore be expressed by abstract numbers. And so in consequence may the velocity; for from the above equation we have

$$\frac{\text{Space}}{\text{Time}} = \text{Velocity,}$$

in which the first member being composed of abstract terms, the second member will also be an abstract number, showing how many units of space are traversed in a unit of time.

And this is the measure of *absolute* velocity when the motion is uniform.

11. Angular Velocity.—This expression relates to rotatory motion, like that of a wheel turning in its bearings; the speed of which may be measured by the linear velocity of any point in the rotating body whose radial distance from the axis is equal to the unit of space. This is called the *angular velocity*, and may be either uniform or variable.

If the angular velocity be uniform, the linear velocity of any point varies directly as its distance from the axis : for the angles are proportional to the times, and the arcs to the radii. Thus if one point be two feet, and another four feet, from the axis, the outer will move twice as rapidly as the inner, since in the same time it describes twice as large a circle.

The speed of a wheel may also be conveniently expressed by stating the number of turns it makes in a given time; which evidently varies as the angular velocity, if the latter be uniform.

The most concise and useful value, however, is the equation

$$\text{Angular Velocity} = \frac{\text{Linear Velocity}}{\text{Radius}}.$$

12. Revolution and Rotation.—A point is said to *revolve* about a right line as an axis, when it describes a circle of which the centre is in, and the plane perpendicular to, that line. When all the points of a body thus revolve, with the same angular velocity and therefore without changing their relative positions, the body itself is said to revolve about the axis.

If the axis passes through the body, as in the case of a wheel, the word *rotation* may be properly used synonymously with *revolution.*

But it frequently occurs in mechanical combinations, as for instance in Watt's Sun-and-Planet Wheels, that a body not only rotates about an axis which passes through it, but at the same time moves in an orbit about another axis. In order to make a distinction between the two motions, we shall in such cases speak of the first as a **Rotation,** and of the second as a **Revolution,** just as we say that the earth rotates on its axis, and revolves around the sun.

13. Continuous Motion.—Motion is in its nature continuous, in the sense that a point cannot move from one position in space to another, without passing through all the intermediate positions, whether its path be rectilinear or otherwise. But in the nature of things it is impossible for a point to go on moving indefinitely in the same direction, unless its path be one that returns into itself, like a circle, ellipse, or other closed curve.

And the possibility of such indefinite continuance is what is implied in the expression *continuous motion,* as technically employed. A wheel turning freely in its bearings affords an example of motion continuous in this sense, which naturally occurs oftener in circular paths than in any others.

14. Reciprocating Motion.—If a point traverses the same path alternately in opposite directions, its motion is called *reciprocating,* whether the path be rectilinear or not. But if the point travel in a circular or other arc, the use of this term will be confined to those cases in which the arc traversed is less than a circumference. For if a wheel make a number of complete turns, first in one direction and then in the other, it is manifestly improper to style such motion *reciprocating,* notwithstanding the recurring reversals in direction.

Reciprocating circular motion, like that of a pendulum, or of a lever swinging on a fixed centre, is also called **vibration.**

15. Intermittent Motion.—When a reciprocating piece has reached the end of its excursion in one direction, there must evidently be an instant of rest, before it begins to return. But it frequently is required that a piece shall remain still for a definite time, after which it again moves, either in the same direction as before or in the opposite. When a piece in its action thus alternates motion with definite periods of rest, it is said to have an *intermittent motion.* If the motions occur alternately in opposite directions, the action may be called an *intermittent reciprocating* motion.

16. **Mechanical Movements.**—The different kinds of motion above specified are to a certain extent interchangeable. That is to say, one kind may be converted into another, by means of various devices, which are called *mechanical movements*.

It is ordinarily the case, that a machine is composed of a number of such movements, or subsidiary combinations of parts, each fulfilling a distinct function in the general operation. They are to the machine what the members are to the body ; but each one, serving a definite purpose in respect to its motion, may be regarded as a little machine, whose action may be studied by itself. For example, the valve gear of a steam engine may be entirely disconnected from the other parts, and its operation investigated without reference to them.

17. **Cycle of Motions.**—When a mechanical combination is set in action, its parts go through a certain series of motions, involving various changes in direction, velocity, or kind of motion, in a regular order. It is usually the case that the parts finally return to their original positions, after which the same motions will recur in the same order, and so on perpetually. The whole series is called the *Cycle*. Under these circumstances the combination is also said to have a **Uniform Periodic Motion.** These terms, however, are used only for want of better ones. Neither the word "Cycle" nor the word "Periodic" have any reference to the time required to go through the series of motions, nor does the word "Uniform" imply that the time occupied is always the same.

18. The terms are intended to convey the idea of regularity of succession and constancy of relation, as obtaining among the motions which make up the series.

To illustrate : One revolution of the crank of a steam engine produces a reciprocation of the piston and a series of different angular positions of the connecting rod, which itself vibrates on a moving axis. And if the speed of the crank be uniform, the velocities of both piston and connecting rod will vary according to a definite law. Now it is clear that the parts will go through the same series of motions, in the same relative order, and with the same variations in velocity as compared with the speed of the crank, at every turn of the latter, whether it go quickly or slowly, uniformly or variably, in one direction or the other.

19. **Phases of Motion.**—This term is used to designate the successive phenomena of varied motion. Thus, one *phase* in the movement of a steam engine would be represented in a diagram showing the relative

positions of the piston, connecting rod, and crank, at the beginning of a stroke ; another if the middle of the stroke had been selected, and so on. Such diagrams, representing different phases of the motion, are often of the greatest utility in conveying clear ideas of the action of complicated mechanical movements. Since the order of recurrence is the same, we may select at pleasure any phase as the beginning of a cycle.

20. Recapitulating a little, we see from the foregoing that every machine, regarded broadly as a contrivance for utilizing power, consists essentially of three classes of parts ; the function of the first being to receive the power, that of the second to transmit and modify the force and the motion, and that of the third to do the work. Evidently, the nature and form of the first class depend directly and largely on the character of the motive power, those of the third class upon the nature of the work to be done, which also to a great extent determines the proper actual velocity of the machine while in practical operation. It is also apparent that these three classes of parts are independent of each other, in so far that any kind of work may be done by any kind of power, and by means of different combinations of interposed mechanism.

21. Now in what follows, we shall have to do only with the second of these classes. Our object is to investigate the laws which govern the determinate motions involved in the action of the machine : and the motions as well as the form of the first class are determined by the manner of action of the motive power, and those of the third by the nature of the work and the manner in which it is to be done. The motion of the one class has then to be transmitted to the other ; and as the given motion of the former may be and usually is different from the required motion of the latter, it follows that, during transmission, the motion must be modified according to specific conditions. These objects are accomplished by the second of the three classes of parts above enumerated ; and it is the province of Pure Mechanism to discuss the methods by which motion may be transmitted, and to investigate the laws which govern its modification during the process.

22. **Elementary Combinations.**—If two pieces be so connected and arranged that a given motion imparted to one compels the other to move in a determinate manner, these two constitute an *elementary combination*. Practically, the motions are made determinate by means of a rigid frame, in relation to which, as well as in relation to each other, the two pieces move. But, obviously, the modification of the motion is best seen by comparing the movements of the two pieces with each other, so that for our purposes it is sufficient to take note of the

motion of which each one is capable, without regard to the means by which it is restrained or limited.

23. Driver and Follower.—That piece of an elementary combination to which motion is supposed to be imparted, is called the *Driver*: and the one whose motion is made compulsory by the action of the other upon it, is called the *Follower*.

A Train of Mechanism consists of a series of parts, composing the whole or a portion of a machine, each of which receives motion from the preceding one, and transmits it to the next in order. The train is, therefore, made up of elementary combinations, and each piece is a *follower* to the one which comes before, and a *driver* to the one which comes after it. As the motion may be modified at each successive step, it is necessary to begin by considering the modifications which may be effected by means of elementary combinations only.

24. Modes of Transmission.—Strictly speaking, if we leave out of the question the agency of attractive or repulsive forces, such as magnetism, one piece cannot compel another to move unless the two are in actual contact.

But in many cases the motion of one piece is communicated to another by the intervention of a third one, under such circumstances that the movements of the latter are of no possible consequence, the proper action of the whole depending entirely on the relative motions of the first and second : and these two may then be properly regarded as forming an elementary combination. We have, then, that motion may be transmitted from a driver to a follower,

1. *By Direct Contact.*
2. *By Intermediate Connectors.*

25. Links and Bands.—Such an intermediate connector must be either rigid or flexible. If it be rigid, it is called a *Link*, and can either push or pull, like the connecting rod of a steam engine : being necessarily pivoted or otherwise jointed to both the driver and the follower.

If the intermediate connector be flexible, it is called a *Band*: for our purposes it is supposed to be inextensible, and it can transmit motion only by pulling.

26. Modification of Motion.—In the action of an elementary combination, the motion of the follower may differ from that of the driver in kind, in velocity, in direction, or in all three. For example, a continuous rotation with uniform velocity may transmit continuous rotation whose velocity is greater or less, uniform or variable, in the same direction or the reverse ; it may transmit rotation intermittently : or the follower may receive a reciprocating motion with a varying velocity in a rectilinear or a curvilinear path. In an elementary combination, as has been pointed out, the path of each piece is determined by its connection with the frame-work of the machine, and it remains for us to ascertain, at each instant of the action, its direction and velocity. If this be done, the question is already settled as to whether the motion has been changed in *kind*. Hence we may properly say that the most important function of an elementary combination is to modify motion in *velocity* and *direction*.

27. The laws which govern this modification are determined by the *comparative* movements of the two pieces. It is apparent on reflection that, at every instant of the action, or, in other words, for every possible position of the driver with respect to the follower, there will exist a certain definite *proportion between the velocities,* and an equally definite *relation between the directions,* of their motions; which will depend entirely upon the pieces themselves and the manner in which they act upon each other, and cannot be affected by the *absolute* directions or velocities.

Consequently, whatever the nature of the combination, the analysis of its action will be complete, if throughout its range we are able to determine, as between the driver and the follower,

1. *The Velocity Ratio.*
2. *The Directional Relation.*

28. Now, the velocity ratio of the two motions may remain the same during the entire action, or it may vary; and this is also true of the directional relation. To illustrate : If two circular wheels gear with each other, turning about fixed axes, it is clear that the *velocity ratio* is constant. If one wheel is twice as large as the other, it will at any instant be turning half as fast, whether the motions be uniform or not. And so of the relative directions of the rotations ; if the wheels are in external gear, they will turn in opposite directions, if they are in internal gear they will turn in the same direction : but the *directional relation,* whichever it may be, does not change. If two elliptical wheels engage, the directional relation, as before, is constant : but the velocity ratio will change as the radii of contact vary. In the case of the piston and crank of a steam engine, neither the velocity ratio nor the directional relation is constant ; supposing the crank to turn at a uniform rate in one direction, the piston travels to and fro with a varying speed.

It is this feature of constancy or the reverse, in these two particulars, which distinguishes the actions of elementary combinations from each other, and forms the true basis for their proper classification.

29. Graphic Representation of Motion.—It has already been suggested, that the action of a combination may often be most clearly illustrated by drawings which show the parts in their proper relative positions at convenient phases of the motion.

But the motions themselves of any points in either piece also admit of perfect graphic representation at any given phase. For though the path of a moving point may be a curve of any kind, yet, as the direction of a curve at any point is that of its tangent at that point, the *direction* of the motion at any instant may be indicated by that of a right line. And the *velocity* being an abstract number, may be properly represented by the length of that line.

30. Geometrical Method of Investigation.—Between the lines thus representing the motions of properly selected points, and other lines closely connected with the moving pieces, definite relations may usually be established, in such a manner that, by means of diagrams thus constructed, the velocity ratio and the directional relation may be ascertained in the particular phase represented, and the law governing the modification of motion throughout the action deduced, by simple geometrical reasoning.

The method here outlined is peculiarly appropriate to this subject, as directly leading to, if not directly involving, the accurate constructions of the movements considered, which are essential in practical operations : and to it we shall adhere throughout. But there are some general principles relating to motion and contact, by a previous study of which the analysis of motion as modified by mechanical devices will be much facilitated : and these will accordingly receive our attention in the following chapter.

CHAPTER II.

COMPOSITION AND RESOLUTION OF MOTION—INSTANTANEOUS AXIS OF
ROTATION—MOTIONS OF TRANSLATION—COMPOSITION OF ROTATION
AND TRANSLATION—SLIDING, ROLLING, AND MIXED CONTRACT.

Composition and Resolution of Motion.

31. Resultant.—If a material point receive a single impulse in any direction, it will move in that direction with a certain velocity. If it receive at the same instant two impulses in different directions, it will obey both, moving in an intermediate direction, and with a velocity different from that due to either impulse alone. Now such a point may receive at the same instant any number of impulses, differing in magnitude and direction. But the point can move only in one direction and with one velocity; this actual motion is called the **Resultant**; and the separate motions, which the different impulses taken singly tended to give it, are called the **Components.** These components being represented by right lines, which may lie in the same or in different planes, the resultant may be found graphically by the following constructions :

32. Parallelogram of Motions.—In Fig. 1, let the point A have two component motions, represented by AB, AC. These two lines determine a plane, in which the resultant must lie. In this plane draw through B a parallel to AC, and through C a parallel to AB. These parallels intersect at D ; and AD is the resultant sought.

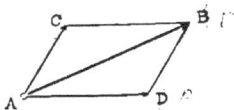

FIG. 1.

This fundamental proposition may be thus stated : *If two component motions be represented in magnitude and direction by the adjacent sides of a parallelogram, the resultant will be similarly represented by the diagonal passing through the point of intersection.*

33. Composition of Motions.—This process of finding the resultant of simultaneous independent motions is called *Composition :* and any number of them may be compounded by repeating it.   Thus in Fig. 2, let there be three components, *AB*, *AC, AE,* all lying in the plane of the paper.   We first compound any two of them, say *AB* and *AC,* as in the figure, giving a resultant *AD,* which is next compounded with *AE,* by which finally we find *AF,* the required resultant.   The process is the same for any number of components, and it makes no difference in what order they are taken.

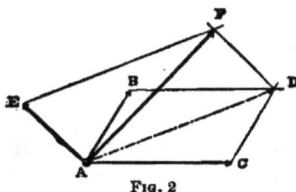

Fig. 2

34. Parallelopipedon of Motions.—The above holds true whether the components all lie in the same plane or not.   In Fig. 3, let *AB, AC, AE,* be three components, the first two lying in the plane of the paper, while the last does not.   Proceeding as before, *AD,* the resultant of *AB* and *AC,* will lie in the plane of the paper also ; but *AF,* the final resultant, found by compounding *AD* with *AE,* will lie in a different plane, determined by those two lines.   Had there been a greater number of components, we should have continued in a similar manner, compounding *AF* with one of them, their resultant with another component, and so on.   But the figure sufficiently illustrates not only the process, but also another fundamental proposition, relating to three components not in one plane, which may be thus expressed :

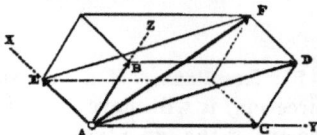

Fig. 3.

If three component motions be represented by the three adjacent edges of a parallelopipedon, the resultant will be represented in magnitude and direction by the body diagonal which passes through the point of intersection.

35. Resolution of Motion.—This is the inverse process to that above explained.   It is obvious that if two or more independent motions can be compounded into a single equivalent motion, that resultant can be again separated, or *resolved,* into its components.   This would be just as true if the resultant thus found had been originally given as an independent single motion.   Consequently any motion may be resolved into two components, either of these into two others, and so on *ad libitum.*

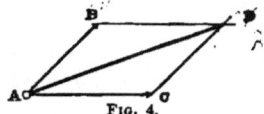

Fig. 4.

And these components may be given any direction at pleasure.   In Fig. 4 let *AB* represent the given motion ; through *A* draw *AC* in any direction, and through B draw BC, cutting AC in any point C. Then completing the parallelogram, it is evident that AB is the resultant of the two motions represented by AC, AD. Of these the direction of the first was assigned; that of the other is not wholly arbitrary, since it must lie in the same plane with AB and AC. But AD may be again resolved, the direction of one component being assumed as before : and this may be repeated until components of the original motion have been found, in as many different directions in space as may be desired.

36. From the above it will be perceived that if the given motion and the required components all lie in one plane, whatever the number of the latter, the resolution can be exactly effected. If they do not, there may be one component in addition to those whose directions are assigned : should these all lie in one plane which does not contain the given motion, this must obviously be the case.

But a motion can be exactly resolved into three components not in one plane, provided that it does not lie in the plane determined by either two.

Thus in Fig. 3, let AF be the given motion, and Ax, Ay, Az, the directions of the three components. Now these three lines last mentioned determine three planes; and by passing through F three other planes respectively parallel to these, a parallelopipedon is constructed, of which AF is the body diagonal, and the edges passing through A are therefore the required components.

37. Normal and Tangential Components.—It is often required to resolve a motion into two components, of which one shall be perpendicular, the other parallel to a given plane : usually a plane tangent to the surface of some moving piece under consideration. In this case the former is called the *normal*, and the latter the *tangential*, component. They are easily found as in Fig. 5. Let A be the moving point, AB representing its motion : draw through A the plane MN parallel to the given plane ; upon this let fall the perpendicular BC, and draw AC. Completing the parallelogram CD, it will be perceived that the

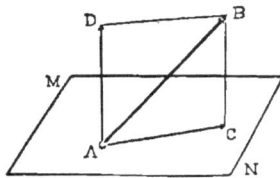
FIG. 5.

tangential component is AC, the orthographic projection of AB upon the plane MN, and that AD, the normal component, is equal and parallel to BC, the projecting perpendicular of the point B.

38. Motions of Connected Points.—If two points be so connected that the distance between them is invariable, they may be supposed to be connected by a right line. The motions of the points at any instant

may be represented by right lines, which may be in the same or in different planes. Each of these may be resolved into two components,

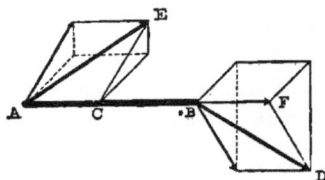

FIG. 6.

one of which is in the direction of the connecting line, the other perpendicular to it. And the simultaneous motions are subject to this condition, that the components along the line, thus found, must have the same magnitude and direction. Thus in Fig. 6, the components AC, BF, along the line AB, must be equal and in the same direction ; otherwise the distance AB would change, which is contrary to the hypothesis. These components are at once found by drawing through E and D, planes perpendicular to AB, cutting it in C and F.

The Instantaneous Axis of Rotation.

39. A right line, moving in any manner in space, may at any given instant be regarded as revolving about some other right line more or less remote. The latter may, from instant to instant, change its position not only in space, but in relation to the line whose motion is under consideration, with reference to which it is therefore called the **Instantaneous Axis.**

40. If the motions of two given points in a right line are known, the motion of the whole line is fully determined. Consequently, if it can be shown that the actual motions of these two points, at any one instant, are the same as though they were revolving with the same angular velocity and in the same direction about any axis, the motion of the line is at that instant equivalent to one of revolution about the same axis.

First, considering the motion of a single point in space : whatever the path, the motion is, at any given instant, equivalent to one of revolution about any right line in the plane normal to the path at the position occupied at that instant by the moving point. Thus in Fig. 7, let P be the moving point, PT the tangent to its path, and MN the normal plane. In this plane draw any right line AB, and PC perpendicular to it. Then if P be supposed to revolve about AB as an axis, it will describe a circle whose plane, being perpendicular to MN, will contain the line PT: and this line being perpendicular to the radius CP at its extremity, will be tangent to

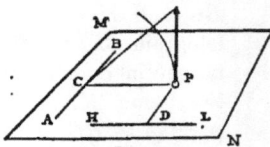

FIG. 7.

the circle thus described. Similarly the actual motion of P may be proved equivalent to one of revolution about HL, or any other right line lying in the normal plane MN.

The angular velocity will depend upon the distance of the moving point from the assumed axis, and is conveniently measured by dividing the given linear velocity by that distance, since in general we have (11)

$$\text{ang. vel.} = \frac{\text{lin. vel.}}{\text{rad.}}$$

Thus if in the figure, PT represent the linear velocity of P at the given instant, we shall have for the equivalent rotations,

$$\text{ang. vel. about } AB = \frac{PT}{PC}$$

and

$$\text{ang. vel. about } HL = \frac{PT}{PD}$$

41. Now, if the motions of two given points on a right line be known, the above reasoning applies to each. Consequently, if we draw two planes, respectively normal to the paths at the positions simultaneously occupied by the moving points, the intersection of these planes will be the only right line about which both points can at the given instant be regarded as revolving. But we have already seen that the motions of two connected points are subject to another condition (**38**), and it remains to be shown that these equivalent revolutions will have the same direction and the same angular velocity when that condition is satisfied.

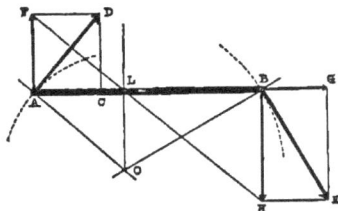

FIG. 8.

42. We will first consider the case in which all the motions of the line upon which the given points are situated, are confined to one plane. In Fig. 8, let AB be the moving line, A and B the given points. Let AD represent the motion of A in magnitude and direction, and suppose B to move in the direction BE; AD may be resolved into the

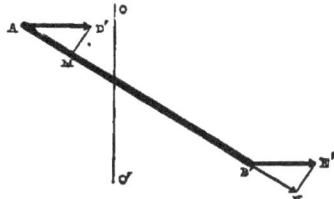

FIG. 9.

two components, AC in the direction of AB, and AF perpendicular to it. The motion of B must have a component BG along AB, equal to AC and in the same direction. Having set this off, draw at G a perpendicular to AB, cutting BE at E. This determines the linear velocity, BE, of the point B, and completing the parallelogram, this is seen to be the resultant of the two components BG, BH. Suppose all these lines to lie in a horizontal plane parallel to the paper ; then AO perpendicular to AD, and BO perpendicular to BE, will be the traces of the two normal planes, and their intersection O will represent a vertical line. Now draw OL perpendicular to AB, then the triangles ACD, ALO, are similar, and also the triangles EGB, BLO ; whence

and
$$\left. \begin{array}{c} \dfrac{AC}{OL} = \dfrac{AD}{OA} \\[2mm] \dfrac{BG}{OL} = \dfrac{BE}{OB} \end{array} \right\} \text{ But } AC = BG \quad \therefore \dfrac{AD}{OA} = \dfrac{BE}{OB}$$

But since AD, BE, are the linear velocities of A and B, the fractions $\dfrac{AD}{OA}$ and $\dfrac{BE}{OB}$ are the angular velocities of the revolutions of those points about O ; which are thus proved to be equal, and their directions are the same by construction.

43. The same pairs of similar triangles will therefore give

$$\frac{CD}{AL} = \frac{AD}{OA} = \frac{BE}{OB} = \frac{GE}{BL}$$

but $CD = AF$, and $GE = BH$; whence

$$\frac{AF}{AL} = \frac{BH}{BL} .$$

Therefore the right line FH must pass through L, the foot of the perpendicular from O upon AB. This fact is important, for the angle between AO and BO may be so acute as to render it difficult to determine the exact point of intersection. In that case, if the relative velocities are known, we have only to find the components perpendicular to AB, which may be laid off upon any scale, so that FH need not cut AB acutely ; and the point L can thus be found with precision.

44. The general case is that in which the motions of the two given points are in different planes ; in illustration of which, Fig. 8, still regarded as a horizontal projection, is to be studied in connection with the vertical projection given in Fig. 9. AD and BR are parallel to the horizontal plane, to which AB is inclined, though parallel to the vertical plane. In Fig. 9, therefore, AB appears as $A'B'$, and AD is foreshortened into $A'D'$: and the actual component $A'M$ along the line is found as in Fig. 6, by passing through D' a plane perpendicular to $A'B'$. Then making $B'N$, the corresponding component of the motion of B, equal to $A'M$, and drawing a plane through N perpendicular to $A'B'$, we have $B'E'$ the vertical projection of the resultant ; from this the horizontal projection BE, which is seen in its true length, is determined, and is evidently of precisely the same length as in the previous case. Now AD and BE may as before be resolved into the components AF, AC, and BG, BH, lying in horizontal planes. The preceding argument depends upon these only, and accordingly applies without change.

45. In general, then, the instantaneous axis of a moving right line may be found if the *directions* only of the simultaneous motions of two points upon it are given : but this may not be possible if those directions are parallel. Thus in Figs. 10 and 11, let AC and BE, the motions of the points A and B at the given instant, be parallel to each other and to the paper, and perpendicular to the moving line AB. In this case the two normal planes coincide ; yet there is an instantaneous axis, in order to locate which we must know also the

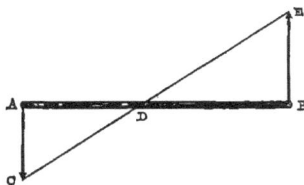

Fig. 10.

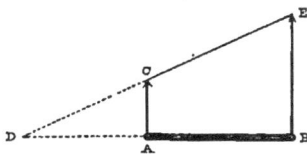

Fig. 11.

relative velocities at least of the two points. Setting these off as AC, BE, of their proper proportionate lengths by any scale, CE will cut AB or its prolongation in D. If we conceive a right line perpendicular to the paper at D, it will lie in the normal planes, and the motion of AB is evidently equivalent to one of rotation about it. And this will as obviously be true, whatever may be the inclination of the moving line to the plane of the paper. Any line in the normal plane, therefore, which passes through D, may be taken as the instantaneous axis.

46. Now, in Figs. 10 and 11, let AB, AC, be regarded as parallel to the plane of the paper, while BE is the projection of a line inclined to that plane, though perpendicular to AB. Then the planes normal

2

to AC and BE will intersect in the line AB itself; which is then its own instantaneous axis, about which we may for the sake of consistency say that it is revolving with infinite velocity. It is, certainly, the only line about which it can be said to have a motion of rotation *only*; it does not follow, however, that it will remain stationary in space, which it will not do: but, as will be shown subsequently, its actual motion will be helical.

47. Motion of Translation.—From consideration of Fig. 11 it will be seen that the more nearly equal AC and BE are, the more remote will be the intersection D: and that if they be exactly equal, the instantaneous axis will be at an infinite distance. Since in that case all the points of AB are at the instant traveling in the same direction with the same velocity, the consecutive position of the moving line will be

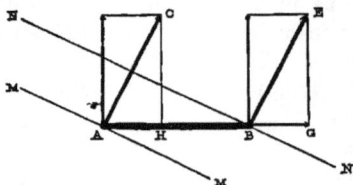

Fig. 12.

parallel to the present one. This will occur whenever the simultaneous motions of the two given points are in parallel lines and in the same direction, whether perpendicular to the moving line or not. Thus in Fig. 12, the normal planes AM, BN, are parallel, although the motions of A and B are inclined to the line AB. It is evident that in this case AC and BE, the actual linear velocities of the two points, *must* be equal, since (**38**) the components AH and BG, along the line, must be equal and lie in the same direction. The motion of AB in Fig. 12 may then be regarded as one of rotation about an infinitely remote line parallel to BN, and lying in a plane parallel to AB: if AC, BE, are made perpendicular to AB, the instantaneous axis will be infinitely remote and parallel to the moving line.

48. The motion of the line, at any instant when it is thus revolving about an axis at an infinite distance, is called **Translation**. It may be continuous; and not only a right line, but any geometrical magnitude, is said to have a motion of translation, when all its points move with the same velocity and in the same direction, in equal and similar paths. Under these circumstances, evidently, all the tangents to these paths, at the positions simultaneously occupied by the moving points, will be parallel: and the right line joining any two of these points will remain parallel to itself during the whole movement. If the paths are rectilinear, the motion is called *straight translation*, or, more commonly, *sliding*: if they are curvilinear, the motion is called circular, elliptical, or helical, translation as the case may be, the name depending on the form of the curve.   A familiar instance of circular translation is afforded by the coupling-rod of a locomotive, shown in Fig. 13 : the points A and B move as indicated by the arrows, in the circles whose centres are D and E, with the same velocity.   And any other point C, in a rigid exten- sion of the bar AB, will move similarly in the equal circle whose centre is F.

49. In relation to a single right line, it is to be noted that a mo- tion of circular translation, if the

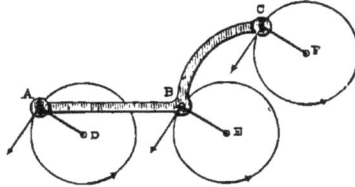

Fig. 13.

planes of the circles be perpendicular to the moving line, is identical with one of revolution about the axis of the cylinder thus generated. Thus, in Fig. 13, the point C may be considered as a line perpendicu- lar to the paper ; and its motion is the same as though it revolved about a parallel axis passing through the point F.

It is also to be observed that the motion of a single right line of a rigid body does not determine the motion of the whole body, which if not otherwise controlled would be free to rotate about that line.   But if the motions of two right lines of the body be given, those of all its other points are thereby fully determined.

Composition of Rotation and Translation.

50. Now, just as a simple motion in a right line can be resolved into two others, so a simple rotation about a fixed axis may be regarded as a resultant of two other motions, viz., a rotation about another axis parallel to and revolving around the first, and a translation in which the paths are circles whose radii are equal to the distance between

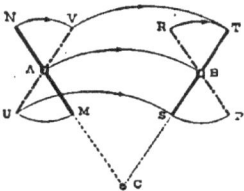

Fig. 14.

these axes ; the angular velocities and the di- rections of both components being the same as those of the rotation which was to be re- solved.

Thus in Fig. 14, let C represent a fixed axis perpendicular to the paper, and let MAN re- volve about this axis into the position SBT. We may also let A represent a right line per- pendicular to the paper, about which MN may rotate into the posi- tion UV, afterward moving by circular translation to the position ST.   Or, MN may move by circular translation into the position PR, afterward rotating about B into the position ST : the path of A being in all cases the arc AB. On either supposition the result is the same as if both motions progress simultaneously and uniformly, which produces the original revolution about C: and the angles MAN, RBT', and ACB, are equal by construction. In Fig. 15, the traveling axis passes through N; now let the points M and A, during translation from position MN to position PT, describe circles whose radii are equal to CN. Then if in the same time MN rotate about N, in the same direction, through an angle

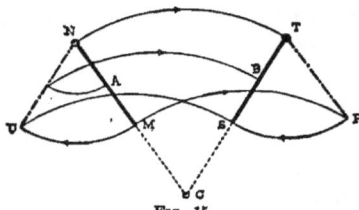

Fig. 15.

$PTS = NCT$, it will come into the same position, ST, as in Fig. 14. In short, any axis parallel to the fixed one may be assumed as that of the component rotation, and thus the revolution around C may be resolved in an infinite number of ways.

51. On the other hand, a motion of circular translation may be considered as a resultant. For (49) any line perpendicular to the planes of the translation, and partaking of that motion, may be regarded as revolving about a fixed axis. Assuming such line as a traveling axis of rotation, the component motions are, *first*: a revolution about that fixed axis, the direction and angular velocity being

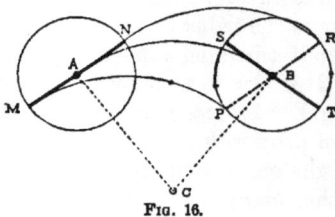

Fig. 16.

the same as those of the translation; and, *second*, a rotation in the opposite direction, but with the same angular velocity, about the assumed traveling axis. Thus in Fig. 16, if MAN revolve about the fixed axis C, until A reaches B, and this be its only motion, it will have the position SBT. If in the same time it also rotates in the opposite direction about A, as shown by the arrows, through an angle TBR, equal to ACB, its new position will be PBR, parallel to the original position MAN; and the resultant motion will be one of circular translation, in which M, N, and all the moving points, describe circles whose radii are equal to AC.

52. If any geometrical magnitude have a motion of rectilinear translation parallel to an axis about which it is revolving, the resultant is a helical motion about that axis. But a motion may be helical with respect to one axis, and yet simply rotatory with respect to another. For instance, the rectilinear generatrix of an oblique helicoid has a continuous helical motion about the fixed axis of the surface;

but in any given position of the moving line, its motion is equivalent to one of *rotation only* about an instantaneous axis, determined (41) by the intersection of planes normal to the paths of any two of its points : and either may be regarded as the actual motion of the generatrix at that instant, the latter being the less complex. In the case of the right helicoid, the intersection of these normal planes is the generatrix itself ; and the motion of that line in space cannot be reduced to any simpler form than that of a *helical* one about an axis, which, in this special case, is at every instant the same, being the fixed axis of the surface.

53. But, in general, if the motions of two connected points are perpendicular to the right line joining them, and are not parallel to each other, the motion of that line will be a helical one about an instantaneous axis : which may be determined, if the relative velocities are given, in the following manner :

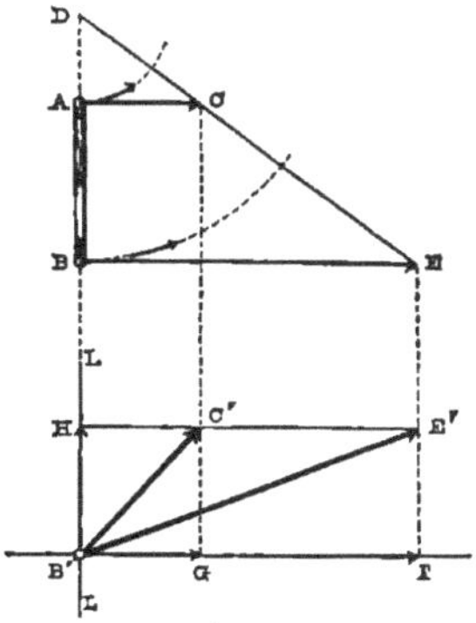

In Figs. 17 and 18, the moving line AB is perpendicular to the vertical plane, and B' is its vertical projection. The motions of the points A and B are seen in their true lengths and directions in the vertical projection as $B'C'$, $B'E'$ (for convenience so drawn that $C'E'$ is horizontal), and their horizontal projections are AC, BE. Through

Fig. 17.

AB pass a plane parallel to $C'E'$, and resolve these motions normally and tangentially with reference to this plane. The tangential components are $B'G$, $B'F$, horizontally projected in AC, BE. Draw EC,

Fig. 18.

and produce it if necessary to cut BA or its prolongation in D, whose vertical projection is B'. Then the motion due to these components, which are horizontal and parallel, will be (45) a rotation about any axis passing through D, and lying in the vertical plane which contains AB. The normal components are vertical and equal, the true length, $B'H$, being seen in the vertical projection. The motion of AB due to these components, therefore is one of vertical translation ; and the actual motion in space is a helical one about a vertical axis LL passing through D.

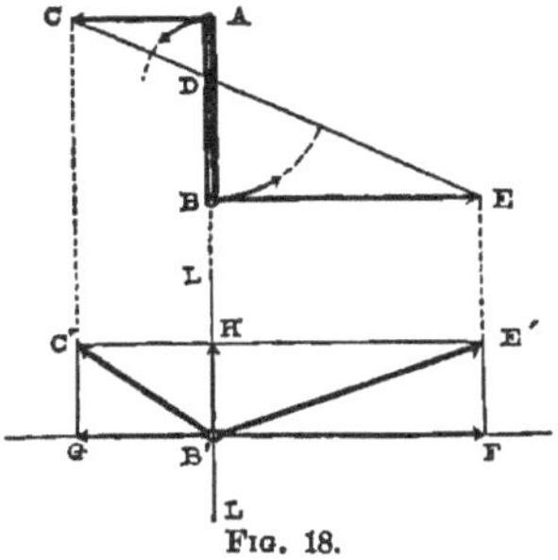

54. Let the points of a material right line be so connected with those of a rigid body that the distances between them cannot change. We may then suppose that line to move, and the body primarily to move in the same manner. If now the body also rotate about the line as an axis, then these two motions may be compounded : and **(49)** if it can be shown that any two lines of the rigid body are at the same instant moving in a rotatory or helical manner about any axis, the whole body may at that instant be regarded as doing the same. Evidently the moving line itself may be taken as one of these two, and any line intersecting it as the other.

The problem of determining the resultant presents several different cases, depending upon the nature of the motions primarily assigned to the rigid body and the line with which it is connected ; and these will be considered separately.

55. I. Let the primary motion be one of revolution about an axis parallel to the line.

The resultant will then be a rotation of the body about an instantaneous axis, which will be parallel to the other two axes, and lie in the same plane.

In Fig. 19, let C represent the moving line, P the axis about which

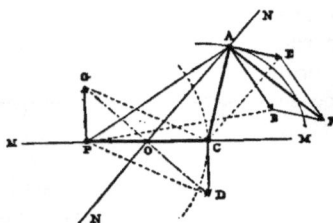

FIG. 19.

it is revolving, both being perpendicular to the paper ; and let CD, perpendicular to CP, and parallel to the paper, represent the linear veloc. ity of any point of the moving line in its revolution around P. Join any point of this line with any point A of the body : then CA represents a line which may or may not be inclined to the paper. Let AE, perpendicular to CA, be the linear velocity of A in its rotation about C; the motion of A due to the revolution about P will be AB, perpendicular to AP, the magnitude being determined by making the angle APB equal to the angle CPD. Both AE and AB, and therefore their resultant AF, the actual motion of A, are parallel to the paper. Consequently the normal planes NN, MM, of which the latter contains the axes C and P, intersect at O in a line perpendicular to the paper, which is the instantaneous axis.

The location of this axis may be found otherwise, thus : regarding CP as a line of the rigid body (which may be inclined or parallel to the paper), the rotation of P around C will be represented by PG, the angle PCG being made equal to the angle ACE: then **(45)** GD will cut CP in O.

56. II. Let the primary motion be one of translation, in which the paths are in planes perpendicular to the line.

In this case also the resultant will be a rotation about an instantaneous axis parallel to the line.

This will be apparent from inspection of Fig. 20, the construction differing from the preceding only in this ; that AB is parallel and equal to CD, the motion of C in its path RS, whose radius of curvature at C is for the purpose of comparison made equal to CP of Fig. 19 : also CD and AE are of the same lengths in both diagrams, in which therefore themo tions of C at

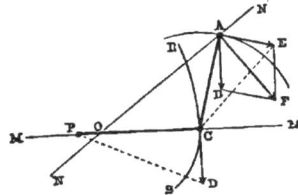

Fig. 20.

the given instant are identical, and the rotations about the moving line are alike.

57. III. Let the primary motion be one of revolution about an axis which intersects the line.

. The resultant will then be a rotation about an instantaneous axis lying in the plane of the other two, and passing through their point of intersection.

In Fig. 21, let CO be the line, OA the axis, both in the vertical plane. Draw CA at pleasure, and suppose it a line of the rigid body. Then the motion of A is due only to the rotation about CO, that of C is due only to the primary revolution about OA, and both are perpendicular to the vertical plane. Let these motions be represented by $A'E$, $C'D$, in the horizontal projection ; and draw DE, cutting $C'A'$ in P', whose vertical projection is P. Then (**45**) CA may be regarded as rotating about any axis lying in the vertical plane and passing through P, and CO as rotating about any axis in the same plane,

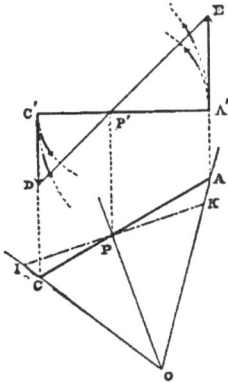

Fig. 21.

and passing through O. Consequently OP is the instantaneous axis of both CO and CA, and therefore of the whole body.

58. Draw IPK perpendicular to OP and consider it another line of

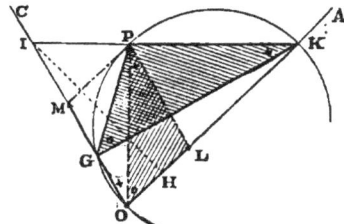

Fig. 22.

the rigid body : then in Fig. 22, draw KG perpendicular and PL parallel to OC, also IH perpendicular and PM parallel to OA, and join PG.

Now let

$$V = \text{ang. vel. of } K \quad \text{around } OC,$$
$$V' = \text{``} \quad \text{``} \quad \text{``} \ I \qquad \text{``} \quad OA,$$
$$v = \text{``} \quad \text{``} \quad \text{``} \ I \text{ or } K \ \text{``} \quad OP.$$

Then

$$\begin{cases} \dfrac{V}{v} = \dfrac{PK}{KG}. \\[2mm] \dfrac{V'}{v} = \dfrac{PI}{IH}. \end{cases}$$

By construction, OPK, OGK are right angles, therefore a circle will go round $OGPK$, whose diameter $= OK$; in which $KOP = KGP$, standing on same arc PK, and $\quad GKP = GOP$,   "   "   "   " GP; but $\quad OLP = GOP$, by reason of parallels PL, OG: $\therefore$ triangles GPK, OPL, are similar.

whence

$$\frac{V}{v} = \frac{PK}{KG} = \frac{PL}{OP} = \frac{OM}{OP};$$

similarly

$$\frac{V'}{v} = \frac{PI}{IH} = \frac{MP}{OP} = \frac{OL}{OP}.$$

That is to say : that if we lay off upon the axes the distances OM, OL, proportional to the angular velocities of the rotation and the revolution about those axes respectively, and complete the parallelogram ML, its diagonal OP will lie in the instantaneous axis, and be proportional to the angular velocity of the resultant rotation about it.

In Figs. 21 and 22, the direction in which the rigid body rotates about the traveling axis, and that in which the traveling axis revolves around

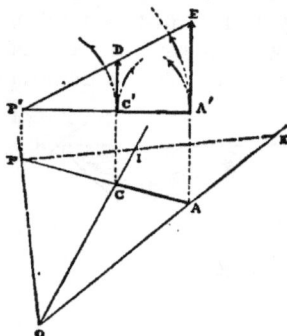

Fig. 23.

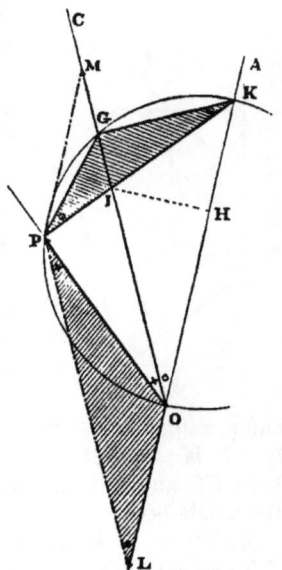

Fig. 24.

the fixed one, are the same.   But the above reasoning applies equally
well if they are opposite ; and Figs. 23 and 24 show the modifications
in the diagrams due to this change in the conditions.   In regard to
which it will be noted that in both cases OM is set off in the same
direction from O ; but OL is measured in one direction if the rotation and the revolution are alike, and in the other if they are unlike.

59. IV. Let the body and the line primarily revolve about another
line in a different plane.

In this case the resultant will be a helical motion of the whole body
about an instantaneous axis lying in a
plane parallel to the other two lines,
and intersecting their common perpendicular.

In Fig. 25, let OA be the fixed axis,
OC the moving one, both parallel to the
vertical plane ; O being the vertical
and DG the horizontal projection of
their common perpendicular.   Considering DG as a line of the rigid body, let
the motion of D be represented by OE
perpendicular to OA, that of G by OB
perpendicular to OC; the axes being

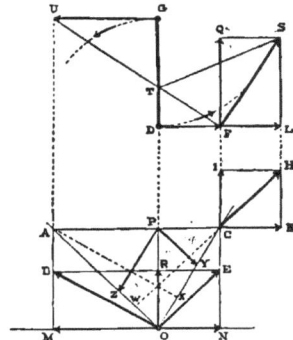

Fig. 25.

for convenience so placed that BE is horizontal.

Resolve these motions normally and tangentially with reference to
a horizontal plane through O : the normal components will be vertically projected in OR, the tangential ones in ON, OM, of which the
horizontal projections are DF, GU: therefore UF is the horizontal
projection of a line joining the points M and N.

Produce MB to cut OA in A, also NE to cut OC in C; then
AC will be parallel to MN. For the triangles OMA, ONE, are
similar : so are also the pair OMB, ONC: and $BM = EN$. Hence
we have

$$\frac{AM}{ON} = \frac{OM}{EN}, \text{ and } \frac{CN}{ON} = \frac{OM}{BM}, \therefore AM = CN,$$

and UF is also the horizontal projection of AC.

60. Moreover, UF is the trace on a horizontal plane through A, of
a plane passing through the point C and the axis OA. The actual
motion of C about that axis, is perpendicular to this plane ; therefore
its horizontal projection FS must be perpendicular to UF. Its vertical projection CH must (see Fig. 9) be parallel and equal to OE;
whence the magnitude of FS is determined.   With reference to the

vertical plane, this motion may be resolved into a normal component
FQ, vertically projected in C, and a tangential one of which the projections are FL, CH.   The latter component being parallel and equal
to OE, may be resolved into the vertical component CI equal to
OR, and the horizontal one CK parallel and equal to ON.   The horizontal projection of CK is FL, equal to DF, and compounding it with
FQ, the resultant FS is a horizontal line, vertically projected in CK.

61. The motions of the two lines DG, OC, then, have each a component of vertical translation equal to OR.   The motion of OC has
for its remaining components, DF for the point O, and FS for the
point C, both horizontal.   The resultant of these is a rotation about
the instantaneous axis OP, whose horizontal projection is T, the intersection of the two vertical planes normal to DF and FS.   The remaining component of the motion of G, is the horizontal line GU, and
the resultant of GU and DF is a rotation of DG about an instantaneous axis determined as in Fig. 10; which axis is also the vertical
line through T.   And the rotations of the two lines about this axis
have the same angular velocity; for, drawing TS, we shall have from
the similar triangles FLS, FDT, in which $DF = FL$, the ratios

$$\frac{TD}{FL} = \frac{TD}{DF} = \frac{TF}{FS};$$

but TFS, TDF, are both right angles, hence those triangles are similar, and the angles STF, FTD, GTU are equal.

Combining this rotation with the translation, the final resultant is
a helical motion of both lines and therefore of the whole body about
OP.

62. In the vertical projection, draw PY parallel to OA, and PZ
parallel to OC, making PY equal to OZ, and the triangles PCY,
ACO, similar.

Let
$$V = \text{ang. vel. of } D \quad \text{around } OA,$$
$$V' = \text{``} \quad \text{``} \quad \text{``} G \quad \text{``} \quad OC,$$
$$v = \text{``} \quad \text{``} \quad \text{``} D \text{ or } G \text{ ``} \quad OP.$$

We shall then have, since ang. vel. $= \dfrac{\text{lin. vel.}}{\text{rad.}}$,

$$V = \frac{OE}{DG}, \ v = \frac{ON}{DT}, \ \therefore \ \frac{V}{v} = \frac{OE}{ON} \times \frac{DT}{DG}.$$

But from sim. triang. ONE, OPA,................$\dfrac{OE}{ON} = \dfrac{OA}{OP}$, and from sim. triang. TDF, TGU.................$\dfrac{DT}{DG} = \dfrac{TF}{UF}$.

$$\therefore \frac{V}{v} = \frac{OA}{OP} \times \frac{TF}{UF} = \frac{OA}{OP} \times \frac{PC}{AC} = \frac{OA}{AC} \times \frac{PC}{OP}.$$

Now from sim. triang. ACO, PCY,.................$\dfrac{OA}{AC} = \dfrac{PY}{PC}$;

$$\therefore \frac{V}{v} = \frac{PY}{PC} \times \frac{PC}{OP} = \frac{PY}{OP} = \frac{OZ}{OP}.$$

And similarly, $\dfrac{V'}{v} = \dfrac{OY}{OP}.$

Draw AX, CW, respectively perpendicular to OC and OA; then by reasoning similar to that used in connection with Fig. 22, it may be shown that

$$\frac{V}{v} = \frac{CP}{CW}, \text{ and } \frac{V'}{v} = \frac{AP}{AX}.$$

63. From which it appears, that in the parallelogram ZY, of which the adjacent sides OZ, OY, are parallel to the axes and proportional to the angular velocities of the revolution and the rotation about those axes respectively, the diagonal OP is parallel to the instantaneous axis and proportional to the angular velocity of the rotation about it. And that the seg-ments AP, PC, into which AC perpendicu-lar to OP and limited by the prolongations of OZ and OY, is divided at P, are propor-tional to those into which DG, the com-mon perpendicular of the two axes, is cut by OP.

In Fig. 25, as in Figs. 21 and 22, the ro-tation about OC and the primary revolution about OA are in the same direction. Should they have contrary directions, the argu-ment would still apply without change; and the modifications in the diagram caused by that change are shown in Fig. 26, which being lettered throughout to correspond with Fig. 25, requires no explanation.

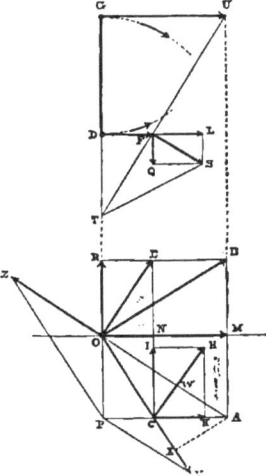

Fig. 26.

64. V. Let the primary motion be one of translation, in which the paths are *not* in planes perpendicular to the line.

The resultant is then a helical motion of the body about an instantaneous axis parallel to the line.

In Fig. 27, let the line OU be vertical, and D its horizontal projection; the motions OE, CH, of the points O and C, being parallel to the vertical plane, and horizontally projected in DF. These motions being equal and parallel, may be resolved with reference to the horizontal plane into the equal vertical components OR, CI, and the equal and parallel horizontal ones ON, CK, both of the latter being horizontally projected in DF.

Let O be the vertical and DG the horizontal projection of a line of the rigid body, and let OB, GU, be the corresponding projections of the rotation of G about OC. Now G has also a motion of translation, of which OE is the vertical and GL the horizontal projection, which

Fig. 27.

may be resolved into a vertical component OR, and a horizontal one, of which the projections are ON, GL. Compounding this latter with GU, the resultant is $GS = GU - LG$; the vertical projection is OM. Hence the resultant of these horizontal components is a rotation of OC and DG about the instantaneous axis whose horizontal projection is T, and vertical projection OC; with which is to be finally compounded the vertical motion of translation, producing the helical resultant motion of the whole body.

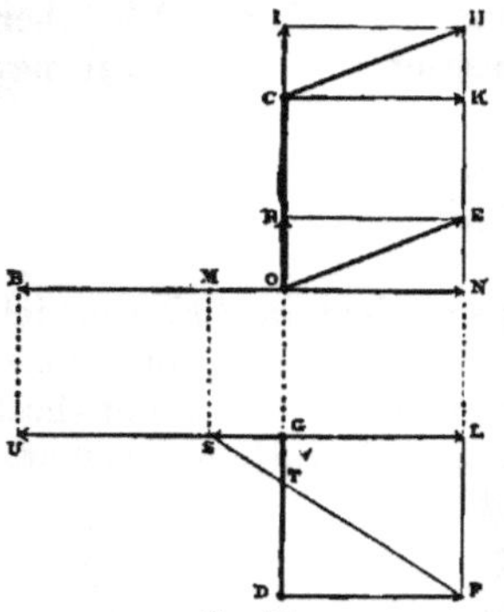

If OE and CH coincide with OC, that is to say, if the line primarily move endlong, the resultant is, obviously, a helical motion about an axis coinciding with the line itself.

65.   VI. Let the primary motion be a helical one about an axis perpendicular to the line.

In this case also the resultant will be a helical motion of the whole body about an instantaneous axis. For the original helical motion is compounded of a rotation about the fixed axis and a translation parallel to it; this rotation being compounded with that about the traveling axis, the resultant is **(57)** a rotation about a third axis, in the plane of the other two, and passing through their common point. The translation may again be resolved into components respectively perpendicular and parallel to the third axis; compounding the former with the rotation about that axis, the resultant is **(56)** a rotation about a fourth axis parallel to the third, which finally compounded with the component of translation parallel to that axis, determines the instantaneous helical motion of the body about the fourth axis.

66. Thus in Fig. 28, let the vertical fixed axis OA, and the horizontal traveling one OC, both lie in the vertical plane, and let OG, CK, be the vertical components of translation of the
helical motion of OC about
OA. Draw in the vertical
plane any line CA, and consider it as a line of the rigid
body; its horizontal projection is $A'C'$. Let $C'D$ be the
rotatory component of the motion of C, in its helical path,
and let $A'B$ be the motion of
A in its rotation about OC:
then as in Fig. 21 we find OP
the instantaneous axis of the

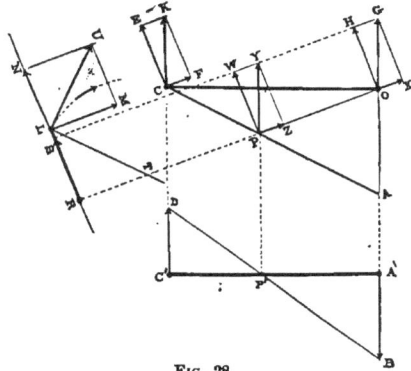

FIG. 28.

resultant rotation. Since P lies in this axis, its only motion is one
of vertical translation, PY, equal to OG and CK. These three motions may be resolved into PW, OH, and CE, perpendicular to OP,
and PZ, OX, CF, parallel to OP. The projections of OC and PC,
on a plane perpendicular to OP, will coincide in RL, and those of
the perpendicular components OH, PW, will coincide in RS. In
this projection, the motion of C will be represented by LU, the resultant of LM, perpendicular to LR and equal to $C'D$, compounded
with LN, the projection of CE. The planes normal to LU and RS
intersect in T, which represents a line perpendicular to this plane of
projection and therefore parallel to OP; this is the instantaneous axis
about which OC and PC are rotating, and the helical motion results
from the final addition of the parallel components OX, CF and PZ.

67. It will be seen that the preceding cases include every possible
motion of a rigid body in space. For if the motion of one of its right
lines be determined, the body can have no other motion except one
of rotation about that line as an axis. And whatever the law by
which the movement of this axis may be governed, its motion at any
instant is either one of translation, or one of simple rotation about
another axis in the same or a different plane, or a helical one about an
axis perpendicular to it. Hence it appears that the most complicated
motion of which any rigid body is capable at any instant, is a helical
one about some right line : that is to say, all its points have at the
instant a uniform rotation about the axis, combined with a uni-

form motion parallel to it, their paths being helices of the same pitch.

Sliding, Rolling, and Mixed Contact.

68. Sliding Contact.—The motions of a piston in a cylinder, of a journal in its bearing, and of a screw in a nut, afford instances in which the contact during the action is purely *sliding*.

Regarding closely the action of a single point in the surface of the piston, the journal, or the screw, we see that it moves over a certain path in the surface of the cylinder, the bearing, or the nut, and that it comes into contact, one after another, with all the points in that path.

In these instances the contact of the two pieces is superficial, the moving surface being identical with the fixed one. But one piece may touch another at discontinuous points, or even in only one point, and yet the same peculiarity may characterize the action.

We may therefore define the condition of **Pure Sliding Contact** to be, *Such relative motion of two pieces that every point of contact in the one, is brought into coincidence with all the successive points in their order, of a line in the other.*

69. Sliding contact in the nature of things as they are, is attended with friction, because there is not any known substance whose surface is absolutely smooth. We might however admit that friction would exist even if all surfaces were perfectly polished and free from asperities. But its existence even then would imply relative motion of two pieces in contact, of such nature that a given point of one should, before quitting contact, be brought into coincidence with more than one point of the other.

Could this be entirely avoided, and the action be made such that no point of either should touch two consecutive points of the other, there would be no friction, which cannot be conceived of as being produced by contact alone.

Practically, this is impossible, owing to imperfections of materials and workmanship : but the abstract condition, compliance with which, if it were attainable, would effect the result, is an exceedingly simple one.

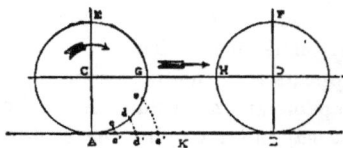

FIG. 29.

70. In Fig. 29, let the tangent *AB* remain fixed ; if the circle now rotate about the fixed centre *C*, the points *c, d, e.*, etc., of its circumference will successively be brought into coincidence with the point *A* of the tangent. Or if the circle be translated without rotating, as shown by the horizontal arrow, the point A of the circumference will come into coincidence successively with the points c', d', e', etc., of the tangent : in either case the action is one of pure sliding contact.

But if we suppose the circle to roll, like a hoop, along the tangent, from C to D, it is readily seen that the point A will instantly quit contact, and the points c, d, e of the circumference will come successively into contact with the tangent at c', d', e'. As it is commonly expressed, the circle measures itself off upon the tangent ; thus if AK be equal to the arc AG, then the point G will come into contact at K, and if AB be equal to the semicircumference AGE, then E will come into coincidence with B, and so on.

No point of the circumference can however be said to *move* in contact with the tangent. The circle has in this case both the motions of rotation and translation ; and the movement of the point A to the left, due to the former, is neutralized by an equal movement toward the right, due to the latter. That point is therefore for the instant at rest ; and the same will evidently be true of the other points c, d, e, etc., as they successively become the points of contact.

71. In Fig. 30, let ML be the common tangent, NN the common normal, of the two plane curves AB, EG, in contact at P. Let o, o', be points on these curves respectively, consecutive to P ; and let EG be fixed.

Then if AB move so that P goes in the direction PL, the effect will be to bring this point P of the upper curve, into coincidence with o' on EG ; but if P goes in the direction PM, the point o of AB will be brought into coincidence with the point P of the fixed curve EG. Now the motion PL may be considered as either a translation, or a rotation about any centre D in NN. If it be the former, evidently the curve AB cannot

Fig. 30.

not be at the same time translated in the opposite direction. But in either case the motion of P can be neutralized by an equal and opposite motion PM, which may be considered as a rotation about any centre (other than D) in NN, as for instance C. The resultant of these two simultaneous motions of AB will be a rotation about P as an instantaneous centre, the effect of which will be to bring the point o into coincidence with o', the points of the two curves which now fall together at P, instantly separating: and if these conditions be continuously maintained, the action will consist of a rolling contact between AB and EG, precisely similar to that in the preceding case. The point of contact of the moving curve is always at rest, and consequently the action is unattended by friction.

72. It is apparent that the same kind of contact will ensue if in Fig. 29 the circle merely rotates about C as a fixed centre, while AB moves in contact with it, if the linear velocities and directions at the point of tangency are the same: so also in Fig. 30, if the points of the two curves which fall together at P, move in the same direction at the same rate, as represented by PM, the consecutive points o, o' will come into coincidence, and the action will be that of rolling contact.

It is also obvious that these two figures perfectly represent the rolling of a cylinder upon a plane, and of one cylindrical surface upon another, the rectilinear elements being perpendicular to the paper. These illustrations sufficiently explain the nature of the action under consideration, which may be defined as follows:

Pure Rolling Contact *consists in such a relative motion of two lines or surfaces, that the consecutive points of one come successively into contact with those of the other in their order.*

73. It will be readily seen that a double curved surface, like that of a ball or an egg, can roll without slipping upon a surface of any kind, plane, single curved, warped, or double curved, convex or concave; provided that the curvature of the rolling surface at the point of contact be the greater. Tangency existing in this case at a single point only, the action in fact consists merely in the rolling of a line of one surface in contact with another line lying on the other surface, and one or both of these lines may be of double curvature.

We are thus led to consider this kind of moving contact between

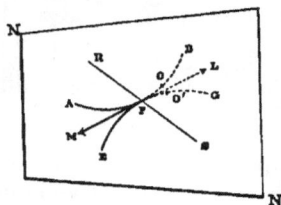

FIG. 31.

lines, in its most general aspect: and in illustration, let NN, Fig. 31, represent the common normal plane of two curves of any kind, AB and EG: ML being their common tangent at P, and o, o' points on those curves, consecutive to P, as in Fig. 30. Considering EG as stationary then, it appears, from the foregoing, that the rolling of AB upon it is effected by communicating to that curve two simultaneous motions, one of which imparts to P the motion PL, and the other imparts to the same point the equal and opposite motion PM. Of these two motions, one may, and the other must, be regarded as a rotation about a line at a finite distance, lying in the plane NN: their resultant is a rotation about an instantaneous axis also lying in NN, and passing through P. But the directions of the axes of the component rotations, and also their distances from P, are arbitrary: hence any line RPS, drawn through P in the plane NN, may be assumed as the instantaneous axis; a rotation about which will bring the points o, o', into coincidence as required.

74. But though the instantaneous axis must pass through P, it is not essential that it lie in the plane NN. For if any line be drawn through P, then both PL and PM may be resolved into components, perpendicular to, and lying in, that line. The latter may be regarded as motions of translation, and, being equal and opposite, they will neutralize each other. The former will also be equal and opposite, and regarding one or both as rotatory, the irresultant will be a rotation about the assumed line through P, as an instantaneous axis.

Thus in Fig. 32, let PR, PS, be the vertical projections of two helices, lying on the surfaces of two cylinders tangent along PB; and let MPL be the horizontal projection of their common tangent, and NN the horizontal trace of their common normal plane. It will be apparent that PL, PM, may be considered as rotations about an axis in NN. But also, with reference to OPO, we may resolve PL into the components PC, PD, and also PM into the com-

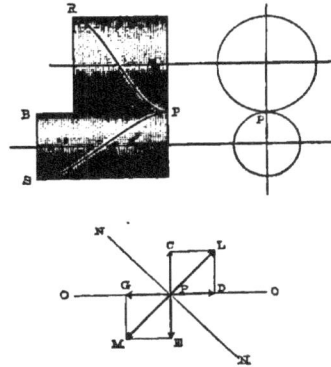

Fig. 32.

ponents PE, PG. Then regarding PD, PG, as motions of translation, they neutralize each other: and PC, PE, may be regarded as rotations about axes lying in the plane of which OO is the trace. These also neutralize each other, leaving P for the instant at rest: and it will readily be seen that as a matter of fact, the instantaneous axis of the upper cylinder, in rolling around the lower, must in the present position be the element of contact, of which PB is the vertical, and OPO the horizontal projection; as also that if one cylinder rolls upon the other, the two helices move in rolling contact.

75. In so far then as the rolling of any one line upon another is concerned, the moving line may at any instant be regarded as rotating about any right line passing through the point of tangency. This, as above pointed out, includes the case of rolling contact between sur-

3

faces which touch each other in only one point. We have now to consider the conditions under which surfaces having more than one point of tangency can roll together without sliding. It has been stated by Prof. Reuleaux,* that helicoids and certain other warped surfaces are capable of moving in perfect rolling contact. From a practical point of view it is of comparatively little consequence whether they are or are not : but the fact that such a statement is made by such an authority, warrants the presentation of the following considerations :

76. 1. The rolling of one surface upon another is a motion of rotation about an instantaneous axis, which must pass through every point of tangency.

2. An axis is a right line ; hence if there be more than one point of tangency, all of them must lie upon a rectilinear element of contact.

3. This rotation about the instantaneous axis may be resolved into two component motions, of which one must be a rotation about another axis.

4. In this component rotation, each point in the moving surface describes a circle with plane perpendicular to and centre in the second axis; of which the radius is a perpendicular from the point to that axis.

5. The other component motion must be such that each point of tangency shall by it be made to move with a velocity equal to that due to the first component rotation, and in the opposite direction.

6. If this second component be a motion of revolution, the axis must lie in a plane normal to the path of every point of tangency, determined as above. Hence those paths must be parallel ; which requires the axis of the first component rotation to lie in the same plane with the common element.

7. If the second component be a motion of translation, all the points of tangency are in consequence moving in parallel directions. Therefore the axis of the first component rotation must lie in a plane containing the common element, and perpendicular to those parallel motions.

8. Consequently the axis of the first component rotation must in all cases and at all times lie in the same plane with the common element : and the motions of the points of contact, due to this rotation, must at any instant be in parallel directions.

77. 9. Now the effect of the first component rotation must be, to bring the consecutive element of the moving surface into coincidence with the present line of contact on the fixed one. But as all the points of the line of contact are moving in parallel directions, by virtue of

---

* Kinematics of Machinery, 1876, pp. 81, 82.

this rotation, the consecutive position (that is to say the consecutive element of the moving surface) must lie in the same plane with the original one : thus determining a plane tangent all along an element.

10. Again the effect of the second component must be, to bring the present line of contact on the moving surface, into coincidence with the consecutive element of the fixed one. But whether this motion be one of translation or of revolution, it imparts parallel motions to all points of the common element, which consequently will lie in a plane containing the consecutive position, and tangent to both surfaces all along an element.

11. Pure rolling contact, then, is not possible between surfaces touching each other in more than one point, unless they are plane or single curved.

12. Nor is it possible between all such surfaces. No argument is needed to show that a cone cannot roll upon a cylinder, nor upon another cone unless the two have a common vertex. The rolling of one single curved surface upon another involves the simultaneous rolling of both upon the common tangent plane. Hence it is an essential condition, that through any point of the common element, it shall be possible to draw two lines, one upon each surface, of which the developments upon the common tangent plane shall coincide.

78. Mixed Contact.—By this is meant a combination of rolling and sliding contact ; the nature of such action will be readily understood by reference to Fig. 29, if we suppose the circle to rotate about the centre, and at the same time to be translated, but in such manner that the points $c, d, e,$ do not come into coincidence with the points $c'\ d'\ e'$. If, for instance, the circle rotate through an arc AG, and in the same time be translated through a space AB or Ad', greater or less than that arc, it is clear that though the action partakes to some extent of the nature of rolling, yet there must be a sliding between the circumference and the tangent, to an amount depending upon the difference between the arc AG, and the distance AB or Ad'. The resultant of the two motions imparted to the circle will be a rotation about an instantaneous axis, which will however not pass through the point of tangency $A :$ and this affords a means of ascertaining whether in any given case the contact motion is of this description or not.

But we may for all the practical purposes of mechanism, disregard this particular combination ; and we shall, in what follows, take note only of the distinction between those contact motions in which the action is purely *rolling*, and those in which it is not ; all the latter being called cases of *sliding* contact.

CHAPTER III.

79. It has been shown, that even the simplest motion, a rectilinear one, may be resolved into components. But if there be any motion employed in mechanical structures, entitled to be called a simple one by reason of the ease with which it can be produced and continuously maintained, that motion is certainly the one of rotation about a fixed axis.

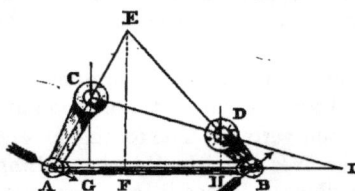

FIG. 33.

Naturally enough, therefore, it is more frequently met with than any other, and the very mention of machinery suggests ideas of levers and wheel-work. Nothing can be simpler than the means by which rotation is determined, nor can the axes have a simpler relation than that of parallelism : and, accordingly, by far the largest and most important class of elementary combinations is that in which both driver and follower rotate about fixed axes which are parallel to each other. In the analysis of their action it will be most convenient to consider first those in which the motion is transmitted by means of a link.

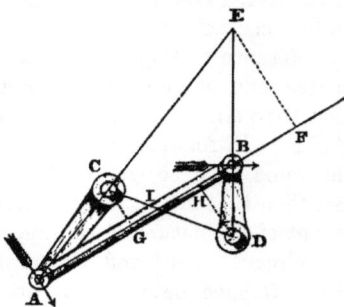

FIG. 34.

Velocity Ratio in Link Motions.

80. In Figs. 33 and 34, let C and D be fixed axes, about which turn the levers AC, BD, whose free ends are connected by the link AB; the axes being perpendicular, and the planes of motion parallel, to the paper.

The motions of A and B are necessarily perpendicular to AC and BD respectively; whence the instantaneous axis, E, is found by prolonging those radii till they intersect.

From C, D, E, let fall the perpendiculars CG, DH, EF, on the line of the link AB or its prolongation. Also draw the *line of centres* CD, cutting the line of the link, prolonged if necessary, in the point I.

Now let

$$v = \text{ang. vel. of } A \quad\quad \text{around } C,$$
$$v' = \text{``} \quad \text{``} \quad \text{``} \; B \quad\quad \text{``} \quad D,$$
$$w = \text{``} \quad \text{``} \quad \text{``} \; A \text{ or } B \quad \text{``} \quad E.$$

Then by similar triangles,

$$\left.\begin{array}{l} \dfrac{v}{w} = \dfrac{AE}{AC} = \dfrac{EF}{CG}, \\[2ex] \dfrac{w}{v'} = \dfrac{BD}{BE} = \dfrac{DH}{EF}. \end{array}\right\} \quad \therefore \dfrac{v}{v'} = \dfrac{DH}{CG} = \dfrac{DI}{CI}.$$

81. This result may be deduced otherwise, thus: In Fig. 35, let the linear velocity of A be represented by AL, perpendicular to AC: then AM is its component in the line of the link, to which BN must be equal. The direction of BO, the motion of B, is perpendicular to BD, and its length is ascertained by drawing NO perpendicular to AB.

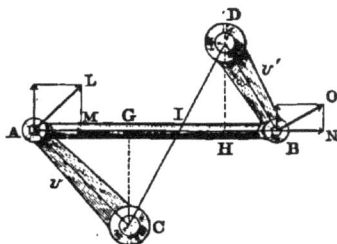

Fig. 35.

Then by similar triangles we have

$$\left.\begin{array}{l} v = \dfrac{AL}{AC} = \dfrac{AM}{CG}, \\[2ex] v' = \dfrac{BO}{BD} = \dfrac{BN}{DH}. \end{array}\right\} \quad \therefore \dfrac{v}{v'} = \dfrac{DH}{CG} = \dfrac{DI}{CI}.$$

We have, then, two simple and convenient values for the velocity ratio, which may be thus expressed:

1. The angular velocities of the arms are to each other inversely as the perpendiculars from their centres of motion upon the line of the link.

2. The angular velocities of the arms are to each other inversely as the segments into which the line of the link cuts the line of centres.

82. Directional Relation.—The perpendiculars *CG* and *DH*, from the centres of motion to the line of the link, are called the *Effective Lever Arms*. When they lie on the same side of *AB*, as in Fig. 33, the arms will turn in the same direction. When they lie on opposite sides of the line of the link, as in Figs. 34 and 35, the rotations will be in opposite directions.

Dead Points.—It is evident that in the combination shown in Fig. 33, the shorter lever *BD* is capable of turning completely round. When a lever thus makes entire revolutions, as is very often the case, it is usually called a *Crank ;* but whether it does or not, it is obviously possible for the system to come into either of the positions shown in Figs. 36 and 37, in which *AB* and *BD* coincide.

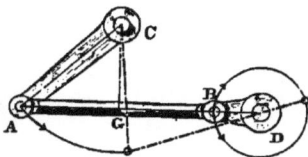

Fig. 36.

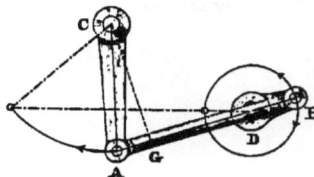

Fig. 37.

When this happens the system is said to be at a *dead point*, because, the effective lever arm *DH* having disappeared, and the motion of *B* having no component in the line of the link, the arm *AC* is momentarily at rest. That arm has a reciprocating circular motion ; and the dead points obviously occur at the extreme limits of its travel. By way of distinction, the crank *BD* is said to be at an *outward* dead point in Fig. 36, and at an *inward* dead point in Fig. 37.

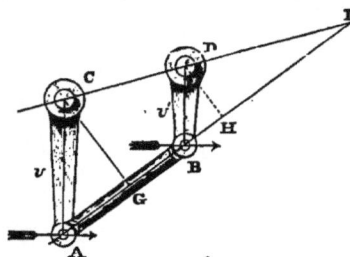

Fig. 38.

83. Momentary Constancy of Velocity Ratio.—In general, the velocity ratio in such combinations is varying, as the ratio between the effective lever arms changes from instant to instant : though if, as in Fig. 13, the levers are equal and parallel, the velocity ratio is constant throughout the action. But it may be *momentarily* constant under other circumstances. Thus in Fig. 38, the arms *AC*, *BD*, are parallel, but not equal. The motions of the points A and B are parallel to each other, and since they are not perpendicular to the line of the link, their linear velocities must be equal, whence we have directly

$$\frac{v}{v'} = \frac{BD}{AC} = \frac{DH}{CG} = \frac{DI}{CI}.$$

Now, the positions of the link consecutive to the present one will be parallel to AB, and cut the line of centres in points consecutive

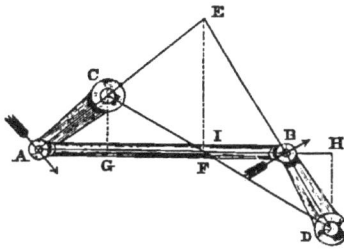
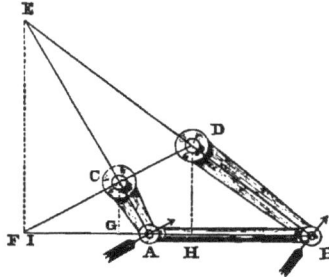

FIG. 39.   FIG. 40.

to and practically coincident with I. Hence during such elementary motion the variation in the velocity ratio will be inappreciable.

84. Again, the line of the link in Figs. 33 and 34 is tangent at F to the circle described by that point about the instantaneous axis E. The consecutive positions of AB therefore intersect each other in F: as may be seen very clearly in Figs. 36 and 37, where F coincides with A, which is momentarily at rest, while BD moves in either direction to its consecutive position.

Now in Figs. 39 and 40, the foot F of the perpendicular upon AB from the instantaneous axis, coincides with I, the intersection of the link with CD the line of centres. In its consecutive positions, therefore, the line of the link will cut the line of centres into practically the same segments as in its present one, and the velocity ratio is momentarily constant in this case also.

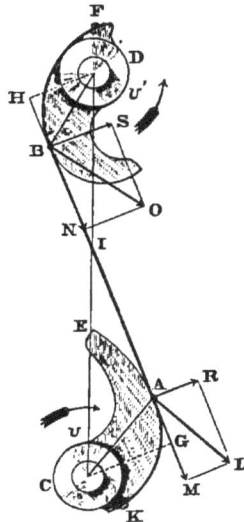

FIG. 41.

Band Motions.

85. Let the two curved pieces whose fixed centres of motion are C and D, Fig. 41, be connected by a flexible and inextensible band or cord, wrapped upon their convex edges, and fastened to them at the points K and F. Let the lower curve turn to the right, as shown by the arrow : the band will then be wound upon AE, and unwound from BF, thus compelling the upper curve to rotate. It is evident that the parts which are wrapped upon the curves being always idle, the effective length of the band is the distance between the points of tangency; which may or may not change during the action.

In the position shown, the points of tangency are A and B, whose motions at the instant must be perpendicular to the contact radii, AC, BD. Their linear velocities, AL, BO, must be such that the components AM, BN, along the line AB, shall be equal, because the band is inextensible ; the components AR, BS, perpendicular to AB, are ineffective, because the band is flexible. Now drawing the perpendiculars CG, DH, from the centres of motion to the line of the band, and also the line of centres CD cutting AB in I, it is evident on comparison that this combination at the instant differs from that of Fig. 35 in no particular except the flexibility of the connector AB. And we shall have, precisely as in that case,

$$\frac{v}{v'} = \frac{DH}{CG} = \frac{DI}{CI} .$$

And also as in that case, the rotations will be in the same direction if CG and DH lie on the same side of AB, and in contrary directions if they lie on opposite sides of that line.

Contact Motions.

86. In Fig. 42, let C and D be the fixed centres of motion of the two curved pieces in contact at P, their common tangent being TT, and their common normal NN: it is evident that if the upper curve turn as shown by the arrow, the lower one will be compelled to rotate. Draw the radii of contact, PC and PD; then the motion of P, considered as a point of the upper piece, will be in the direction PA, perpendicular to PC, while the direction of the motion of P on the lower curve must be PB, perpendicular to PD.

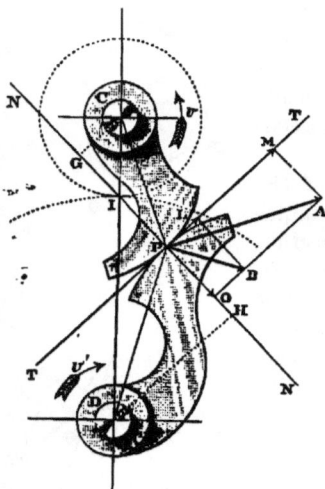

FIG. 42.

Resolve PA into the components PM, PO; of these the former obviously does not tend to move the lower curve, and it is also evident that the magnitude of PB must be such that its normal component shall be PO; for were it greater than PO, the lower curve would quit contact with the upper, and if it were less the two curves would intersect, both of which are contrary to the hypothesis. Draw CG, DH, perpendicular to NN; then the triangles PCG, APO, are similar, as also are PDH, BPO. Draw CD cutting NN at I; then ICG, IDH are similar; and letting

$$v = \text{ang. vel. around } C;$$
$$v' = \text{``} \quad \text{``} \quad \text{``} \quad D.$$

we have

$$\left. \begin{array}{l} v = \dfrac{PA}{PC} = \dfrac{PO}{CG} \\[2ex] v' = \dfrac{PB}{PD} = \dfrac{PO}{DH} \end{array} \right\} \therefore \frac{v}{v'} = \frac{DH}{CG} = \frac{DI}{CI}.$$

Which may be thus expressed:

1. The angular velocities of two pieces in contact are to each other inversely as the perpendiculars let fall upon the common normal from the centres of motion.

2. These angular velocities are to each other inversely as the segments into which the common normal cuts the line of centres.

87. Condition of Constant Velocity Ratio.—This is at once deduced from the second of the above values, viz;

$$\frac{v}{v'} = \frac{DI}{CI}:$$

for $DI + CI = CD$, which is constant, therefore I must be fixed, and the two segments at all times the same.

In other words, the curves, in order to maintain a constant velocity ratio, must be such that *their common normal shall always cut the line of centres in the same point.*

88. Condition of Compulsory Rotation.—A rigid link can both push and pull, so that the direction of the driver's motion is arbitrary. A band is capable of pulling only: while on the other hand one piece can drive another in direct contact with it only by pushing. In Fig. 43, the lower piece is the driver, turning as shown by the arrow. Let x, z, be points consecutive to and on opposite sides of P, the former being in advance. Then CP is greater than Cx, but less than Cz; and it is clear that in order to produce compulsory rotation, the

2. { curves and their motions must be such *that the contact radius of the driver shall be on the increase.*

89. Rate of Sliding.—The tangential components PM, PL, represent the linear velocities with which the curves are at the instant sliding on the common tangent. In Figs. 42 and 43, these components lie in the same di-

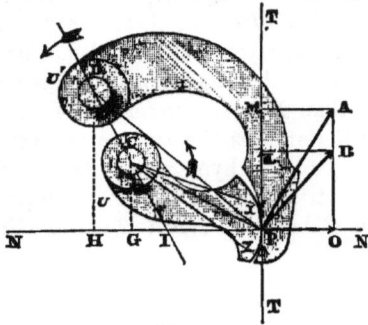

Fig. 43.

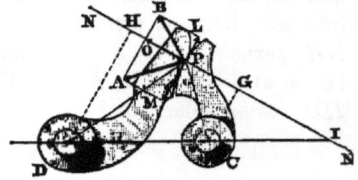

Fig. 44.

rection from P, and consequently their difference LM represents the rate at which the two curves are sliding upon each other. In Figs. 44 and 45 these components fall in opposite directions, and the rate of sliding is represented by their sum.

Directional Relation.—From these various diagrams, which are lettered to correspond throughout, we also perceive that the directional relation depends upon the positions of the perpendiculars CG, DH, with reference to the common normal; if they lie upon the same side of NN, the rotations will be in the same direction; if on opposite sides, the rotations will be in contrary directions.

90. Condition of Rolling Contact.—In order that there may be no

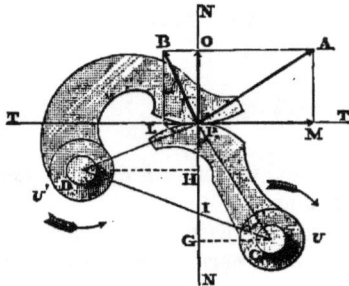

Fig. 45.

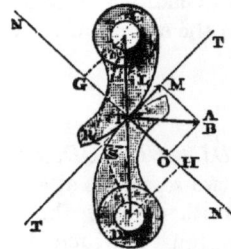

Fig. 46.

sliding between the two moving pieces, the tangential components PM, PL, must have the same magnitude and the same direction, as in Fig. 46. The normal components PO being the same in both motions, the resultants PA and PB must also be alike in magnitude and direction ; and these being perpendicular to the contact radii PC, PD, the latter must also coincide in one right line, which can be no other than CD.

That is to say : the condition of rolling contact is, *that the point of contact shall lie upon the line of centres.*

The curves may be of such a nature that this condition is continuously satisfied, the point of contact traveling along the line of centres, and the velocity ratio varying accordingly.

On the other hand, that point may travel across the line of centres, the action taking place partly on one side and partly on the other ; in which case the velocity ratio may or may not vary ; but whether it does or not, there will be more or less of sliding between the curves, except at the instant when the point of contact crosses that line.

91. Similarity of Action in all the Modes of Transmission.—It will be observed that the common normal in contact motions has a very striking resemblance to the lines of the link and the band, previously discussed. In fact, we may select any two points in this normal, and, drawing right lines from them to the centres of motion, form thus a pair of arms and a link, which combination will at the instant have precisely the same action as the original contact pieces.

We perceive, in short, that the motion of the driver is transmitted to the follower in a right line, whether they be in contact or not ; and that the action of all elementary combinations in which the two pieces rotate about fixed parallel axes, is governed by these laws :

I. The angular velocities are to each other inversely as the perpendiculars from the centres of motion upon the **Line of Action**; or, inversely as the segments into which the line of action divides the line of centres.

II. The rotations have the same direction if the centres of motion lie on the same side of the line of action ; and contrary directions if they lie on opposite sides.

III. The rate of sliding, in contact motions, is the difference of those components of the actual motions of the common point about the two centres respectively, which are perpendicular to the line of action, when they lie on the same side of that line ; and their sum if they lie on opposite sides.

92. We may now proceed either to examine separately the action of given combinations, or to ascertain, by the aid of the principles thus far explained, the forms and proportions which may or must be given to the elements in order to satisfy assigned conditions. The latter is plainly the more systematic and fruitful course ; and wheels whose axes are parallel, enter so largely into the composition of machinery of all descriptions, that the fact is a sufficient reason for at present confining our attention to the transmission of motion by direct contact; considering first those cases in which the action is purely rolling, and taking up afterward those in which the contact is of the mixed variety.

CHAPTER IV.

1. *Velocity Ratio Constant.*

93. If the teeth of a pair of ordinary circular spur wheels be indefinitely reduced in size and increased in number, they will ultimately become mere lines, or rectilinear elements of cylinders of revolution, as shown in Fig. 47.

These are technically called the *Pitch Surfaces* of the wheels. Evidently they are *capable* of moving in perfect rolling contact about their axes, the linear velocities of

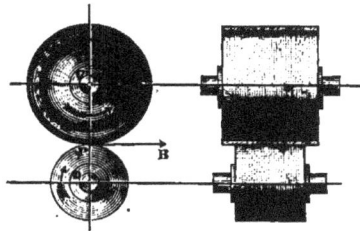

Fig. 47.

their circumferences being the same, and the angular velocities inversely proportional to the radii, or to the original numbers of the teeth. Which agrees with the deductions of the preceding chapter ; the condition of rolling contact is (**90**), that the point of tangency shall always be upon the line of centres : and the condition of a constant velocity ratio is, (**87**), that the common normal, which passes through that point, shall always divide the line of centres into the same segments. Hence in order to satisfy both conditions at once, the contact radii must be constant ; or in other words the contact curves must be circles whose centres are in the axes.

94. But it is important to note that although circles are the only curves which can move in rolling contact about fixed axes, yet one has no tendency to *drive* the other. The condition of compulsory rotation is, (88), that the contact radius of the driver shall be on the increase; which is here impossible.

And again, in Fig. 47, let the upper circle turn as shown by the arrow, the motion of the contact point being represented by *PB*; then this motion, being tangent to both circles, has no normal component at all. Consequently the driver tends merely to slip upon the follower, not to cause the latter to move. And it is a well-known fact, that two polished rollers cannot be used to transmit rotation with a constant velocity ratio: the more nearly they approach theoretical perfection as cylinders, the farther they depart from practical perfection as wheels, and if motion be transmitted at all, it is by reason of adhesion between them.

95. Friction Gearing.—Nevertheless such toothless wheels are quite extensively used in practice, constituting what is known as *Friction Gearing*. The axes are pressed together with considerable force, and in order to secure the greater adhesion, the peripheries in contact are usually made of different materials: thus if one wheel be of iron, the other may be made of wood, or have its rim covered with leather or india-rubber. There is, of course, always a possibility that such wheels may slip upon each other; in point of fact they usually do, and the amount of slip, moreover, varies under different conditions of pressure and speed. There is, then, no certainty that a constant velocity ratio will be preserved, and they cannot be employed where that is imperative; but there are many cases in which it is not.

96. Grooved Friction Gearing.—Another arrangement of friction gearing, very generally used in hoisting machines, especially for mining purposes, is shown in Fig. 48; the two wheels having a series of angular grooves turned in their peripheries, so that by pressing them together an effective degree of adhesion is secured: both wheels are usually made of cast iron. The action is evidently attended with considerable sliding in any case, and the amount may vary between wide limits: but for such work a constant velocity ratio is not requisite, and the possibility of slipping is a practical advantage, as it reduces the shock if the machine be suddenly started or stopped, thus diminishing the risk of breakage. The velocity ratio when in full work is substantially the same as that between two

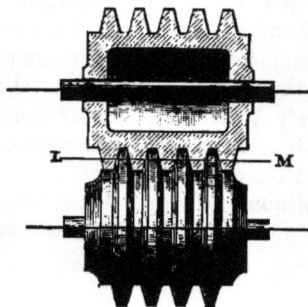

FIG. 48.

pitch cylinders whose line of tangency is midway between the tops and bottoms of the grooves, as LM in the figure.

2. *Velocity Ratio Varying.*

97. If the contact curves are not required to maintain a constant velocity ratio, then, since the point of tangency must still be always upon the line of centres, the contact radii will vary. But their sum, evidently, remains constant, if the point of contact lies between the centres of motion ; or if those centres lie on the same side of that point, the difference of the contact radii must be constant.

And by reference to Fig. 46, it will be seen that if PR, PS, be two equal arcs of the curves, the latter must be of such form that

$$CR + DS = CP + DP = CD.$$

We proceed now to show how some curves may be constructed, which will satisfy the above condition, and the manner in which various changes in the velocity ratio may be effected by means of them.

98. The Logarithmic Spiral.—In Fig. 49, let the parallels YU, TW, be cut at any angle by TB, and perpendicularly by the other parallels CD, IM, BN, etc. About C as a centre, describe an arc with radius $CE = AM$; and about P as centre, describe an arc with radius PA, cutting the first in E, thus constructing the triangle CPE. With centres C and E, and radii CG, EG, respectively equal to BN and AB, describe arcs intersecting in G. Continuing in a similar manner, the result is the formation of the broken line RPV.

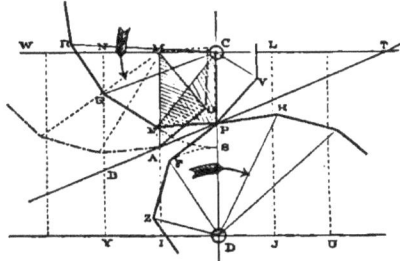

Fig. 49.

Now let us imagine C to turn first about P until PE coincides with PA, and then about A until EC coincides with AM; then RPV will have the position shown in dotted lines, the triangle EPC appearing as AOM. After which in a similar manner G may be brought to coincide with B, the point C meantime advancing to N, and so on.

99. The polygon RPV thus travels with a sort of imperfect rolling action, measuring off its perimeter on TB, since PE, EG, are by construction equal to PA, AB, respectively ; also $PV = PH$, etc.

Now if the subdivisions on TB be made indefinitely minute, so that A and H are points consecutive to P, then RPV will become a curve to which *TB* is tangent at *P*. This curve will not, however, pass through the angles of the polygon here shown; since the *arc EP*, for instance, must be equal to *AP*, the angle *ECP* will in the curve be less than it now is.

By reason of the parallels *LH*, *CD*, etc., the angles *LHT*, *CPH*, *MAP*, etc., are equal; hence the curve must be such that its tangent at any point shall make a constant angle with the radiant from *C* to that point. It is apparent from the mode of derivation that the radiants from *C* will increase as we go from *CP* to the left: and these conditions determine the curve to be the logarithmic or equiangular spiral, of which *C* is the pole and *TB* the tangent at *P*.

Supposing this curve to replace *RPV*, the rolling action will be perfect, *C* traveling in a right line to *M* and *N*, while *E* and *G* go to *A* and *B*.

100. The same process being repeated below *TB*, with *D* as the pole, we shall have a polygon of the same description as *RPV*, but in a reversed position. We have in this second one, *PF* equal to *PA*, and *DF* equal to *IA*. In the upper one, *PE = PA*, and *CE = AM*; hence if the two rotate in opposite directions about *C* and *D* as fixed centres, *E* and *F* will come together at *S* on the line *CD*. So in like manner will *G* and *Z*, or *V* and *H*, meet upon *CD*.

It is apparent that these two broken lines will at the limit become parts of equal, similar, and opposite equiangular spirals. The curves cannot be constructed in this manner, except approximately: but this mode of derivation exhibits in the clearest light the fact that they will move in perfect rolling contact about fixed axes passing through their poles. Referring the reader to the Appendix for detailed instructions relating to the accurate delineation of these and other curves, we proceed to consider the conditions which may be assigned in some ap-

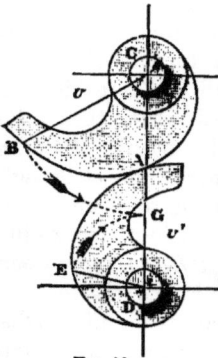

FIG. 50.

plications of this spiral in mechanism.

101. Since the curve does not return into itself, continuous rotation cannot be transmitted from one shaft to another by means of a single pair of the spiral arcs. But for reciprocating circular motions they are well adapted; of which two cases are illustrated in Figs. 50 and 51.

In the former, let the centres *C* and *D* be given; and suppose it to be required, that while the upper piece rotates through a given angle, the velocity ratio shall vary between two assigned limits.

Divide *CD* at *A* into two segments, whose ratio is one limit, and at *G* into two others whose ratio equals the other limit. Set off the given angle *ACB*, and make *CB* equal to *CG*. We have then two given radiants *CA*, *CB*, and the pole *C*, to construct the spiral *AB*. The other sector is a portion of the same curve ; of which one radiant is *AD*, and the other *ED*, is equal to *GD* : whence the angle *EDA* is determined by simply cutting the spiral by arcs described about the pole *D*, with those given radii.

Letting v = ang. vel. of Upper Sector,

v' = " " Lower "

we have in the position shown,

$$\frac{v}{v'} = \frac{AD}{AC};$$

and when *B* and *E* meet at *G*,

$$\frac{v}{v'} = \frac{GD}{GC}.$$

102. It may be required that the two shafts shall turn through equal angles. In that case one limit may be assigned to the velocity ratio, but the other will of necessity be the reciprocal of the first.

For, as shown in Fig. 51, we must have *CB* = *AD* ; also the angles *BCA*, *EDG*, are to be equal, whence, from the nature of the curve, *ED* = *AC*, and the velocity ratio will vary between the limits,

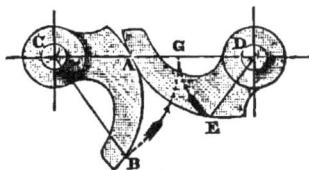

Fig. 51.

$$\frac{v}{v'} = \frac{AD}{AC}, \quad \frac{v}{v'} = \frac{GD}{GC} = \frac{AC}{AD};$$

of which the one is determined when the other is assigned.

103. Lobed Wheels.—In the transmission of continuous rotation, it may be required that a certain number of determinate changes in the velocity ratio shall take place during the revolution of one of the shafts. If the same number of changes is to occur during the revolution of each shaft, the equiangular spiral may be employed as follows :

In Fig. 52, let *C*, *D* be the given centres, and let the angles *ACB*, *ADE*, each be equal to 60°. Make *DE* = *AC*, and *CB* = *AD* : construct the equal spiral arcs *AE*, *AB* ; draw *AH*, *BF*, similar curves

4

symmetrically placed with reference to AD and BC. The velocity ratio in the position shown is

$$\frac{v}{v'} = \frac{AD}{AC};$$ when B and E meet at G, the ratio will be

$$\frac{v}{v'} = \frac{AC}{AD},$$ as in Fig. 51.

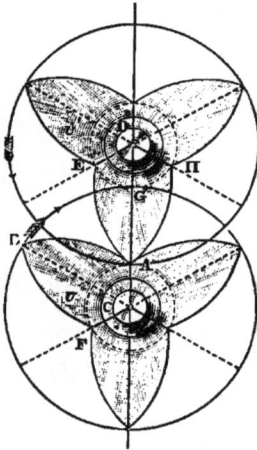

Fig. 52.

The angles EDH, ACF, each including one-third of a circumference, the pair of trilobes may be completed as shown. They will roll together, the velocity ratio varying from maximum to minimum and back again, three times in each revolution of each shaft.

In this construction the limits are necessarily reciprocals of each other, so that only one of them is arbitrarily assignable.

The angles ACB, ADE, may include any aliquot part of the circle whose denominator is even; and in this way pairs of wheels may be made with any desired number of lobes, which will roll together. In Fig. 53 those angles are each equal to 180°, the result being a pair of unilobes; in Fig. 54, they are made equal to 90°, for constructing bilobed wheels.

It is to be observed that the wheels laid out as above will work to-

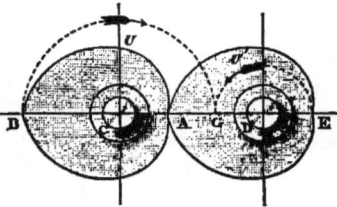

Fig. 53.

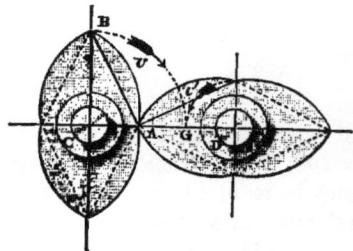

Fig. 54.

gether only in equal and similar pairs, trilobe with trilobe, one unilobe with another one, and so on.

104. The Rolling Ellipses.—Two equal and similar ellipses, of whatever eccentricity, will also work in per-
fect rolling contact, each revolving
about one of its foci as a fixed centre ;
the distance between the centres being
equal to the major axis.

In Fig. 55, let C and D be the fixed
centres, A and B the free foci, and E
a point of contact. It is to be proved
that E will always lie upon the line of
centres CD.

From E draw lines to both foci of
each ellipse. Then since the common
tangent FG makes equal angles with
all these lines, we have $AEG = BEF$,
and $DEG = CEF$, therefore AEB,
CED, are both right lines.

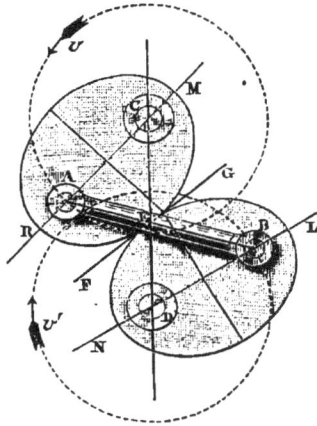

Also, $EC + EA = EC + ED = EB$
$+ ED$. Consequently $AB = CD$, and $EC = EB$.

Fig. 55.

The latter equality shows that the elliptical arcs EL, EM, measured from any point of tangency to those extremities of the major
axes which will come together, are equal. The rolling action is therefore perfect.

105. From the fact that AB is constant, we are enabled, when the
rolling ellipses are fixed on the ends of shafts which overhang their
bearings, to connect the free foci by means of a link, as shown in the
figure, which is sometimes advantageous in practice, as will subsequently appear, even when the ellipses are provided with teeth. These
free foci, obviously, describe circles of which the radius is AC.

106. During each revolution of either ellipse, the velocity ratio varies once from maximum to minimum and back again. The limits
are, evidently, mutually reciprocal, and determined by the ratio of
the segments into which either focus cuts the major axis. Hence if
one limit be assigned, and the distance between the axes given, it is
only necessary to divide that distance into two parts whose ratio is the
assigned limit : we have then the major axis and the foci, whence the
ellipses are determined.

107. Transformation of Rolling Curves—Contraction of Angles.—
If any two curves are capable of rotating in rolling contact about
fixed axes, producing a given variation in the velocity ratio during
definite angular movements, they may be transformed into two others,
also rolling in contact about the same axes, and producing the same

variation in the velocity ratio, during angular movements which are greater or less than the original ones in any assigned ratio. This is effected by a process called the *Contraction of Angles*, since usually the angular movement of the derived curves is less than that of the original ones during the same variation in the velocity ratio.

In Fig. 56, let A be the point of contact of the rolling curves AH, AI, which turn with given angular velocities about C and D: then at the end of a definite time the points I and II will meet at G. In Fig. 57, let the lengths of the radiants be the same as in Fig. 56, but the angles included between them only half as great. Let these new curves turn with the same angular velocity as

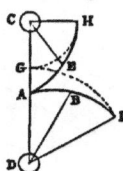

FIG. 56.   FIG. 57.

the original ones : then in half the time, H and I will also meet at G. All the angles being reduced in the same proportion, while the radiants remain the same, it follows that intermediate points, as E, B, which meet on CD in the first pair, will do the same in the second ; which, therefore, will move in pure rolling contact.

108. Lobed Wheels Derived from the Ellipse.—By applying the above process to the rolling ellipses, pairs of wheels may be constructed which will roll together, the velocity ratio varying between the same limits and according to the same law as in the ellipses themselves, but the maxima and minima recurring two, three, or any desired number of times in each revolution.

The operation is illustrated in Fig. 58, in which C is a focus of the ellipse whose major axis is AB. For convenience of construction, a circle is described around C, of which the upper half is divided into any number of parts at I, II, III, etc., and the lower left-hand quadrant is divided in a similar manner, at 1, 2, 3, etc. Radii are drawn to the points of division, the upper ones cutting the ellipse at D, E, F, etc.: and the segments intercepted between

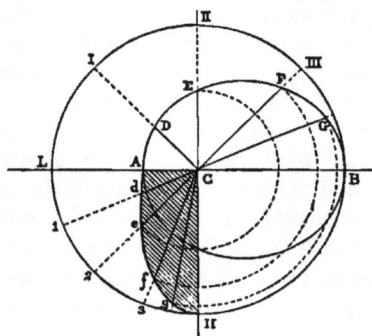

FIG. 58.

the focus and the curve are set off from C on the corresponding radii in the quadrant below, as $Cd = CD$, on $C1$, $Ce = CE$, on $C2$, and so on, CB going to CH.

109. Regarding the ellipse as a unilobe, is is evident that the curve AeH, thus determined, is one fourth of the outline of a bilobed wheel,

which when completed will roll in contact with an equal and similar one, as shown in Fig. 59. By contracting the angles to one third instead of to one half, as shown in Fig. 60, the contour of a trilobe is determined, the complete wheel with its rolling mate being shown in Fig. 61; and in like manner wheels with any number of lobes may be constructed.

110. But these wheels, like those previously constructed from the equiangular spiral, will work only in equal and similar pairs, bilobe with bilobe, trilobe with trilobe, and so on : consequently they can only be used when the number of changes in the velocity ratio is to be the same during each revolution of each shaft. Now, the above process may be so modified that from a pair of rolling ellipses a dissimilar pair of wheels may be derived, as for instance, a bilobe rolling with a trilobe, or

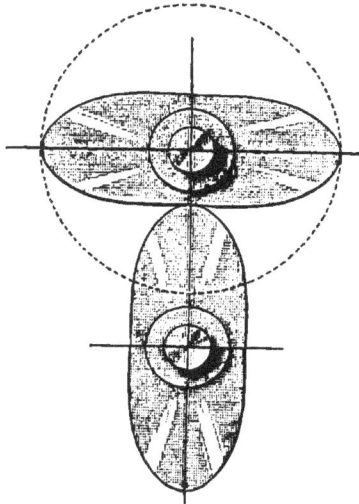

Fig. 59.

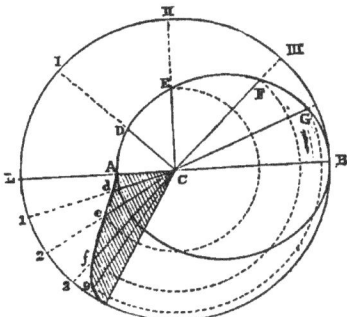

Fig. 60.

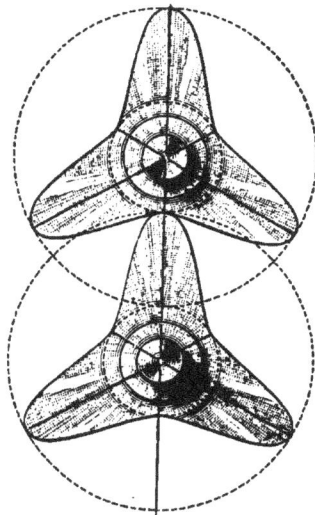

Fig. 61.

a unilobe with a wheel of two or more lobes ; so that the numbers of changes in the velocity ratio will be different in the two revolutions.

The eccentricity of the original ellipses, however, is not entirely arbitrary, but will vary between limits determined by the conditions of each particular combination.

The modification just mentioned consists merely in this, that a part only, instead of the whole, of the elliptical outline, is transformed into a new curve. It will be readily apprehended that equal arcs of two rolling ellipses may subtend at the fixed foci unequal angles, the ratio of whose magnitudes shall be that of two integers : and that when this is the case, those arcs may be transformed as above into portions of the contours of lobed wheels.

111. For example, in Fig. 62, let C be the fixed and A the free focus of the ellipse whose major axis is HP ; let D be the fixed focus of its rolling mate, and B and E, two points which will meet at G, the elliptical arcs PB, PE, being equal. Now if the angles BCP, EDP, are to each other as 3 to 2, then by contracting them in the same proportion, the former may be reduced to 90° and the latter to 60°. Thus the elliptical arc PB will be transformed into PI, one fourth of the outline of a bilobe, and PE into PM, one sixth of the contour of a trilobe, and these new curves will roll together.

Let O be the free focus of the second ellipse, then, drawing OE and AB, it is evident that the angles EDP, BAC, are equal.

If, therefore, on the indefinite right line HD, we construct any two angles BAC, BCD, whose magnitudes have the ratio of 2 to 3, and produce the lines which limit them, till they intersect at B; then from the ellipse whose foci are A and C, and major axis is equal to $AB + BC$, a bilobe and a trilobe may be derived, which will roll together.

The velocity ratio in the position shown, is

$$\frac{v}{v'} = \frac{PD}{PC} = \frac{CH}{PC} :$$

Fɪɢ. 62.

and when B and E meet at G, it will be

$$\frac{v}{v'} = \frac{MD}{IC} = \frac{ED}{BC} = \frac{AB}{BC}.$$

112. In Fig. 62, the angles BCP, BAC, are both obtuse; but the construction holds true if they be made acute, as in Fig. 63. But whereas they were contracted in the first case, they have in the second case to be expanded in order to produce the required contours of the lobed wheels;

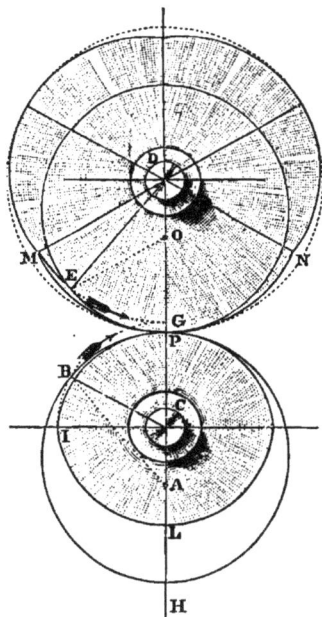

Fig. 63.

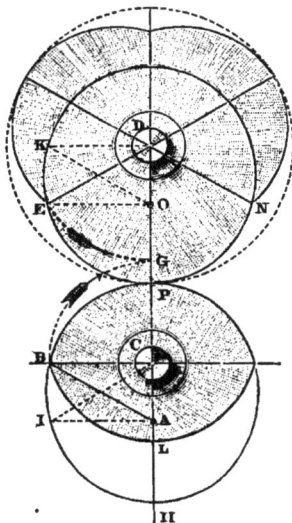

Fig. 64.

a circumstance which points directly to the conclusion that there must be one ellipse so proportioned that it is not necessary to do either. This is shown in Fig. 64, where the angle BCP being 90°, and EDP, or its equal BAC, being 60°, the equal elliptical arcs PB and PE form at once portions of the outlines of the bilobe and the trilobe respectively.

From these same ellipses another pair of dissimilar multilobes can be constructed. Erecting the perpendiculars AI, DK, and drawing CI, OK, the elliptical arcs PBI, PEK, are equal. But the angles ICP, KDO, measuring 120° and 90° respectively, are to each other in the ratio of 4 to 3: therefore, contracting them to one half, as in

Fig. 65, we have portions of the contours of a trilobe and a quad-rilobe which will work in rolling contact.

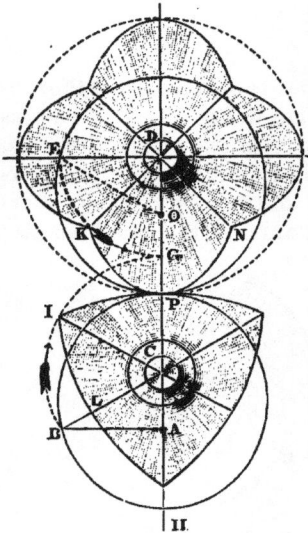

Fig. 65.

113. In general, then, we may proceed as follows: If it be required to construct a wheel with m lobes, to roll with another having n lobes, supposing m to be the greater number: Assuming any two points A, C, on the indefinite right line HD, Fig. 66, construct the angles BCD, BAD, with magnitudes whose ratio is that of m to n, the lines AB, CB, intersecting at B. Then the sum of those two lines will be the major axis, and the points A and C the foci, of an ellipse from which the wheels can be constructed as above explained.

Other pairs of angles having the same ratio may be constructed, determining the points I, G, etc.: these points will lie in a curve, AID, and it is evident that if from any point of this curve, lines be drawn to A and C, they will form with HD, angles in the required ratio, and their sum may be taken as the major axis of the proposed ellipse: as for instance, $HP = EA + EC$, if E be the point selected.

114. The curve itself may be constructed either by determining a sufficient number of points as above described, or by aid of the following property. The centre of the circle circumscribing any one of the triangles whose base is AC, and vertex in the curve, as for example ABC, must lie on the

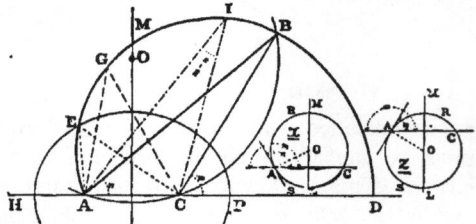

Fig. 66.

line LM perpendicular to and bisecting AC.

Now let the angle $BCD = m$,

  then " " $BAC = n$,

  and " " $ABC = m - n$;

          and in the circumscribing circle we shall have

$$m - n \text{ measured by } \tfrac{1}{2} \text{ arc } AC,$$
$$n \qquad `` \qquad `` \tfrac{1}{2} `` \ CB,$$
$$m \qquad `` \qquad `` \tfrac{1}{2} `` \ ACB:$$

that is to say,

$$\text{arc } CB : \text{arc } AC :: n : m - n.$$

If then we had taken the centre O at pleasure on LM, and, describing the circle through A and C, had set off the arc CB in the above ratio to the arc AC, the point B would have been thus determined.

115. Now, this curve springs from the point A; for when m becomes 180°, as shown in the two small diagrams marked x and z, AC will divide the circle whose centre is O, into segments such that CRA, above AC, is to ASC below that line, as n is to $m - n$: and the curve will be tangent at A to that circle.

The limit D of the curve will be reached when m becomes zero : that is to say when the radius of the circle ACB becomes infinite, in which case the circle itself will coincide with the right line ACD. We shall then have

$$CD = AC \left(\frac{n}{m - n} \right) .$$

But

$$AD + CD = 2\, CD + AC$$

$$= 2\, AC \left(\frac{n}{m - n} \right) + AC$$

$$= AC \left(\frac{2n}{m - n} + 1 \right)$$

or,

$$AD + CD = AC \left(\frac{m + n}{m - n} \right) :$$

whence

$$\frac{AC}{AD + CD} = \frac{m - n}{m + n} .$$

We have, therefore, this limit to the eccentricity of the ellipses from which dissimilar wheels with assigned numbers of lobes can be derived, viz : that the distance between the foci, divided by the major axis, must be greater than the difference of the numbers divided by their sum.

116. Each wheel derived from the ellipse has thus far been supposed to consist of two or more lobes. But the construction last discussed will hold good when n is to m in the ratio of unity to any integer ; therefore by expanding the elliptical arc subtending the angle n, until the latter equals 180°, it will be seen that a unilobe may be formed, which will roll with a wheel having any given number of lobes. This is sufficiently illustrated by Figs. 67 and 68, which represent respectively a unilobe rolling with a bilobe and one rolling with a trilobe.

In the former case, it will be observed that since the angles ABC,

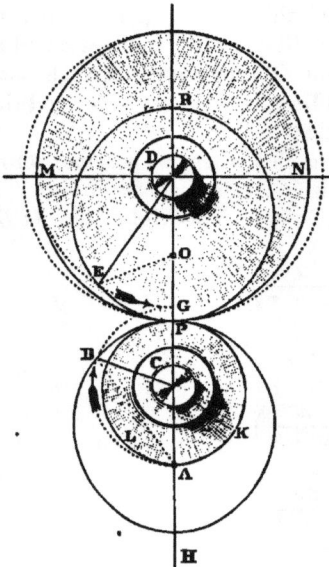

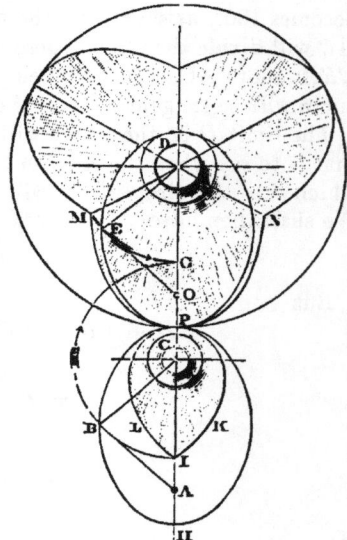

FIG. 67.   FIG. 68.

BAC, are each equal to $\frac{1}{2} BCP$, the curve, AID of Fig. 66, in Fig. 67 becomes the semicircle ABG, of which the centre is C and radius AC, the distance between the foci of the primitive ellipse. The elliptical arc PB, becomes by expansion the curve PLA, forming half the outline of the unilobe ; while the equal arc PE is expanded into PM, bounding one quadrant of the bilobe.

117. Interchangeable Multilobes.—It appears, then, that from a given ellipse there can be constructed pairs of similar wheels having any number of lobes ; and if the eccentricity be great enough, several different pairs of dissimilar wheels may be derived from the same ellipse. But these wheels will work only in pairs as constructed, and are not interchangeable ; thus, the bilobe which rolls with a trilobe, is not the same as the one which works with a unilobe, though derived from the same ellipse, and if a unilobe and a trilobe be derived from the same primitive ellipse, they will be different from either of the others.

There are, however, two methods by which a series of lobed wheels may be constructed, which are interchangeable, any one working in perfect rolling contact with any other one of the series.

118. In the first of these methods, which was discovered by the Rev. Hamnett Holditch, the multilobes are derived, each from one of a series of different ellipses. Their rolling properties do not appear to admit of geometrical demonstration, but the process of constructing them is simple and practical.

A series of ellipses is first made, having the same foci, but with minor axes which are to each other as 1, 2, 3, 4, etc. Then the primitive ellipse, whose minor axis is 1, will roll with a bilobe constructed as in Fig. 58 from the ellipse whose minor axis is 2, with a trilobe derived as in Fig. 60 from the one whose minor axis is 3, etc., and these with each other indifferently.

119. In Fig. 69, let O be the centre, A and C the foci, and HP the major axis of the primitive ellipse. Draw CG perpendicular to HP, and cut it at B by an arc with centre O and radius OH, then setting off BD, DE, etc., each equal to CB the semi-minor axis, describe arcs about O through D, E, F, etc., thus determining OK,

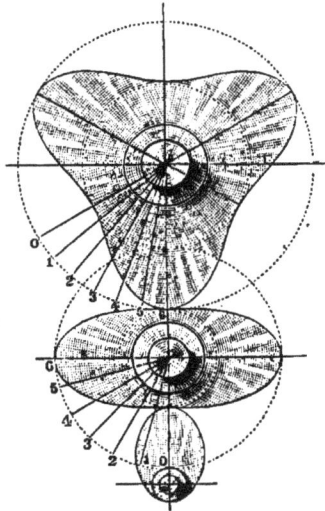

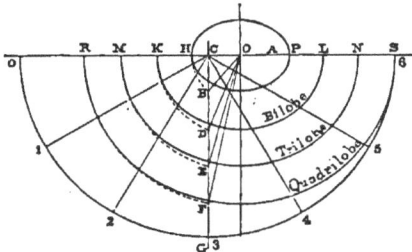

Fig. 69.

Fig. 70.

OM, OR, the semi-major axes of the ellipses from which the bilobe, trilobe, and quadrilobe are to be derived. In Fig. 70, the first three of a series are shown in action, the unilobe, or primitive ellipse, being the same as in Fig. 69. Since the difference between the greatest and the least radii is the same in all the wheels, being equal to the distance between the foci, it will be seen that if one multilobe be given, the original ellipse may be found by reversing the process, and wheels with any desired numbers of lobes constructed, which will roll with the one given.

120. The other method of constructing a set of lobed wheels capable of rolling with each other indifferently, was devised by the author, and the fundamental curve is the equiangular spiral. The process is illustrated in Fig. 71, where C is the pole of the spiral SMN. Drawing through C any right line, cutting the curve at A and B, the portion ASB is one half the outline of a unilobe which will roll with a similar and equal one, as already shown. In order to construct a bilobe which shall roll with this unilobe, it is necessary to find two radiants at right angles to each other, whose difference is the same as that between CB and CA. Produce BA, and intersect its prolongation at G by a circle through B, with centre at the pole; and draw CF perpendicular to BA, cutting this circle

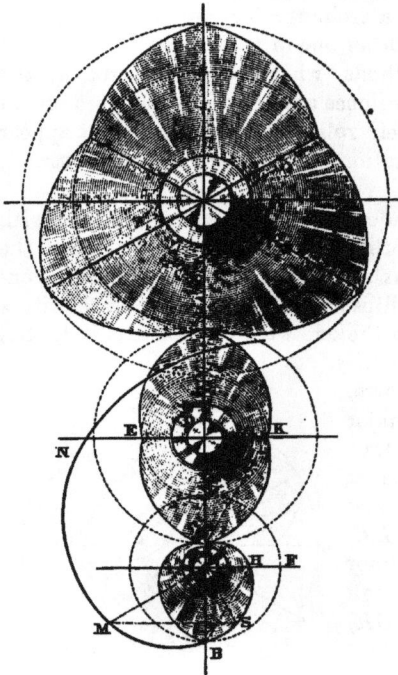

FIG. 71.

at F and the spiral at H.

Then

$$CB - CA = AG,$$
and
$$CB - CH = HF.$$

If, now, two radiants be found, which are to CB and CH respectively, in the same proportion that AG bears to HF, they will be perpendicular to each other, and their difference will be equal to AG.

121. This may be done graphically, as it involves merely the construction of a fourth proportional to three given lines, thus,

$$AG : HF :: CB : CM,$$
$$AG : HF :: CH : CS.$$

The arc SBM of the spiral thus found is one quadrant of the contour of the required bilobe; the triangle ADK is therefore made similar and equal to MCS, and the completion of the wheel requires no explanation.

For the trilobe, draw any two radiants including an angle of 60°, ascertain their difference, and, comparing it with AG, proceed as above to increase or diminish the assumed radiants, as the case may be, in the same proportion, until the difference is equal to AG; the lengths thus determined will be those of the greatest and least radii of the trilobe; and in a similar manner wheels of any number of lobes may be constructed.

122. Since in any system of interchangeable multilobes, the difference between the greatest and least radii must be constant, while their actual lengths vary with the number of lobes, the limits of the variation in the velocity ratio will not be the same for different pairs; and obviously the law which governs the rates of variation as well as the differences between those limits, depends upon the peculiarities of the fundamental curve—upon the eccentricity of the primitive ellipse or the obliquity of the original spiral.

123. Irregular Lobed Wheels.—The non-circular wheels thus far described, whether consisting of one lobe or many, are symmetrical; and moreover, the whole contour of each is made up of curves of the same kind. Neither of these things is so of necessity; and as illustrating the possibilities of varied motion in rolling contact, we give a few examples of irregular wheels composed of different curves.

In Fig. 72, let C and D be the poles of the equal and similar logarithmic spiral arcs ALB, AME; then these arcs will roll together, B and E meeting at G. Now, AB may be taken as the major axis, and C as one focus, of the semi-ellipse ANB, which will roll with the equal and similar one AOE.

In each revolution, then, the velocity ratio will vary between the reciprocal limits

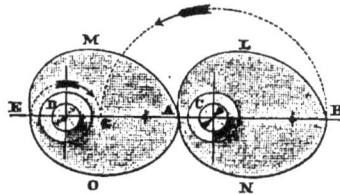

FIG. 72.

$$\frac{v}{v'} = \frac{CB}{CA}, \quad \frac{v}{v'} = \frac{CA}{CD}:$$

but by reason of the different natures of the curves, the rates of variation during the decrease, will not be the converse of those during the increase.

124. In Fig. 73, let C be the pole of the logarithmic spiral arc AFB. With centre C, and radius equal to $\frac{1}{2}AB$, describe an arc cut-

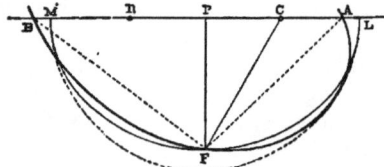

Fig. 73.

ting the spiral at F; draw FA, FB, FC, also FP perpendicular to AB, and on BA produced set off PL, PM, each equal to FC. Then because $CF - CA = CB - CF$, the spiral arcs AF, FB are equal. Consequently the chord FB is greater than the chord FA, whence PB is greater than PA. That is to say, P is not the middle point of AB; but it is the middle point of ML by construction, therefore PL is greater than PA, and PM is less than PB.

The spiral arc AFB, turning round C as a fixed centre, will roll with an equal and similar one, the distance between their centres of motion being equal to AB. Set off $PR = PC$; then R and C are the foci of the ellipse whose major axis is ML, and semi-minor axis PF: which also will roll with an equal and similar one, turning round the same centre C, the distance between centres being equal to ML, which by construction is equal to AB.

125. From these data we can construct, as in Fig. 74, two dissimilar unilobes, whose contours are com-
posed partly of the spiral and partly of the elliptical arcs. We have first the semi-ellipse FLO, of which C is the focus and P the centre, PL being the semi-major axis; this rolls with the equal

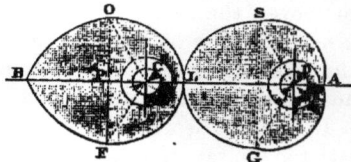

Fig. 74.

semi-ellipse GLS: then the spiral arcs OB, FB, corresponding to FB of the preceding diagram, and these roll with the equal arcs SA, GA, which correspond to FA of Fig. 73.

During the revolution, the velocity ratio varies between the limits

$$\frac{v}{v'} = \frac{DL}{CL}, \quad \frac{v}{v'} = \frac{AD}{BC},$$

which, it will be observed, are not reciprocals, as in the unilobes previously described.

126. In Fig. 72, the wheels, though similar, are not symmetrical; in Fig. 74 they are symmetrical but not similar: and finally, those represented in Fig. 75 are neither the one nor the other. Above the line of centres, each wheel is formed of a single arc of a logarithmic spiral, AMB rolling with ANE. Below that line, we have BH, which rolls with EF, each being a quadrant of the bilobe derived, as in Fig. 58, from the ellipse whose major axis is AB, C being one focus ; AL, which rolls with AI, these curves being parts of quadri-lobes derived in the same way from the same ellipse : and LH, rolling with IF, these two being equal arcs of another logarithmic spiral.

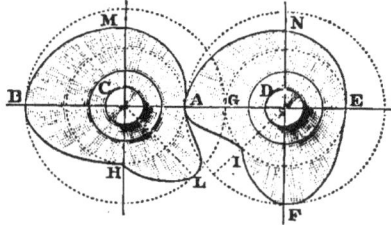

Fig. 73.

127. It may be added, that there are a number of other curves capable of rolling in contact about fixed axes. But it is the province of the mathematician rather than of the mechanician to investigate their properties, which do not admit of simple geometrical demon-stration ; nor are they so easily constructed and adapted to use in me-chanical devices. It may be safely said that in general the number of changes in the velocity ratio, and the limits of its variation, are of greater importance than the precise law according to which it varies between those limits : and it is believed that the principles involved in the examples already given will be found to meet most if not all the requirements of practical mechanism.

128. These various curves, whose action has been discussed as though they were mere lines rolling with each other in their own plane, are of course practically to be considered as the bases, or trans-verse sections, of cylindrical surfaces, tangent to each other along an element. Strictly, therefore, the expression that "the point of con-tact lies always on the line of centres," means that "the element of tangency lies always in the plane of the axes," which is one condition, as shown in Chap. II., of pure rolling contact between two surfaces.

And equally of course, the consideration of these surfaces is only the step preliminary to the investigation of the surfaces of teeth to be formed upon them : for, the motion of the followers being com-pulsory only so long as the contact radii of the driver are increasing, continuous rotation can not be transmitted by means of these rolling surfaces themselves, with any certainty that the desired velocity ratio, whether constant or varying, will be exactly maintained.

CHAPTER V.

ROTATION BY ROLLING CONTACT, AXES NOT PARALLEL.

1. **AXES INTERSECTING.**—Velocity Ratio Constant or Varying. The Rolling Cones, or Pitch Surfaces of Circular and Elliptical Bevel Wheels. Conical Lobed Wheels.

2. **AXES IN DIFFERENT PLANES.**—Velocity Ratio Constant. The Rolling Hyperboloids, or Pitch Surfaces of Skew-Bevel Wheels.

1. *Axes Intersecting.*

129. Velocity Ratio Constant.—It follows directly from the deductions at the conclusion of Chap. II, that the only surfaces capable of rolling in line contact about two fixed axes which intersect each other, are cones, the common vertex being the point of intersection, and the common element lying in the plane of those axes. And in order to maintain a constant velocity ratio, the transverse sections of these cones must be circles. Thus, in Fig. 76, let AB, AC, be the axes, and AP the common element, all lying in the plane of the paper : then in the rotation indicated by the arrows, the motion of the common point P will be perpendicular to the paper. Draw PD perpendicular to AB, PE perpendicular to AC, and let

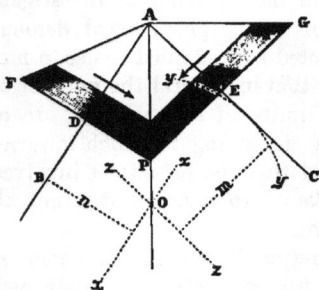

Fig. 76.

$$v = \text{ang. vel. about } AB,$$
$$v' = \text{``} \quad \text{``} \quad \text{``} \quad AC;$$

then at the instant we shall have

$$\frac{v}{v'} = \frac{PE}{PD} :$$

and in order that this ratio may be constant as required, the contact radii PD, PE, must be constant also. That is to say, the plane bases of the cones, PDF, PEG, perpendicular to the axes, must be circles; and the cones themselves are generated by the revolution of the common element AP about each axis in succession.

130. These then are the pitch surfaces of the ordinary conical or bevel wheels. In practice it is of course sufficient to use comparatively thin frusta of the cones, as indicated in the figure: in the great majority of cases they are provided with teeth, but are to some extent used without, constituting what is called Bevel Friction Gearing.

As above deduced, the angular velocities are always inversely proportional to the perpendiculars let fall from any point of the common element upon the axes, whatever the relative positions of those three lines. But it is easy to see that those positions may be very different from what they are in Fig. 76; and it is of interest to consider what results may follow from various assigned or assumed conditions.

131. First, however, it may be remarked that in practice two cases may present themselves for solution. Thus, the axes AB, AC, being given, as in Fig. 76, suppose it to be required to construct a pair of wheels so proportioned that m revolutions about AB shall produce n revolutions about AC. Draw xx parallel to AB, at a distance from it measuring n parts on any convenient scale of equal divisions, and zz parallel to AC, at a distance from it equal to m parts on the same scale. These two lines intersect in o; and APo is evidently the common element of a pair of cones which will roll together with the required velocity ratio.

Or, the cone APF being given, it may be required to find another which will roll with it, the velocity ratio being also assigned as above. In that case draw xx as just explained, cutting AP in o; then about o as centre, describe an arc yy with radius equal to m parts on the scale, and the axis AC of the required cone will be tangent to this arc as shown.

132. Now in Fig. 77, let the axes AB, CD, cut each other obliquely at V, the angle of inclination being given and the velocity ratio assigned. Drawing xx and zz as de-

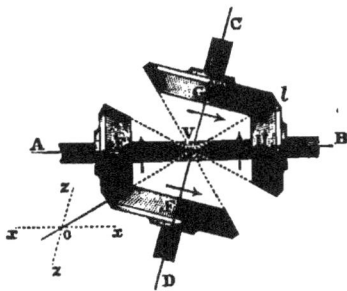

FIG. 77.

scribed in the preceding article, Vo is determined and the cones constructed.

It is to be noted in this case, that these surfaces are tangent along the single line $o\,Vl$; and that if one shaft be carried past the vertex, two pairs of frusta may be used at once, cut from opposite nappes of the cones, as E, F, and G, H. These two pairs may be equidistant from the vertex and therefore exactly alike, as shown; or either pair may be placed nearer the vertex if desired, as they roll together without any interference, with the same velocity ratio, and as indicated by the arrows, with the same directional relation.

133. In this instance the lines xx, zz, are drawn within the *acute* angle formed by the intersecting axes. But in Fig. 78, they are

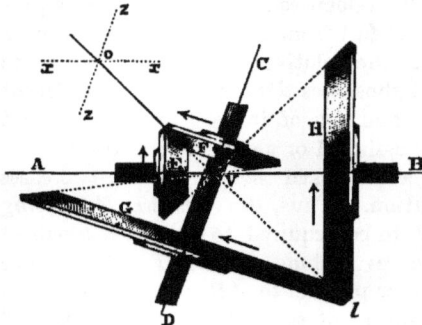

drawn within the *obtuse* angle, the inclination of the axes and the velocity ratio being the same as before. The result is a different pair of cones; the velocity ratio is the same as in Fig. 77, by construction; but the directional relation is changed.

Also, the opposite nappes of these cones intersect each other in such a way, that though under the conditions here assumed one shaft can be carried past the vertex, and two

Fig. 78.

pairs of frusta employed at once if desired, they cannot be placed at equal distances from the vertex: the wheels, therefore, will not be alike, but one pair must be larger than the other in order to avoid interference.

134. But it may not be possible to use two pairs at once, whatever

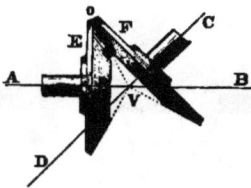

Fig. 79.

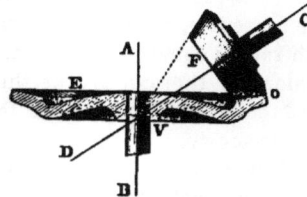

Fig. 80.

their distances from the vertex. Thus in Fig. 79, the cones being constructed so as to produce an assigned velocity ratio and also an assigned directional relation, it is clear that neither shaft can be carried past the other wheel.

Again, it is to be observed that in all these cases the tangency is external: and it obviously always will be when the common element lies within the acute angle formed by obliquely intersecting axes. But when it lies within the obtuse angle, it may happen, as in Fig. 80, that one cone shall touch the other internally. Or, the common element may be perpendicular to one of the axes, as in Fig. 81, the pitch cone degenerating into a plane.

135. Thus far the axes have been supposed to cut each other obliquely; and whatever the forms of the pitch surfaces, it will be noted that in each they are tangent along one line only. We have

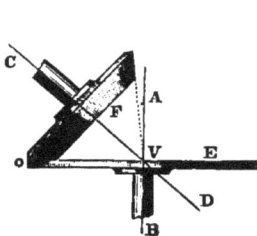

Fig. 81.

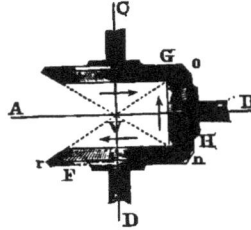

Fig. 82.

yet to consider the case in which the axes are perpendicular to each other, as in practice they are, more frequently than otherwise. And, as shown in Fig. 82, we are confronted by the singular circumstance that the pitch cones are tangent along *two* elements, *or, mn.* The frustum *H* can then roll simultaneously with the frusta *F* and *G*,

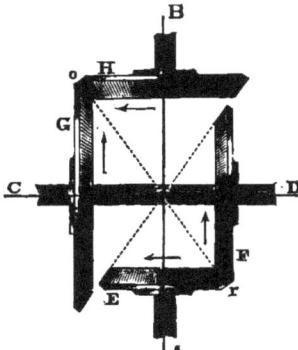

Fig. 83.

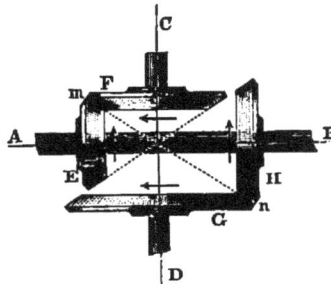

Fig. 84.

cut from opposite nappes of the other cone, but these two are tangent to *H* along different lines, and consequently they will turn in opposite directions. Equal and opposite frusta, like those in Fig. 77,

cannot, of course, be simultaneously employed : but either of the
shafts can be carried past the vertex, and the unequal pairs of wheels
cut from the opposite nappes, being tangent along the same line, will
have the same directional relation.   And, as shown also in Figs. 83
and 84, this directional relation is optional, since either of the two
common elements of the cones may be selected as the line of pragmatic contact.

136. In the case of internal tangency, equal frusta of the opposite
nappes of the hollow cone may be combined in a single wheel, as

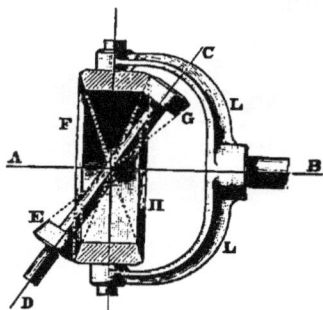

FIG. 85.

shown in Fig. 85.   Here the frusta
F and H are formed in opposite sides
of a solid ring, which is carried by
arms LL, so curved as to avoid
interference with the wheel G, which
engages with H, and is fixed on the
same shaft with E, engaging with
F.   By this arrangement,[*] the forces
transmitted act upon the large wheel
in the manner of a couple, in opposite directions, on opposite sides of
and equidistant from the axis AB ;
thus relieving the bearings from side
pressure.

137. We find, then, that whatever the angle included between the
axes, it is always possible to construct two pairs of cones, rolling together with the same velocity ratio, but having different directional
relations, and we are at liberty to employ whichever may best serve
the purpose to be accomplished.

The cones can always be determined when the axes, velocity ratio
and directional relation are given, by the process described in (**131**).

Another method, differing slightly in detail, is shown in Figs. 86
and 87, which also illustrate the following property, viz : that if distances be set off on the axes from the point of intersection, directly
proportional to the angular velocities, and the parallelogram completed of which these are the adjacent sides, the common element
will lie in the direction of the diagonal.

Thus let v' the angular velocity about AB' be represented by VL,
and v', that about CG, by VM, then VP will be the common element.

For, draw PD perpendicular to AB, PE perpendicular to CG, and
join DE.   Now VP divides the parallelogram ML into two triangles

---

* Which was first suggested by Mr. O. A. Benton.

which are similar to each other; and also to *DPE*, because a circle will go round *VEPD*, in which the angles *DEP*, *DVP*, stand on the

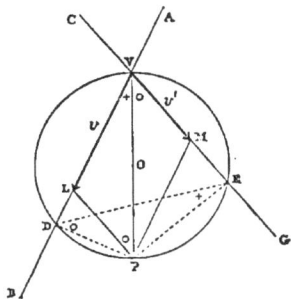

FIG. 86.

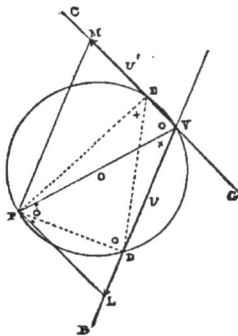

FIG. 87.

same arc, and are therefore equal. For a like reason the angle *EDP* is equal to *EVP*, which is equal to *VPL*. Then if *VP* be the common element, we have

$$\frac{v}{v'} = \frac{PE}{PD} = \frac{VL}{LP} = \frac{VL}{VM}.$$

If the angle between the axes be acute, it may be better to set off *VL* representing *v*, and then to draw *LP* parallel to *CG*, and of a length representing *v'* on the same scale; thus avoiding the uncertainty arising from attempting to locate *P* by an acute intersection.

In these two diagrams the inclination of the axes, and the velocity ratio, are the same, but the directional relations are unlike. Consequently *VM* has the same ratio to *VL* in both cases, but is set off so that in one diagram the angle *MVL* is acute, while in the other it is obtuse.

138. Velocity Ratio Varying.—In Fig. 88 let *E*, *F*, be two pins fixed in the sphere whose centre is *O*; let *EFP* be a loop of fine inextensible thread, passing around the pins and a marking point at *P*. Then as *P* is moved along, the thread being kept always taut and in contact with the sphere, it will trace the curve *APBH*, which is the *spherical ellipse*.

Evidently, *EF*, *EP* and *FP* will always be arcs of great circles; and since the part *EF* of the thread is always the same, and always idle,

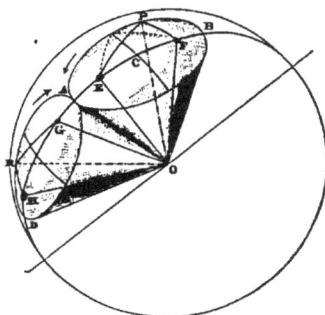

FIG. 88.

while the total length is invariable, this curve may be defined as the locus of the vertices of a series of spherical triangles, of which the base is common and the sum of the other two sides constant. This sum will be equal to the major axis AB; now let AD be the major axis of an equal and similar curve tangent to the first at A, of which G and H are the foci : then BD will be a continuous arc of a great circle, and EH will be equal to AB or AD.

Next, let these two curves be taken as bases of cones whose common vertex is O, the centre of the sphere : then they will be tangent to each other along the element OA, and they will roll in contact about OE and OH as fixed axes. For, let AP, AR be any equal arcs of the two spherical ellipses ; then we shall have

$$PE = RG, \qquad PF = RH,$$

whence,

$$PE + RH = PE + PF = EH;$$

consequently the points P and R will meet on the arc DAB, or in other words, the common element will always lie in the plane of the axes.

139. We have, then, a pair of non-circular cones, capable of rolling in contact about fixed intersecting axes, with a varying velocity ratio, and forming the pitch surfaces of what may be properly called Elliptical Bevel Wheels, since their bases, as thus constructed, are spherical ellipses. But though these curves are easily traced upon the surface of a sphere in the mechanical manner above described, it will be found practically more convenient to employ others, which may be constructed as follows. Referring to Fig. 88, it is seen that the sides of the spherical triangles are the measures of the faces of a series of trihedral angles whose common vertex is O. Join O with the centre C of the spherical ellipse ; then a plane perpendicular to OC will cut the edges of these pyramids in points readily determined, and the problem of finding the plane base of the cone resolves itself into the construction of a number of trihedral angles, of which one face is common and the sum of the other two constant. The base thus found will not be a true ellipse, but it will closely resemble one, and evidently will be symmetrical about two lines at right angles to each other.

140. The processes of constructing one of the trihedral angles in accordance with the above conditions, and of determining by means of it a point in the required curve, are illustrated in Figs. 89 and 90.

The lines OE, OC, OF, of Fig. 89 correspond to those similarly lettered in Fig. 88 ; they all lie in the horizontal plane, and OE is perpendicular to GL the ground line. EOF, then, is the common face of the trihedral angles, bisected by OC. Assume one of the two remaining faces, as EOA, producing OA to cut GL in A. Subtract this assumed face from the given constant sum, thus determining the third, which lay off in the horizontal plane as FOB, making OB equal to OA.

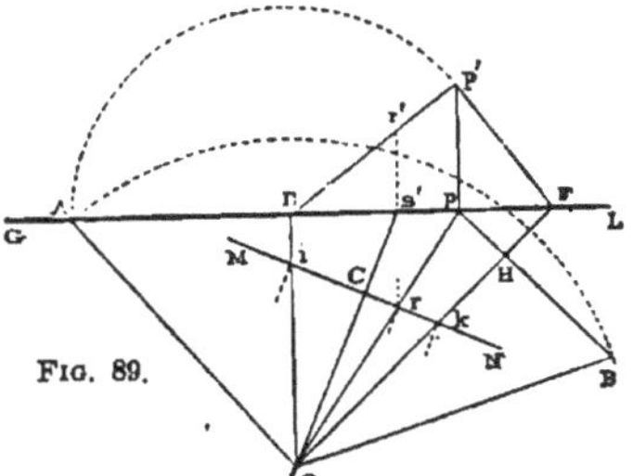

FIG. 89.

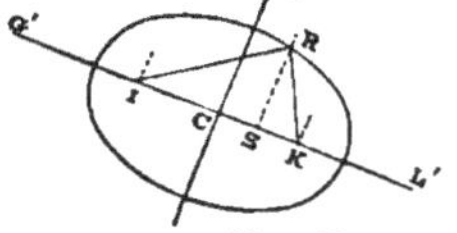

Draw BH perpendicular to OF, produce it to cut GL in p ; erect pp' perpendicular to GL, and of

FIG. 90.

a length equal to the altitude of a right-angled triangle of which HP is the base and HB the hypothenuse : then p is the horizontal and p' is the vertical projection of the point where B will fall in the vertical plane if OB is revolved about OF.

Also, if OA be revolved about OE, A will describe a circle in the vertical plane ; which will pass through p', because $OA = OB$; hence OP is the horizontal and Ep' is the vertical projection of the third edge of the trihedral angle required.

141. Now let MN be a plane perpendicular to OC ; it will cut this third edge in a point whose horizontal projection is r, and $r's'$ will be the vertical projection of a perpendicular let fall from this point to the horizontal plane.

In Fig. 90, which is a projection upon this new vertical plane MN, the horizontal plane is seen as $G'L'$, and the point r appears at R, the distance RS being equal to $r's'$ of Fig. 89 : the line OC appears as the point C, and i and k, the points in which OE and OF are cut by MN, are seen as I and K.

In this projection the required plane base of the cone is seen in its true form ; as many points as may be necessary to determine its contour with precision, being found exactly as R was, by the construction of other trihedral angles.

142. In Fig. 91, OE and OH are the fixed axes of a pair of frusta cut from cones of this description ; the extremities of the major axes of the bases are shown in contact at A, the plane of the paper in this figure corresponding to the plane $ODAB$ of Fig. 88.

The action is obviously akin to that of two ellipses rolling together in their own plane; letting fall upon the fixed axes the perpendiculars AM, AN, the velocity ratio in the present position is $\dfrac{AM}{AN}$, and at the end of a half revolution it will be $\dfrac{AN}{AM}$.

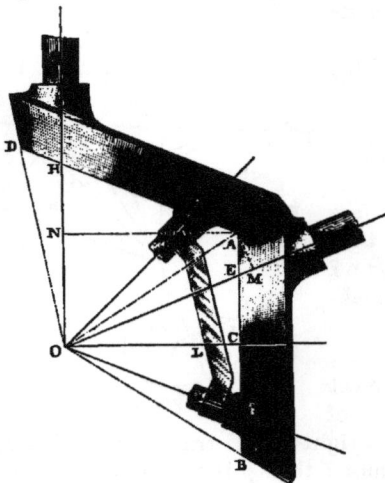

Fig. 91.

This combination also resembles the elliptical spur-wheels in the respect that if these conical frusta overhang their bearings, as here shown, a link, L, may be used to connect what it is proper to call their free foci: the axes of the pins by which this link is pivoted to the wheels, of course converging in O the common vertex.

143. Any reciprocal limits may be assigned to the velocity ratio; and if the angle between the axes be also given, the pitch cones may be constructed as follows.

Let the assigned values of the velocity ratio be $\dfrac{m}{n}$, $\dfrac{n}{m}$; and let OE, OH, Fig. 92, be the given axes. Draw xx, zz, parallel to the axes, and at distances from them which are to each other in the ratio of m to n; these lines intersect at P. Draw PO, and lay off the angle POB equal to EOH; draw OC bisecting POB, and from any point A of PO, draw AB perpendicular to OC, cutting OB in B. Then AB may be taken as the major axis of the plane base of

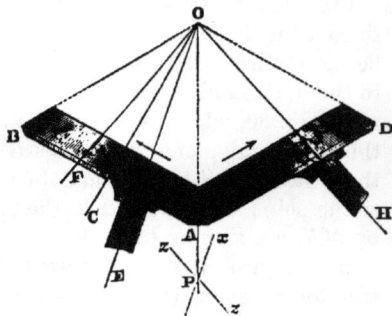

Fig. 92.

the cone, which from these data can now be constructed in the manner previously explained; for making the angle BOF equal to AOE, we have EOF the common face, and AOB the constant sum of the other two faces, to be used in determining the trihedral angles, whose edges are the elements of the cone.

144. In Fig. 93, the inclination of the axes is the same as in Fig. 92, but the lines xx, zz, are so drawn that their intersection P falls

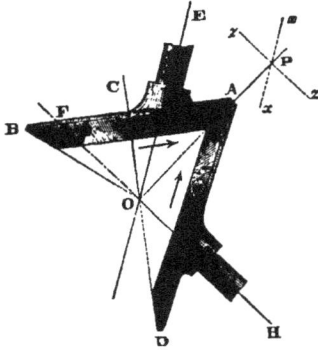

Fig. 93.

within the obtuse instead of in the acute angle : the result being that the directional relation is different.

From the very nature of the case, an engaging pair of elliptical bevel wheels must be equal and similar, and always in external contact. And it will be seen that, since the angle subtended at the vertex by the major axis of the base, is equal to that included between the fixed axes of rotation, neither cone can degenerate into a plane, until that angle becomes 180°, and the axes coincide : much less can either be hollow.

We have then always the choice between two pairs of wheels, having different directional relations but the same action in respect to the velocity ratio.

If the axes be perpendicular to each other, as in Fig. 94, the pitch cones will, like the circular ones under the same condition, have two common elements, either of which may be selected as the line of pragmatic contact.

145. Conical Lobed Wheels. —From these elliptical bevel wheels, it is possible to construct conical wheels with various numbers of lobes, by a process of contraction or ex-

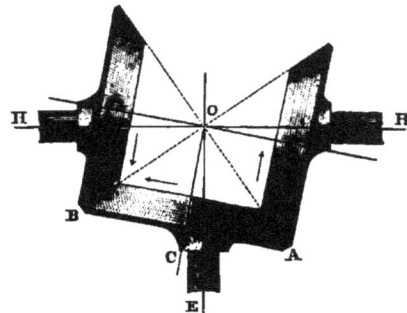

Fig. 94.

pansion of angles analogous to that applied to the plane ellipses.

Let the elements cut from one of a pair of these pitch cones by a series of radiating planes through the fixed axis, be revolved about that axis until the dihedral angles between the planes are expanded or contracted in any desired proportion. Any part of the elliptical cone may be thus transformed into a new conical surface, which will roll in contact with the one formed by treating in like manner the corresponding portion of its rolling mate.

The bases of these lobed cones will be similar to the outlines of the

multilobes derived from the plane ellipses. By applying the process
of contraction to one half the perimeters of the spherical ellipses, pairs
of similar wheels, with any desired number of lobes, may be constructed analogous in form and action to those shown in Figs. 59
and 61.

146. It is also practicable to construct conical lobed wheels which
will roll together in dissimilar pairs, unilobe with bilobe, trilobe with
quadrilobe, and so on.   Referring to Fig. 88, it will be seen that the
dihedral angles *PFE*, *RHG*, are equal. Hence if the spherical ellipses
be such, for example, that the dihedral angles *RGA*, *RHG*, are to
each other as 3 is to 2, the former may be contracted to 90°, and the
latter will by contraction in the same proportion be reduced to 60° ;·
the result being the formation of a bilobe rolling with a trilobe, the
bases resembling those of Fig. 62.

It is sufficient merely to state that by constructing a series of spherical triangles, as was done with the plane triangles in Fig. 66, a spherical curve, analogous to the one there shown, may be drawn, by the
aid of which we can determine the limits of the eccentricity of the
spherical ellipses from which it is possible to derive such dissimilar
pairs with assigned numbers of lobes. It may be said, and no doubt
with truth, that the difficulty of making such wheels would prevent
their being used under any ordinary circumstances. But it is equally
true that only extraordinary conditions would require them to be used :
and should cases occur in which it would be desirable, it is worthy of
note that very many of the combinations of non-circular cylindrical
wheels described in the preceding chapter, may be replaced by combinations of conical wheels nearly identical in their action.

147. Of these, it is probable that the ones derived as above suggested
from the spherical ellipse, would be found practically preferable, as
being the ones most easily constructed, and also as allowing a wider
range in the selection of the limits of variation in the velocity ratio.
But the equiangular spiral has also its spherical analogue, which may
be made the base of a conical surface, capable of rolling in contact
with an equal and similar one, the fixed axes of rotation passing
through the poles of the bases and the centre of the sphere upon which
they lie.

If from any point on the sphere arcs whose lengths are in geometrical progression be set off successively on equidistant meridians passing
through the point, the curve drawn through the extremities of these
arcs will be the one required.

148. This will be readily seen by the aid of Fig. 95. In the side
view let *OC* be a vertical radius of the sphere, and *AB* a plane tan-

gent to it at C. Let C in this plane be the pole of the logarithmic spiral unilobe shown in dotted outline, as AMN in the horizontal projection.

The radiants of this spiral which include equal angles, are in geometrical progression ; and planes passing through them and also through OC, cut equidistant meridians from the sphere. On each meridian set off from C an arc equal in length to the corresponding radiant of the spiral : the result will be the spherical curve shown in full lines ; the vertical projection being $A'P'B'$.

Now let $A'RE$ be the vertical projection of an equal and similar curve, of which D is the pole : the vertical plane being that of the great circle containing OC, OD, and also the arcs $A'C'B'$, $A'DE$. It is then obvious that these curves will be tangent to each other at A'; also, that if $A'P$, $A'R$ be equal arcs of the two, then P and R will meet in the plane of the axes, if the curves revolve about OC and OD respectively, as shown by the arrows. For the angular distances of these points from

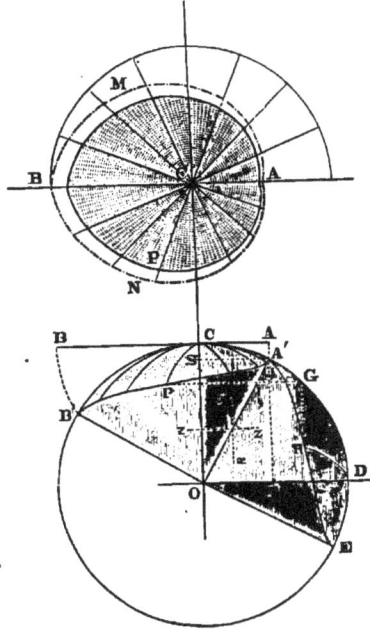

FIG. 95.

that plane are the same as in the original spirals, which are known to roll together, and by construction the sum of the arcs CP, DR, is equal to the arc CD.

149. It will be seen, then, that if these curves be made the directrices of cones whose common vertex is O the centre of the sphere, those surfaces will work in rolling contact about OC and OD as fixéd axes ; and that the velocity ratio will vary from maximum to minimum and back again once in each revolution, this combination being analogous to that of the two logarithmic spiral unilobes shown in Fig. 53.

In fact the plane original AMN is identical with one of those unilobes. But the actions of the two combinations, though very similar, are not identical. Draw $A'S$, $A'T$, perpendicular to OC and

OD, the limiting values of the velocity ratio in Fig. 95 are

$$\frac{A'S}{A'T}, \quad \frac{A'T}{A'S};$$

while in Fig. 53 they are

$$\frac{AC}{BC}, \quad = \frac{\text{arc } A'C}{\text{arc } B'C},$$

and

$$\frac{BC}{AC}, \quad = \frac{\text{arc } B'C}{\text{arc } A'C}.$$

By comparing in a similar manner the values of the velocity ratio for intermediate positions, it will be found that neither the limits nor the laws of variation are precisely alike in the two cases.

150. If the angle COD between the axes be given, and the values of the reciprocal limits of the velocity ratio be assigned, the cones may be constructed, by processes closely resembling those employed in preceding cases.

Thus, it is evident that OA' must pass through the intersection of two lines xx, zz, drawn parallel to OC and OD respectively, and at distances from them which are to each other in the proportion of $A'S$ to $A'T$; then $CB' = A'D$, and the arcs $A'C$ and CB' being rectified give AC and CB, whence the original spiral AMB may be reconstructed, and the spherical curves derived from it as just described.

Like preceding combinations, too, this might be discussed, with reference to the effects of dividing the obtuse instead of the acute angles, of placing the axes of rotation at right angles, and in short, of assuming any special conditions. It will also readily be seen that the bilobes, trilobes, etc., derived from the equiangular spiral and rolling in contact about parallel axes, have their analogues in conical multilobes whose axes intersect. All these matters, however, we shall leave the reader, if so disposed, to pursue farther at his leisure : the examples already given illustrating sufficiently the principles involved in the construction of this class of combinations, which, it is proper to add in conclusion, we believe to be a new one.

2. Axes in Different Planes.

151. Velocity Ratio Constant.—If a right line revolve about an axis in a different plane, the surface generated is the hyperboloid of revolution. Any normal to this surface will intersect the axis; it will also be perpendicular to both generatrices through the point of normalcy, since these two elements determine the tangent plane.

If then a series of normals be drawn through different points of the revolving line, they will lie in planes perpendicular to that line, and therefore parallel to each other. They are, consequently, elements of one generation of a hyperbolic paraboloid, of which the directrices are the axis and the revolving line, and the plane directer is perpendicular to the latter.

Any plane parallel to those directrices will therefore cut the series of normals in points which will lie in one right line, an element of the second generation of the hyperbolic paraboloid.

Now the line thus determined may be taken as a new axis; and by revolving around it, the same line which generated the first hyperboloid will generate another. These two surfaces of revolution, having, at every point of a common rectilinear element, a common normal and a common tangent plane, will be tangent to each other all along that element.

152. Since these hyperboloids are warped surfaces, perfect rolling contact between them, as has been shown, is not possible under any circumstances whatever.

But they are capable of rotating in contact about fixed axes, with a constant velocity ratio; and the sliding between them is quite different from that between two tangent cones or cylinders whose perimetral velocities are not the same, in the respect that it is wholly in the direction of the common element. And these hyperboloids, like the cylinders and the cones, are practically used as the pitch surfaces of toothed wheels. In view of these facts, it is proper to consider their action in this place, notwithstanding the imperfection in their rolling.

153. The form of the surface, and the manner of constructing it, are shown in Fig. 96. As the inclined line AB revolves about the vertical axis, every point in it describes a

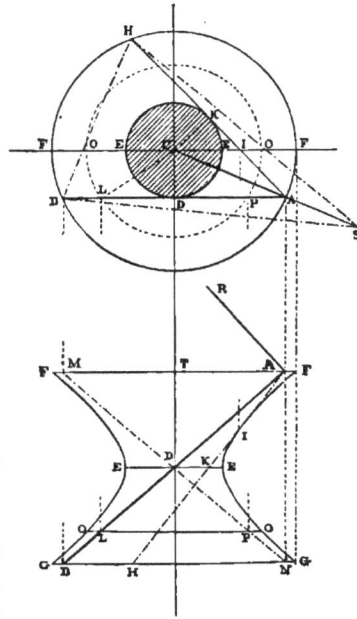

Fig. 96.

circle in a horizontal plane, whose radius is seen in its true length in the horizontal projection. In that projection, the axis appears as the point C, and CD is the common perpendicular of the axis and the revolving line.   Then the point D describes the *circle of the gorge*, EDE; A describes the circle of the upper base, FAF; B, that of the lower base, GBG, in this case equal to FAF, because A and B are equidistant from D: and any intermediate point L describes a circle OO, whose radius is CL.   In this way any number of points in the meridian outline $GOEF$ may be determined.

The same surface may be generated by the revolution about the same axis, of another right line MDN; MN and AB having the same horizontal projection, and being equally inclined to the plane of rotation, but in opposite directions.

For the paths of A and M, also those of B and N, coincide, and D is common to both lines: consequently any two points, one on each line, equidistant from D, as for example L and P, will describe the same circle.

154. Through any point of the surface, then, two rectilinear elements, or companion generatrices, may be drawn; whose projections on a plane perpendicular to the axis will be tangent to that of the gorge circle on the same plane.

Thus AH, tangent at K in the horizontal projection to the circle EDE, is the second generatrix through A.   In this figure, the vertical plane contains the axis; AH pierces this plane at I, as determined from the horizontal projection: and in the vertical projection AH is tangent at I to the hyperbolic outline, of which E is the vertex, and AB and MN are the asymptotes.

These two lines, AB and AH, determine the plane tangent to the hyperboloid at A; the horizontal trace of this plane is therefore parallel to BH; and its vertical trace is parallel to AB, since that line is parallel to the vertical plane. Consequently AC, perpendicular to BH, is the horizontal, and AR perpendicular to AB, is the vertical, projection of the normal to the surface at A.

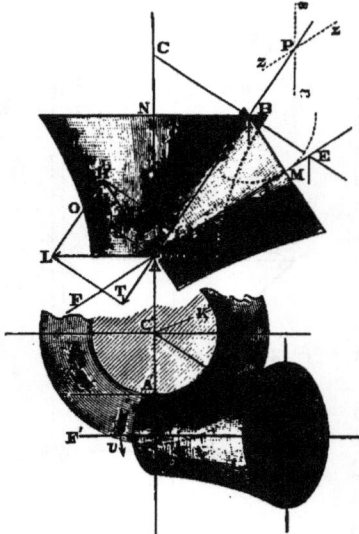

Fig. 97.

155. Now in Fig. 97, two hyperboloids are shown in contact, the axes AC and FE, and the com-

mon element AB, being parallel to the vertical plane. In the horizontal projection the vertical axis appears as the point C', the inclined one as $F'E'$, and the common element as $A'B'$, which at A'
cuts the common perpendicular of the axes, here seen in its true
length as $C'I'$, into segments $A'C'$, $C'I'$, which are the radii of the
two gorge circles. Also $C'B'E'$ is the horizontal, and CBE is the
vertical, projection of the common normal to the surfaces at B. In
the vertical projection, the upper bases and the gorge circles will
appear as lines perpendicular to the axes : the latter intersecting at A,
and the former, which in the figure passes through B, cutting the axes
at M and N.

156. Let these hyperboloids revolve about their axes in the directions indicated by the arrows. The point A of the inclined surface
will at the instant move in the direction of the tangent to the gorge
circle at that point : which being parallel to the vertical plane, this
motion may be represented in magnitude and direction by AH in the
vertical projection : and it may be resolved into the components, AS
in the direction of AB, and AO perpendicular to that line.

Considering the inclined hyperboloid as the driver, the rotation of
the vertical one will not be compulsory, because the contact radii are
constant.

But the common element may here, as in the cases of the cylinders
and the cones, be regarded as the line of contact of teeth formed upon
the two surfaces, so that AO would be the normal, and AS the tangential component of AH; because AB would necessarily be a line
of the plane tangent to both teeth at A. Therefore the resultant
motion of that point of the vertical hyperboloid (whose direction is
tangent to the larger gorge circle), must be of such magnitude, AL,
as to have the same normal component AO, the other component
being AT in the line AB.

157. Now let $v =$ ang. vel. about FE,
$$v' = \text{ “   “   “ } AC;$$
then
$$v = \frac{AH}{A'I'} \left.\begin{array}{c} \\ \\ \end{array}\right\}$$
$$v' = \frac{AL}{A'C'}$$
$$\therefore \frac{v}{v'} = \frac{AH}{AL} \times \frac{A'C'}{A'I'} \cdot \quad \cdots \cdots \quad (1).$$

But from similar triangles AHL, ACE,
$$\frac{AH}{AL} = \frac{AE}{AC};$$

and by the principles of projection,

$$\frac{A'C'}{A'I'} = \frac{B'C'}{B'E'} = \frac{BC}{BE}.$$

Substituting in (1),

$$\frac{v}{v'} = \frac{AE}{BE} \times \frac{BC}{AC} \quad . \quad . \quad . \quad . \quad . \quad (2).$$

From similar triangles ABC, ABM,

$$\frac{AE}{BE} = \frac{AB}{BM};$$

and from similar triangles ABC, ABN,

$$\frac{BC}{AC} = \frac{BN}{AB}.$$

Substituting in (2), we have, finally, for the velocity ratio,

$$\frac{v}{v'} = \frac{BN}{BM}.$$

That is to say, the angular velocities are to each other in the inverse ratio, not of the perpendiculars let fall from any point of the common element upon the axes, but of the projections of those perpendiculars upon a plane parallel to both axes and the common element.

158. Supposing the axes to be given in position, then, and the velocity ratio assigned, the inclination of the common element is found exactly as in the case of intersecting axes: two lines, xx, zz, are drawn parallel to AC and AE, at distances from them which are to each other in the inverse ratio of the angular velocities about those axes respectively; and AB must pass through their intersection P. Then drawing CE perpendicular to AB, it is the vertical projection of the normal, therefore the horizontal projection of E must be at E' on the horizontal projection of FE; and projecting B to B' on $C'E'$, we draw $B'A'$ parallel to $F'E'$, thus determining the horizontal projection of the common element and also the radii of the gorge circles, $A'C'$ and $A'I'$: whence the surfaces may be constructed as before described. Or, if for example the vertical surface be given, and it be required to find the other, the velocity ratio being also assigned: then, knowing BN and its ratio to BM, the length of the latter may be found, and with that as radius, an arc is described about B as centre. The vertical projection FE of the second axis is drawn through A, tangent to this arc. The vertical projection of the normal at B, is CB perpendicular to AB, and it cuts FE in E, whose horizontal projection must therefore be at E' on the prolongation of $C'B'$, and $F'E'$ parallel to $A'B'$ is the horizontal projection of the axis of the required surface.

159. In Fig. 97, the common element lies between the axes, and the two surfaces touch other externally. But if one be larger than the other, the smaller one may be placed within it, and that in such a position as to touch the larger along a line of the concave surface. Thus, in Fig. 98, the same pair of hyperboloids as in the preceding figures, are shown in this new relation : the demonstration of the velocity ratio applies, it will be seen, without any change whatever, but the directional relation in the two cases is different. The surfaces are tangent along an element of the same generation as before, but it now lies on the same side of both axes.

It will be observed that in the case in which the vertical hyperboloid and the velocity ratio are given, FAE is drawn tangent to the same circle, described about B

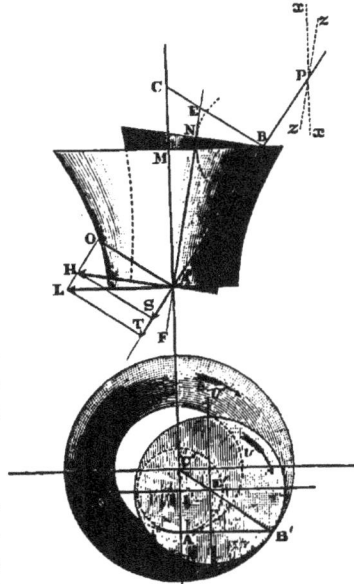

Fɪɢ. 98.

with radius BM, but on the side opposite to that selected in Fig. 97.

Also, that in the other case, where the two axes and the velocity ratio are given, the lines xx, zz, are so drawn that their intersection P lies in the obtuse angle formed by the axes, instead of in the acute one as in the preceding figure.

160. In the above determination of the velocity ratio, the motions of the coincident points of the two gorge circles were selected merely because they were the most convenient ones for the purpose. It is evident that the angular velocity of every point in either surface about its axis must be the same. The motion of the line AB of the vertical surface is one of revolution about AC; and AL being the linear velocity of A, every point in AB must have a component motion along that line equal to AT, and in the same direction, as shown in Figs. 8 and 9. And in revolving about FE, the motion of every point of AB must have a component in the direction of and equal to AS, since AH is the linear velocity of the point A in the inclined hyperboloid.

The action of the combination, therefore, consists in rolling, combined with a sliding in the line of the common element, which is represented by ST, the sum or the difference, as the case may be, of AS and AT.

The rate of sliding, then, is at any instant the same at every point of contact ; but as the linear velocity of each point depends upon its distance from the axis of the surface upon which it lies, it is evident that the proportion of the sliding component to the perimetral velocity will be greatest at the gorge, and diminish as the point under consideration recedes therefrom.

161. Although in Figs. 97 and 98, the hyperboloids, in order to save space, are limited by the gorge planes, it is obvious that they may be extended to any distance on both sides of those planes, and will be tangent from end to end. But practically, as in bevel gearing, comparatively thin frusta only of the pitch surfaces are used ; and their location is optional within certain limits in some special cases hereafter to be noted. Thus in Fig. 99 the surfaces are in contact all along the line mn ; and we may use either of the three pairs of frusta A and B, C and D, E and F, or any two or all three pairs at once : and the steadiness of the motion will practically be greater when two pairs equidistant from the gorge planes are employed.

162. **Transverse Obliquity.**—The plane tangent to the hyperboloid at any point is determined (154) by the companion generatrices through the point. If a meridian plane be passed through the same point, it will cut the tangent plane in a line tangent to the hyperbolic outline at that point : and the angle included between this line and either generatrix is the measure of what is called the *transverse obliquity*, or *skew*, of the teeth which must be used in practice.

As before mentioned, teeth which work in line contact ultimately reduce to rectilinear elements of their pitch surfaces. But when they are of sensible magnitude, there is one instant in the action of each engaging pair, when their line of tangency coincides with the common element of the pitch surfaces. When those are cylinders or cones, the common element lies in the plane of the axes : this is impossible in the case of the rolling hyperboloids, but the less the departure from that condition, the more advantageous will be the action. And this transverse obliquity, which measures the inclination of the generatrix to the meridian plane at any point, is obviously greatest at the gorge, and diminishes as the element is extended, which is another reason for employing, when the circumstances of the case admit of it, frusta at some distance from the gorge planes.

Referring to Fig. 96, it will be seen that the transverse obliquity at the gorge is measured by the angle TDA in the vertical projection. That at A is half the angle between AB and AH; to construct it, describe with radius the true length of AB (seen in the vertical projection), arcs about B and H in the horizontal projection, where BH appears in its true length. These arcs intersect at S on the prolongation of CA ; and CSB or CSH is the required angle.

163. Now since, when the axes are given in position, the angle between their projections on a plane parallel to both is divided, in order to produce any assigned velocity ratio, precisely as though the axes intersected and the pitch surfaces were cones, it will be seen that in every case the problem admits of two solutions. That is to say ; with any given pair of axes, it is possible to construct two pairs of hyperboloids, having the same velocity ratio, but different directional relations.

The cone is, in fact, but the limiting form of the hyperboloid, in which the radius of the gorge circle becomes zero : and all the peculiarities which have been pointed out as resulting from special conditions with intersecting axes, will be found to have almost their exact counterparts under analogous circumstances when the axes lie in different planes.

164. Thus, when the projections of the axes cut each other obliquely, we may divide the acute angle, as in Fig. 97, and the resulting hyperboloids are externally tangent : a comparison of Fig. 99, which represents the same pair of surfaces, with Fig. 77, will clearly show the close resemblance between the two combinations.

If the obtuse angle be divided instead, one hyperboloid may be internally tangent to the other, as in Fig. 98 ; which condition of things is at once seen to correspond to the case of a cone rolling within a hollow one, as shown in Fig. 79.

But whether the axes intersect or not, it does not follow that the phenomenon of internal tangency will always occur when the obtuse angle between their projections is selected. This, in relation to conical surfaces, was illustrated in Figs. 78 and 79 : and the truth of it in regard to the surfaces now under consideration will be evident on examination of Fig. 100, in which the relative positions of the axes, and the velocity ratio, are precisely the same as in Figs. 97 and 99, the only difference being in the directional relation. This, as above stated, is due to the division of the obtuse instead of the acute angle ; but the tangency of the pitch surface is still external, as it always will be if the acute angle be divided.

165. And this figure calls attention to another point of similarity between the cone and the hyperboloid. The latter, although a sur-

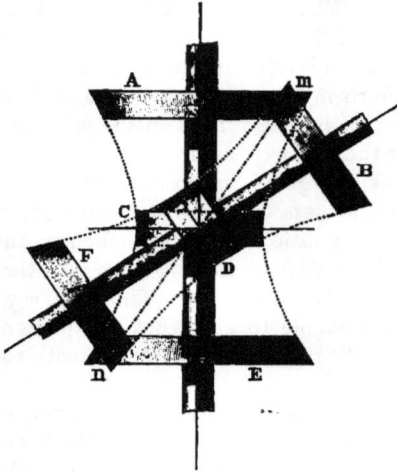

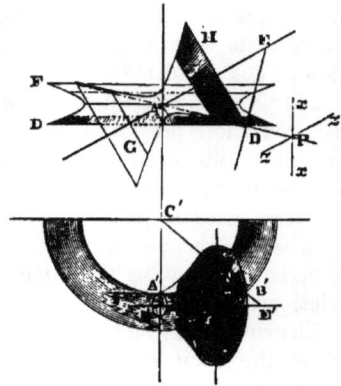

Fig. 99.   Fig. 100.

face of one nappe, is divided by the gorge plane into two parts, whose relations are very like those of the opposite nappes of the cone. In Fig. 100, it will be observed that the shaded frusta, *H* and *D*, might be extended to the gorge planes. Not beyond, however, because the extension *G* of the inclined surface will intersect the frustum *D* of the vertical one, and *F*, the extension of the latter surface, will intersect *H* : a condition of things very similar to that in Fig. 78.

Equal and opposite pairs of frusta, then, cannot be used in this case : nor yet can we employ a pair like the central one of Fig. 99, of which the mid-planes are the gorge circles. But if *H* and *D* be very near the gorges, and very thin, it will be possible to carry the inclined shaft past *D*, and to use at the same time another pair, *G* and *F*; if they be sufficiently far from the gorge planes to clear the first pair.

166. Again, it may happen that when the projection of the common element lies in the obtuse angle, it shall be perpendicular to the projection of one of the axes. This case, shown in Fig. 101, presents the remarkable feature that the pitch hyperboloids retain their limiting forms, the one remaining a cone, the other a plane.

It will be seen that the common element AB intersects the inclined axis FE, by revolving around which it generates the cone, ABL: while in revolving about the vertical axis, the points A and B describe the circles whose radii are $C'A'$, $C'B'$, in the horizontal projection.

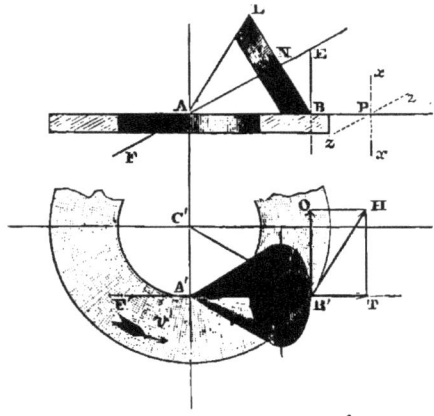

The determination of the velocity ratio will be most conveniently made by considering the motions about the axes of the two points which fall together at B. The motion of the point which belongs to the plane hyperboloid, may be represented by $B'H$, perpendicular to $C'B'$ in the horizontal projection; and it may be resolved into the components, $B'T$ in the line of the common element, and $B'O$ perpendicular to it.

Fig. 101.

Now $B'O$ is tangent to the circle described about the inclined axis by the point which belongs to the cone, and therefore represents its resultant motion, which has no component along $A'B'$.

167. Then as before, let

$$v = \text{ang. vel. about inclined axis,}$$
$$v' = \text{"} \quad \text{"} \quad \text{"} \quad \text{vertical "},$$

and we have

$$\left.\begin{array}{c} v = \dfrac{B'O}{BN} \\[2ex] v' = \dfrac{B'H}{B'C'} \end{array}\right\} \quad \therefore \frac{v}{v'} = \frac{B'O}{B'H} \times \frac{B'C'}{BN} \quad \cdots \cdots \quad (1).$$

But from similar triangles $B'OH$, $B'A'C'$, we have,

$$\frac{B'O}{B'H} = \frac{A'B'}{B'C'}, \; = \frac{AB}{B'C'},$$

since $A'B' = AB$:

Substituting in (1),

$$\frac{v}{v'} = \frac{BA}{BN}.$$

As in the other cases, then, the velocity ratio depends entirely upon the projections of the axes and the common element on a plane parallel to all three, and the singular circumstance results that the angular velocity of the cone is not affected by the lateral separation of the axes or the resulting variation in the diameter of the perforated disc with which it works in contact.

The sliding component $B'T$ depends wholly upon the revolution of AB about the vertical axis, and since it must be the same at every point of the moving line, its ratio to the linear velocity is greatest at the vertex of the cone, diminishing as we recede from that point.

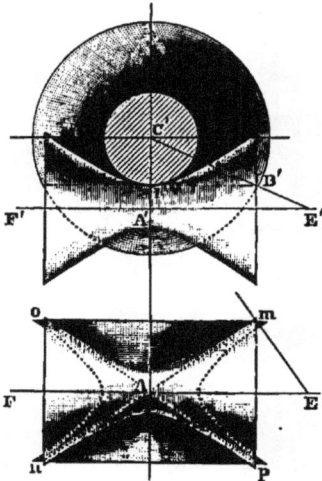

Fig. 102.

168. Finally, in the case in which the projections of the axes intersect each other perpendicularly, as in Fig. 102, it is seen that the pitch hyperboloids are tangent to each other along two elements, as were the cones under the same circumstances.

These elements, mn, op, are necessarily the companion generatrices, and evidently tend to establish the same velocity ratio, but different directional relations. Either of them may be made the line of pragmatic contact, and at the limit, as shown in Fig. 103, a frustum B of one hyperboloid may work in contact with two frusta, A and C, on opposite sides of the gorge plane of the other. But this combination could practically be used only as an arrangement of friction gearing, the wheels being parts of the pitch surfaces only, for teeth of sensible magnitude must be disposed in the direction of a generatrix, common to the engaging frusta; now B and C are tangent along one line, B and A along another, and the teeth of B cannot slope both ways at once.

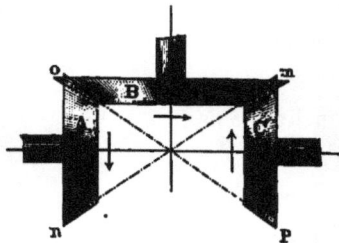

Fig. 103.

169. Evidently this double tangency prevents in this case also the use of frusta whose mid-planes are the gorge circles, as well as of equal and opposite pairs. But pairs on opposite sides of the gorge planes

may be used, having the same common element, and therefore the same directional relation, if they be placed at different distances from those planes.   Either line of contact may be selected, so that as in the case of the cones the directional relation is optional.   The resulting combinations are shown in Figs. 104 and 105 ; comparing these with

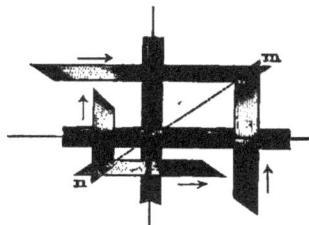

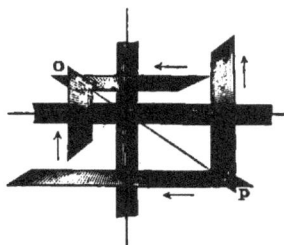

Fig. 104.

Fig. 105.

Figs. 83 and 84, the analogy between the cone and the hyperboloid, considered as the pitch surfaces of wheels, will be most clearly seen.

170. It may be remarked that frusta of these hyperboloids can be employed in the manner of friction gearing.   When this is to be done the frusta are preferably placed at some distance from the gorge planes ; for the curvature of the hyperbola diminishes so rapidly as it recedes from the vertex, that it very soon becomes almost inappreciable. Consequently, if at the mid-planes of the frusta thus located cones be drawn tangent to the hyperboloids, frusta of these cones may practically be employed.   It is hardly necessary to repeat that since the transmission of the motion depends wholly upon the adhesion between the surfaces in contact, no absolute dependence can be placed upon the constancy of the velocity ratio, and it may be noted that, since in order to secure such adhesion, it is necessary to press the surfaces together with considerable force, an amount of friction is thus caused in the bearings, so great that it may be questioned whether such gearing has in many cases any advantage other than that of original simplicity and possibly in freedom from noise at high velocities, over toothed wheel-work ; which latter we will now proceed to discuss.

CHAPTER VI.

1. *The Different Varieties of Gearing Classified.*

171. It has been pointed out, that when a definite velocity ratio is to be maintained, the pitch surfaces previously discussed are unsuitable for the transmission of rotation, on account of their liability to slip upon each other ; and we propose now to consider the forms of the teeth which in practice must be employed in order to prevent this slipping.

But it is desirable first to gain clear ideas of the general nature of the various kinds of toothed wheels in use, and of the peculiarities upon which their classification is based.

172. Not only may the axes of a pair of engaging wheels have different relative positions, but the teeth themselves may be of different kinds, and act upon each other in different ways ; for example, the mode of action of a pair of screw-wheels is quite dissimilar in its intrinsic nature from that of a pair of skew-bevel wheels, although the relative positions of the axes may be the same in each case. There are, in consequence, six varieties or classes of toothed gearing to be met with in practice, viz :

1. Spur Gearing.   4. Twisted Gearing.
2. Bevel Gearing.   5. Screw Gearing.
3. Skew Gearing.   6. Face Gearing.

173. Regarding this matter from a new point of view, it is seen that the teeth of engaging wheels act upon each other by contact whatever their number. If then that number be indefinitely increased, the size being correspondingly diminished, the teeth will ultimately become, in general, mere lines, or elements of surfaces in contact. The relative motions of these surfaces will be the same as those of the wheels from which they are thus derived, their forms and disposition depending on the nature of the class of gearing to which those wheels originally belonged : these are the pitch surfaces, whose action in some combinations has already been investigated. For the purpose of comparison, and to illustrate the distinctive features, a pair of wheels of each class, and also their pitch surfaces, are shown in Figs. 106 to 111, inclusive.

174. In the first three classes, the engaging teeth, which are bounded

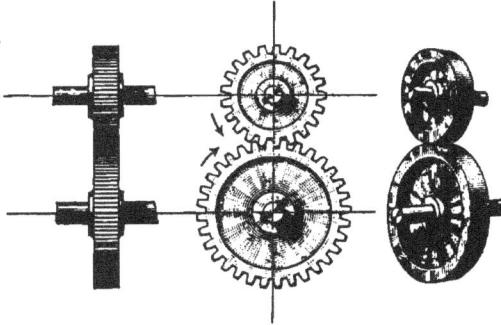

Fig. 106, spur gearing.

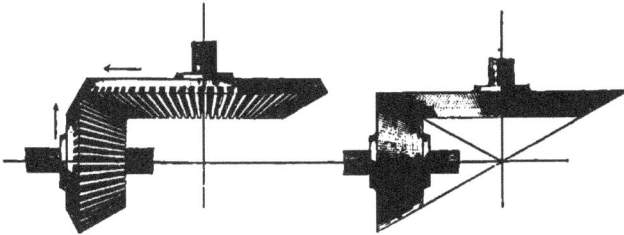

Fig. 107, bevel gearing.

by ruled surfaces, touch each other along right lines, and by the process above indicated they are reduced to *rectilinear* elements of the pitch surfaces, which by the mode of derivation must be tangent all along a right line. As already stated, the axes of spur wheels are parallel, and their pitch surfaces are cylinders ; the axes of bevel wheels intersect, and their pitch surfaces are cones whose common vertex is the point of intersection : the axes of skew wheels lie in different planes, and their pitch surfaces are hyperboloids.

175. Let us now suppose one of a pair of engaging circular wheels, belonging to either of these three classes, to be uniformly twisted on its axis, each successive transverse plane being rotated through a greater angle than the preceding one; then the other wheel of the pair will receive a corresponding twist, as will be readily understood by the aid of Fig. 109. It will hereafter be shown that the twisted

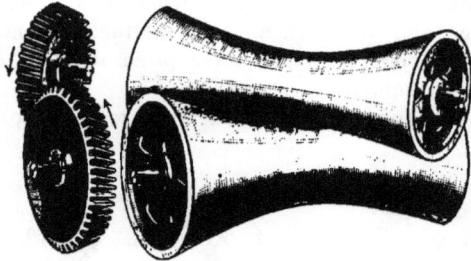

FIG. 108, SKEW GEARING.

FIG. 109, TWISTED GEARING.

wheels thus formed will gear together as well as before, and in substantially the same manner. The teeth are now distorted into surfaces of a helicoidal nature, and by the above process of indefinite subdivision, they reduce to helical lines.

But it is to be noted that these lines lie upon surfaces which are tangent along a right line, whether the axes are parallel, intersecting, or neither. And it will also be seen that whatever of screw-like action may be involved in their motions, tends only to cause pressure in the direction of the common element of the pitch surfaces, and has nothing to do with the transmission of rotation.

176. In these respects there is a marked distinction between these twisted wheels, and those belonging to the next class, of Screw Gearing properly so called; although they are frequently confounded with each other. In all the latter, the teeth, it is true, are also of helicoidal form, and reduce to helical lines : but these helices lie upon cylinders whose axes are in different planes, and the pitch surfaces therefore touch each other in a single point only. Moreover, as illustrated by the familiar combination of the "worm and wheel," it is the screw-like action *alone* of one wheel upon the other, by which the rotation is transmitted in that class of gearing.

177. Face Gearing is not much used in modern machinery ; the name is derived from the fact that the wheels were usually formed with teeth consisting of turned pins projecting from the faces of circular disks, as shown in Fig. 111 : a mode of construction well adapted to wooden mill work, and to that only. In the case illustrated here, the axes are perpendicular to each other ; but turned pins may

FIG. 110, SCREW GEARING.

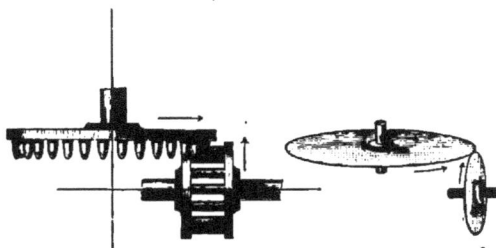

FIG. 111, FACE GEARING.

be inserted in other surfaces than planes, and in this way such wheels can be made to work together when the axes have different relative positions. All these may be properly said to belong to the same class ; of which the distinguishing features are, that whatever the relation of the axes or the general forms of the wheels, the teeth are circular in their transverse sections, touch each other in a single point, and ultimately become *points* in the circumferences of circles which are in contact. The reason of this last peculiarity is, that an increase of the number involves a diminution of the length as well as of the diameters of the teeth, so that at the limit they vanish altogether : whereas, in the other classes of gearing, the length of the teeth is not affected by any variation in the height or thickness, and they reduce to lines. Face wheels, then, have no *pitch surfaces* properly so called, although in constructing them, surfaces of some kind must be provided in which to secure the teeth or pins.

178. The following table exhibits in a convenient manner the peculiar features of the different kinds of gearing above mentioned : the teeth, of whose linear elements the forms are given in the last column,

being supposed to be of sensible magnitude, in order that the circular sections of those in the sixth class may be kept in view.

CLASS OF GEARING.	RELATIVE POSITIONS OF AXES.	PITCH SURFACES.	ELEMENTS OF TEETH.
1. SPUR.	Parallel.	Cylinders.	Rectilinear.
2. BEVEL.	Intersecting.	Cones.	Rectilinear.
3. SKEW.	In Different Planes.	Hyperboloids.	Rectilinear.
4. TWISTED.	Any.	Either.	Helical.
5. SCREW.	In Different Planes.	Cylinders.	Helical.
6. FACE.	Any.	None.	Circular.

2. The Teeth of Spur Wheels.

Epicycloidal System.

179. Generation of the Tooth Outline.—In Fig. 112, let C, D be the

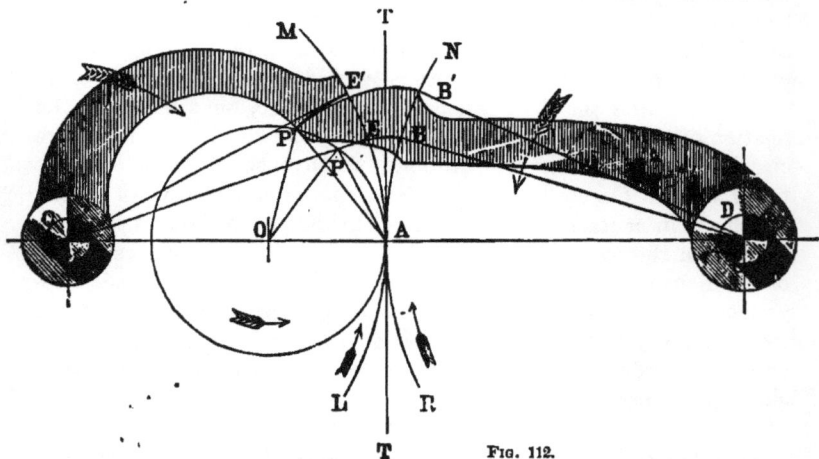

FIG. 112.

centres of the pitch circles LM, RN. Tangent to these at A, is a smaller circle whose centre is O. Suppose all the centres to be fixed, then the three circles can move in rolling contact, with equal linear velocities. Set off from A the three equal arcs AB, AE, AP. Suppose a marking point fixed originally at A, in the circumference of the smaller circle; then while this travels to P, it will trace, with reference to RN, the curve BP, and with reference to LM, the curve EP.

Now the relative motions of the circles are precisely the same as though the smaller one, which carries the marking point, had rolled upon the outside of RN, and upon the inside of LM, regarding these two as fixed base lines: the curves are, therefore, an epicycloid and a hypocycloid respectively; and AP is their common normal at P, because on whichever of the two fixed lines we regard the small circle to be rolling at the instant, the point of contact A is the instantaneous axis (**73**), so that the motion of P in either curve is perpendicular to AP. If the tracing point go on to P', the arcs AP', AE'', AB', being equal, the resulting curves $B'P'$, $E'P'$, are, clearly, but extensions of the first pair, and AP' is their common normal.

180. We perceive, then, that the curves thus simultaneously generated are tangent to each other at some point, throughout the generation; that the point of tangency is always in the describing circle: and that the common normal always passes through the fixed point A, upon the line of centres.

Consequently these curves are correct outlines for parts, at least, of teeth; if the curved lever CE turn, as shown by the dotted arrow, it will drive the other before it, the point of contact following the arc $P'PA$, until E' and B' meet at A', and as the common normal always cuts the line of centres at the same point, the velocity ratio will be constant.

181. It is to be noted, that this condition would have been satisfied, had the tracing point been fixed, not in the circumference of the describing circle, but at a greater or less distance from its centre. And it will also be readily perceived, that the tracing point need not be carried by a *circle* at all; any other describing curve might have been used, provided that it were capable of rolling in contact with the pitch circles; the point of contact would not travel in a path coinciding with the describing curve, but the generated curves would always have a point of tangency, and the common normal would always have passed through A. And conversely, any two curves of which these two things are true, can be generated in the manner above described. In general, then: *The tooth-outlines which act in contact, must be such as can be simultaneously traced upon the planes of rotation of the two wheels while in action, by a marking point which is carried by a describing curve moving in rolling contact with both pitch circles.*

By using various describing curves, then, an infinite number of
tooth-outlines may be generated, all of which geometrically satisfy
the conditions. But many of them are of impracticable forms ; of
those which are not, none have been more extensively employed than
the Epicycloid and the Involute, to which therefore we shall at present confine our attention ; and we now proceed to the practical operations of "laying out the teeth" of a pair of wheels in outside gear.

182. Circular Pitch.—Supposing the distance between the centres of
a pair of wheels to be given, and the velocity ratio to be assigned ; the
first step is to divide the line of centres into segments having the
given ratio, thus determining the radii of the pitch circles. The circumference of each circle is next to be divided into as many equal
parts as its wheel is to have teeth.

The *pitch* of the teeth is the length of the circular arc measuring
one of these subdivisions ; or, in other words, it is the distance measured on the pitch circle, occupied by a tooth and a space. This pitch
arc, it is obvious, must be the same on each wheel ; although the
teeth *may* be smaller, and the spaces larger, on one wheel than upon
the other. The numbers of the subdivisions, then, are proportional
to the diameters of the pitch circles ; and a fractional tooth being
impossible, the pitch must be an aliquot part of each circumference.

The pitch as above defined is sometimes called the *Circular* pitch,
in distinction from what is called the *Diametral* pitch, which will be
explained hereafter.

183. Face and Flank.—The part of a tooth-outline which lies outside its pitch-circle, as $B'P'$ in Fig. 112, is technically called the *face*
of the tooth ; and the part which lies within the pitch circle, as
$E'P'$ in the same figure, is called the *flank*. Usually, each tooth has
both ; but wheels can be made, and sometimes used to great advantage, in which one of a pair has faces only, the other one having only
flanks ; we will consider this case first.

184. Arc and Angle of Action.—The angle through which a wheel
turns, while one of its teeth is in contact with the engaging tooth of
another wheel, is called the *angle of action :* and the arc of the pitch
circle by which it is measured, is called the *arc of action*. The latter
must, evidently, be at least equal to the pitch arc, in order that each
tooth may continue in gear until the next one begins to act ; and it
ought in practice to be considerably greater.

185. Backlash.—The size of the tooth depends, partially at least,
upon the pitch, since, as above stated, the pitch arc is to be divided
between a tooth and a space. Practically, of course, the teeth of both
wheels are made of the same thickness ; hence, were perfect work-

manship attainable, the tooth and the space might be made exactly
equal. But since it is not, the space must be made a little wider than
the tooth ; the difference is called *backlash*, and should be as small as
it is practicable to make it.   For our present purposes we may neg-
lect it altogether, and make the thickness of the tooth just one half
the pitch.

A Pair of Wheels.—Limiting Case.

186. In Fig. 113 the pitch and describing circles being drawn as in

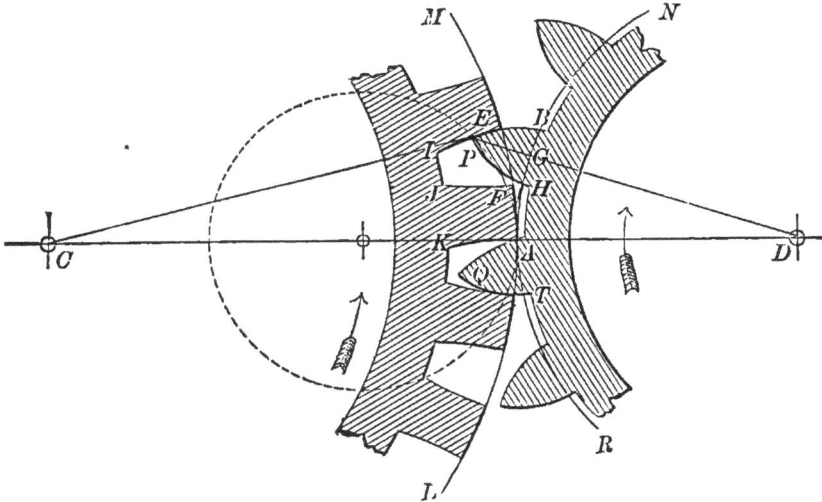

FIG. 113.

the preceding figure, let AB, AE, be the pitch arcs, and AP an equal
arc on the describing circle.   Then the *face* for the tooth of RN cannot be less than BP, since if made of just that height, as shown, contact is ending at P at the very instant when the next tooth begins to
act at A. Bisect AB in H, which gives the thickness of the tooth ;
and draw through H a reversed face similar to BP.

The conditions are purposely so chosen that this reversed face passes
through P : the case is, therefore, a barely possible one, the tooth
being pointed, and just high enough to make the angle of action equal
to the pitch angle.

We found that the face must be of the height BP, in order to secure this angle of action : drawing PD, which cuts the pitch circle
in G, we see that in this case BG is just half the thickness of the

tooth. Had BG been greater, GH must have been less, so that the reversed face through H would not have passed through P, but between P and G, and the case would have been impracticable, without reducing the pitch and giving both wheels more teeth.

But if BG had been *less* than half the thickness of the tooth, the tooth itself could have been made higher by extending the face above P, or it might be left of the height BP, which would have given it some thickness at the top, as in the next figure.

187. Clearance.—The *acting* flank is EP ; but in order to let the teeth of the other wheel pass, the hypocycloid is continued to I, making the depth of the space a little greater than PG ; the difference is called *clearance*, and a similar provision is made in the other wheel by cutting in radially, as shown at A, H, B, a little below the pitch circle. The tooth of LM is completed by bisecting the pitch-arc AE at T, and drawing the curves AK, FJ, similar to EI.

188. A Practical Case.—Limiting cases like the preceding are to be avoided in practice. A pointed tooth is bad, as being weak and liable to wear at the top. And even if it be not pointed, the angle of action should be greater, as otherwise the least wear at the top reduces the face below the requisite height, and causes one tooth to quit correct driving contact before the next one properly begins to act. We say *correct driving contact*, for it will be seen that if in Fig. 113 we suppose all the teeth to be removed except the pair in contact at P, the face EB would push the flank IE out of its way, even if the height were slightly reduced, but the two acting curves would not be tangent to each other, and the velocity ratio would not remain constant, but the speed of the wheel C would diminish.

A reasonable case is shown in Fig. 114; the arc of action in this instance is $1\frac{1}{2}$ times the pitch, and drawing the radial line PS, we find BG to be much less than $\frac{1}{2}$ BH, thus enabling us to give the tooth a substantial thickness, PK, at the top.

189. Approaching and Receding Action.—In Figs. 113 and 114, the action takes place wholly on one side of the line of centres. If RN be the driver (the directions being as shown by the arrows), the action begins at A and ends at P, the point of contact continually *receding* from the line of centres, in which case AB, AE, are called **arcs of recess**, or of receding action. If LM drive (in the opposite direction), the action begins at P, ending at A : the point of contact is always *approaching* the line of centres, and AB, AE, are then called **arcs of approach**, or of approaching action.

It has been found by experience that the friction is greater and more injurious in the latter case than in the former : when such wheels as those under consideration are used, therefore, the one whose teeth have faces only should always drive.

190. But even then there is one drawback, which will be understood by reference to Fig. 112. The longer the arc of action, the longer the face of the tooth, and the greater the *obliquity* of the line of action, that is, its inclination to TT, the common tangent of the pitch circles. Now the pressure, as well as the motion, is transmitted in the line of action, and the greater its obliquity, the greater will be the component of pressure in the line of centres, tending to cause friction and wear in the bearings.

The amount of sliding also increases more rapidly as the point of

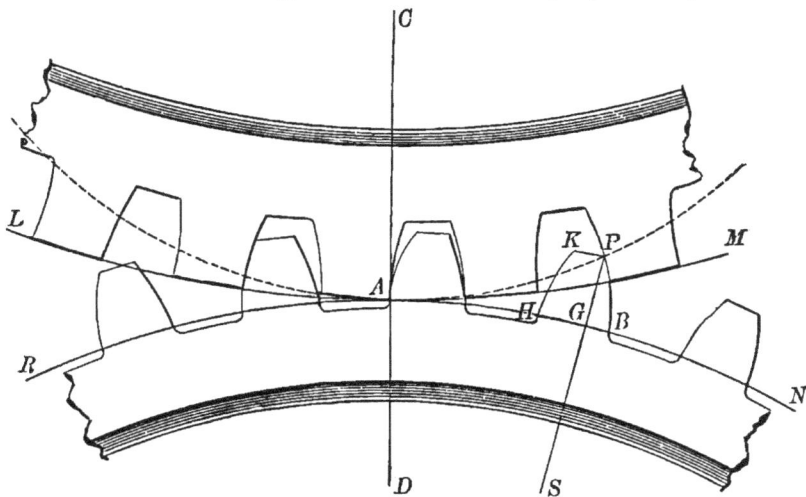

Fig. 114.

contact recedes from the line of centres, so that upon the whole such wheels are better suited for use in light mechanism where the teeth can be made small and numerous, and smoothness of action is important, than for the transmission of heavy pressures.

Teeth with both Faces and Flanks.

191. It will now readily be seen that by using another describing circle on the other side of the point of contact of the pitch circles, thus giving both faces and flanks to the teeth of each wheel, two things will be accomplished : a given angle of action may be secured with shorter faces, and therefore with less sliding, and this angle will be divided into an *angle of approach* and an *angle of recess*, thus enabling us to use either wheel as the driver.

If a wheel has both to drive and to follow, the arcs of approach and of recess may be made equal or nearly so, but if one wheel of a pair is always to be the driver, it may be desirable to make the arc of recess the greater, in order to reduce the amount of the more detrimental friction.

192. The construction is shown in Fig. 115 ; all that relates to the face *BP* for *RN*, and to the flank *EP* for *LM*, is precisely the same as in Fig. 113, and the lettering being made so far to correspond in

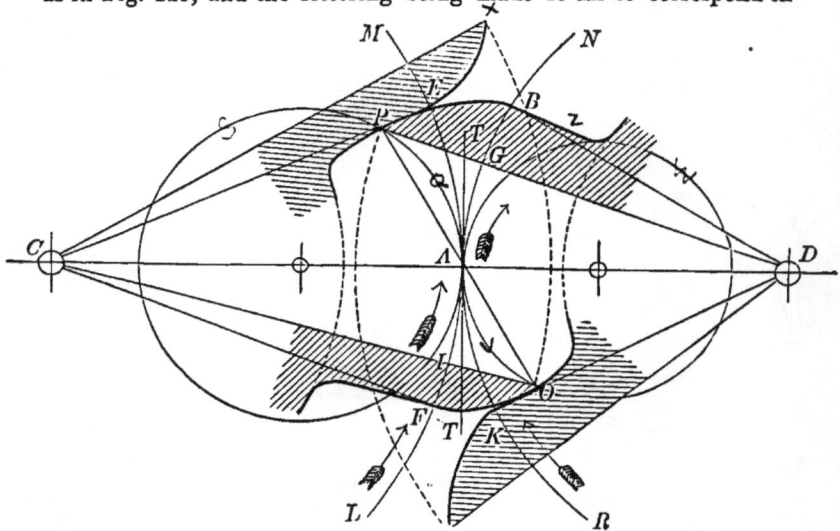

Fig. 115.

the two diagrams, no further explanation is needed in reference to those curves.   To complete the teeth, another describing circle is used on the opposite side of the pitch circumferences, which generates the face *OF* for *LM* and the flank *OK* for *RN*.

If we assume the arc of action on that side of *CD*, as for instance *AF* or *AK*, the possibility of securing it with a given number of teeth is at once ascertained by making the arc *AO* equal to *AF*, and drawing *OC* to cut *LM* in *I*.   If *FI* be less than half the thickness of the tooth which is required by the given pitch, or *equal* to it, the construction is possible, the tooth in the latter event being pointed ; if greater, it is impracticable.

Should it prove to be feasible, we have only to draw the epicycloid *OF*, which joined to *EP* completes the outline of the tooth for *LM*, and the hypocycloid *OK*, joining the latter to *BP*, which finishes the outline of the tooth of *RN*.

That is to say, these are the whole of the *acting* outlines ; the flanks are, as already explained, extended to a greater depth in order to give *clearance*.

193. The operation will be readily traced, as in the diagram the acting side of a tooth of each wheel is drawn in two positions, showing the state of affairs at the beginning and at the termination of the action respectively. Supposing RN to drive, the action begins at O, the driver's flank pushing the face of the follower, and the point of contact moving in the arc OA, until the points K and F meet at A. The face of the driver then urges the follower's flank, the point of contact now traveling in the arc AP, and at P the action ends.

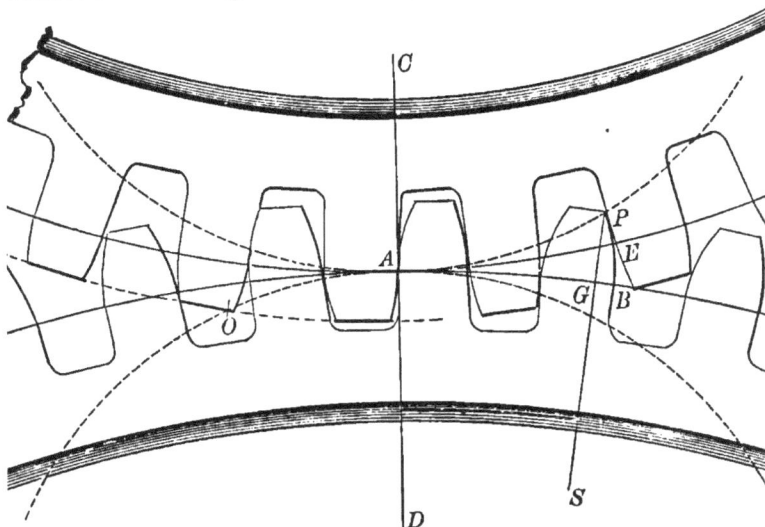

FIG. 116.

We see, then, that the angle of approach depends upon the length of the follower's face, and the angle of recess upon that of the driver's face : and if these lengths be assumed or given, the angles are easily found. For instance, had the length of the follower's face been assigned, as ES, then a circle described about C through S cuts the describing circle of that face in O, determining AO the length of the arc of approach, to which AF and AK are to be made equal.

194. A Practicable Example.—The diagram, Fig. 115, is drawn without regard to practical proportions, the only object being to illustrate the construction clearly ; but in Fig. 116, we have shown a feasible case. The cut is half size, and the conditions assigned are as follows :

Distance between centres, 27 inches.

Wheels to have 63 and 45 teeth respectively.

The smaller wheel to be the driver.

Whole arc of action to be $2\frac{1}{4}$ times the pitch.

Angle of recess to be $\frac{1}{3}$ greater than angle of approach.

We have, then,

$$63 : 45 :: 7 : 5, \quad 7 + 5 = 12, \quad \tfrac{27}{12} = 2\frac{1}{4},$$

also

$$2\frac{1}{4} \times 7 = 15\frac{3}{4}, = \text{radius of larger pitch circle.}$$
$$2\frac{1}{4} \times 5 = 11\frac{1}{4}, = \quad \text{``} \quad \text{`` smaller ``} \quad \text{``}$$

Again,

$$2\frac{5}{8} = \tfrac{21}{8},$$

which is to be divided into parts in the ratio of 3 to 4 ;
whence,

$$3 + 4 = 7, \quad \tfrac{21}{8} \div 7 = \tfrac{3}{8} ;$$

and

$$\tfrac{3}{8} \times 3 = 1\frac{1}{8} \text{ times the pitch} = \text{angle of approach,}$$
$$\tfrac{3}{8} \times 4 = 1\frac{1}{2} \quad \text{`` } \quad \text{`` } \quad \text{`` } = \quad \text{`` } \quad \text{`` recess.}$$

Interchangeable Wheels.

195. Inasmuch as the face and the flank which act upon each other are generated by the same describing circle, it makes no difference

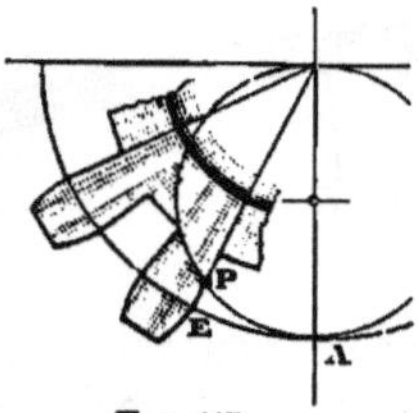

Fig. 117. 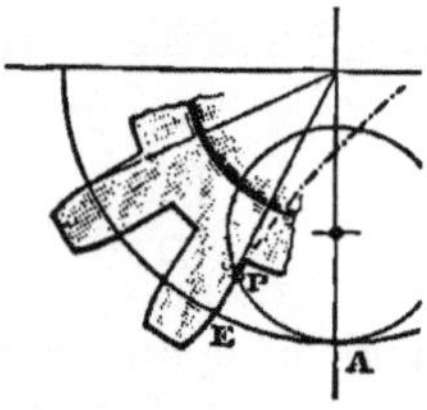   Fig. 118. 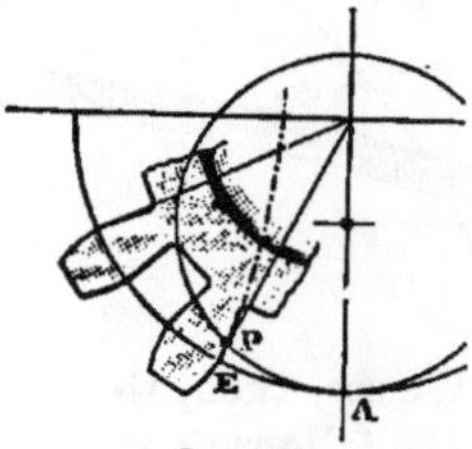   Fig. 119.

whether the diameter of the one which traces the other face and flank be the same or not, in laying out a single pair of wheels, and in Fig. 115, the describing circles are of different diameters. But for the very reason just stated, it is clear that if we wish to make a number of wheels, any one of which will gear with any other one, we must use the same describing circle for all the faces and all the flanks.

196. Size of the Describing Circle.—In making such a set of interchangeable wheels, the question at once arises, how large should the describing circle be ?

The answer to this depends upon properties of the hypocycloid, which are illustrated in the next three diagrams. In Fig. 117, the describing circle is half as large as the pitch circle; and the generated curve degenerates into a right line, so that the tooth, having radial flanks, is comparatively weak at the root. In Fig. 118, the describing circle is smaller, and the flank curves away from the radius, outwardly with respect to the body of the tooth, as it recedes from the pitch circle, giving a stronger form. In Fig. 119, on the other hand, the describing circle is larger, and the flank curves in the opposite direction, rendering the tooth both weak and difficult to make.

The safe practical deduction would seem to be, that the diameter of the describing circle should not be more than half that of the pitch circle of the smallest wheel of a set. Still, it will be found that if it be made five-eighths instead of one-half that diameter, the curvature of the flanks will not be so great, with the customary proportions of height to thickness of the teeth, as to make the spaces any wider at the bottom than at the pitch circle; the teeth can therefore be made as usual by means of a milling cutter, and a describing circle of the size last mentioned has been employed with excellent results.

197. In special constructions, as, for instance, in laying out a single pair of wheels for which cutters are to be made expressly, good results for general purposes may be attained by the use of two describing circles, the diameter of each being three-eighths that of the pitch circle within which it rolls.

However, with a given arc of action, the face is shorter, and the obliquity of the line of action less, the larger the describing circle. Consequently in very delicate mechanism, as, for example, in watch-work or clock-work of the finest grades, the advantages thus gained may make it advisable to use teeth of the form shown in Fig. 119, notwithstanding the difficulty of making them, and their inherent weakness; the latter may be to some extent obviated by using large fillets at the junction of the sides and bottom of the spaces, which is quite admissible, because, as already shown, the *acting* flank is comparatively short, and the exact outline of the clearing space is of no consequence, so long as the space is great enough.

198. Rack and Wheel.—A *rack* is simply an infinitely large wheel. The curvature of a circle diminishes as the radius increases, and disappears when the radius becomes infinite. Thus the *pitch line* of a rack is only a straight tangent to the pitch circle of the wheel with which it works, and the line of centres becomes a perpendicular to this pitch line, passing through the centre of the wheel.

The rack will travel through a distance equal to the circumference
of the pitch circle of the wheel during one revolution of the latter,
whatever the num-
ber of teeth, and
in the same propor-
tion for any frac-
tion of a revolu-
tion. The pitch of
the rack teeth,
therefore, is found
by rectifying the
pitch arc of the
w h e e l, whatever
that may be, and

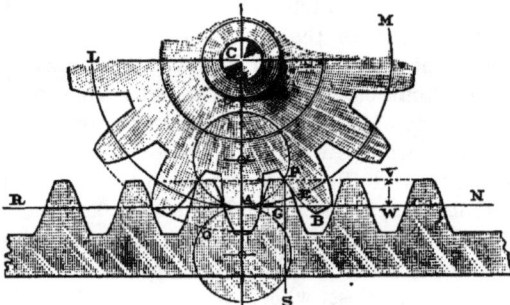

FIG. 120.

setting off that length upon the pitch line. The construction is
shown in Fig. 120; the two describing circles are here made of the
same diameter, and it is clear that if the same circle be used to generate the faces and flanks of a set of wheels, any one of them will gear
with the rack if the pitch be also the same.

199. Evidently, both faces and flanks of the rack teeth are cycloids,
being generated by the rolling of a circle upon the pitch line. If the
length of the face be assumed, as WV for instance, a line parallel to
RN, through V the highest point, cuts the pitch circle in P, thus determining AP, to which AB must be made equal, and fixing the part
of the action which will take place on the right of CD. Or if AB
be assigned, we make AP equal to it, thus ascertaining the necessary
length of face. In either case, PS is now to be drawn perpendicular
to the pitch line, which it cuts at G; and as in the preceding constructions, BG cannot be greater, and should be less, than half the
thickness of the tooth as determined by the pitch. The part of the
action which will take place on the left of CD, depends upon the
length of the face of the wheel tooth, and is ascertained as in the
cases previously explained.

200. If in constructing the teeth for a pair of wheels, we employ
two describing circles, the diameter of each being one-half that of the
pitch circle within which it rolls, the teeth of each wheel will have
radial flanks. A similar course may be pursued in laying out a rack
and wheel, as shown in Fig. 121. The describing circle whose diameter is AC, generates the radial flanks of the wheel-teeth and the
cycloidal faces of the rack-teeth.

The diameter of the pitch-circle of the rack being infinite, one-half
of it is also infinite, and RN is therefore the describing line for the

faces of the wheel-teeth, which consequently are involutes of the pitch-circle *LM*, and for the flanks of the rack-teeth.   The latter being

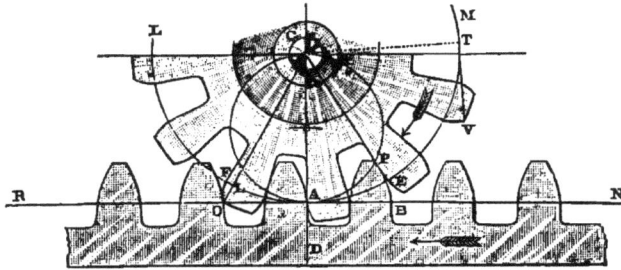

Fig. 121.

radii of the infinite circle *RN*, would, strictly, be simply straight lines perpendicular to it, as shown in the teeth on the right of *CD* ; and they were formerly made so.   But when these parts of the teeth are in action, the point of contact moves in the describing line, that is to say, in the pitch line *RN* itself.

The fact is, that the *acting* flank of the rack-tooth degenerates into a point ; thus a marking point carried by *RN*, in moving from *A* to *O*, traces on the plane of the wheel, which turns as indicated by the arrow, the involute *FO*, the arc *AF* being equal to *AO* : while, having no motion relatively to the plane of the rack, it marks on that plane simply the point *O*.   The action is consequently bad, this point being subjected to excessive wear : but the flank of the rack tooth may be made, as shown on the left of *OD*, an arc of a circle whose centre is on *RN*, and radius equal to the radius of curvature of the face of the wheel-tooth at its highest point.   This radius is found thus : let *V* be the highest point of the involute tooth ; draw through *V* a tangent to the pitch-circle *LM*, lying, as shown, on the concave side of the involute, and find the point of tangency *T* : then *VT* is the required radius of curvature.

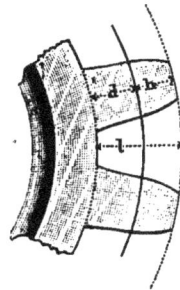

Fig. 122.

201. Arbitrary Proportions.—It is not necessary in all cases, to pay particular attention to the relative amounts of the approaching and the receding action.   And it is a very common practice to make the whole radial height of the tooth a certain fraction of the pitch ; the part without the pitch circle being a little less than that within, by which the clearance is provided for.   In Fig. 122, these parts are marked *h* and *d* respectively. *l* being the whole height.   Three of these arbitrary proportions, as they may be called, which have been extensively adopted, are as follows :

1. $l = \frac{1}{4}$ pitch ; $h : d : : 11 : 13$. $(b = \frac{1}{15}$ pitch.$)$
2. $l = \frac{1}{3}$ pitch ; $h : d : : \ 4 : \ 5$. $(b = \frac{1}{20}$ pitch.$)$
3. $l = \frac{7}{10}$ pitch ; $h : d : : \ 3 : \ 4$. $(b = \frac{1}{11}$ pitch.$)$

The whole angle of action, as well as the ratio of the approach to the recess, will of course vary according to the numbers of the teeth, in the use of any such system : but either of these rules will give satisfactory results for most purposes, the wheels acting as drivers or followers indifferently, provided that there are at least twelve teeth upon the smallest wheel. Less than that should not be used unless it is necessary : sometimes, however, the use of lower numbers cannot be avoided, in which event it will often be requisite to extend the faces of the pinion's teeth beyond these limits ; and the proper length should be determined as before explained.

202. It was also formerly the custom to make the backlash a definite fraction of the pitch, which we have added above in parenthesis as usually given in connection with each of the preceding rules. But although these values may have been proper in many cases, as allowing for imperfections of workmanship in wooden mill-work or when the wheels were simply to be cast, it is certain that they are in many cases too large, if the teeth are to be cut with the slightest pretension to accuracy. Nor does there appear to be any reason why the backlash should vary *directly* with the pitch ; on the contrary, it seems almost self-evident that the coarser the pitch, the smaller will be the proportion borne to it by any unavoidable error. From this point of view, it appears more reasonable to say that the backlash should vary *inversely* as the pitch ; and perfectly safe to insist that it ought in every case to be as small as the skill of the workman will enable him to make it with the facilities at command.

Since theoretically the teeth may be in contact on both sides at once, we have in the diagrams entirely disregarded the backlash. If it were introduced, the constructions, evidently, would be modified only in this respect, that the "thickness of the tooth as required by the pitch," instead of being exactly half the pitch arc, would be less than that by just the amount of backlash allowed.

3. Annular Wheels.

203. In Fig. 123, the smaller pitch circle lies within the greater, which it touches internally ; the teeth of the outer wheel being therefore formed on the concave circumference of an annular rim. But neither the mode of generating the tooth-outlines, nor the nature of

their action, are in any way changed; two describing circles are shown, each of which generates a face for one pitch circle and a flank for the other,—and in short, a comparison of this diagram with Fig. 115, which is lettered similarly throughout, will show that the two are identical in all particulars relating to the construction.

It will also be ob-
served that the contour
of the annular wheel,
in respect to the forms
of the acting curves, is
identical with that of
an ordinary spur wheel
having the same pitch
and describing circles,
the tooth of the one cor-
responding precisely to
the space of the other.

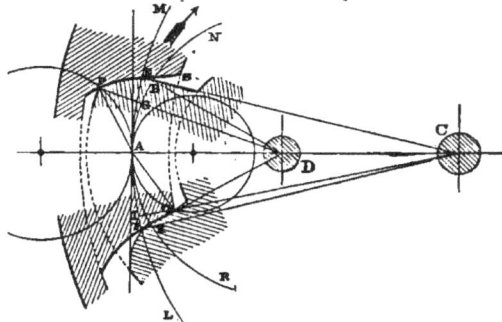

FIG. 123.

The describing circles in the figure are of different diameters : but they might have been equal, and it will readily be seen that the spur wheels which are thus made interchangeable with each other (195), may also be made interchangeable with annular ones. But in the construction of inside gearing the diameters of the describing circles are limited by considerations which relate entirely to the peculiar condition of internal tangency between the pitch circles. The manner in which these limits are determined, will be most clearly seen by first regarding the tooth-outlines as generated in another way.

FIG. 124.

204. Intermediate Describing Circle.—In Fig. 124, let C be the centre of the inner pitch circle, D that of the outer, and MOL a curve lying between the two circumferences, tangent to both at A, and capable of rolling in contact with them. Let the arcs AK, AF, of the pitch circles, be equal to each other, and during the rotation indicated by the arrow, let a marking point be carried from A to O, by the describing curve : it will in its progress, evidently, trace upon the planes of rotation, the face OK for the pinion, and the face OF for

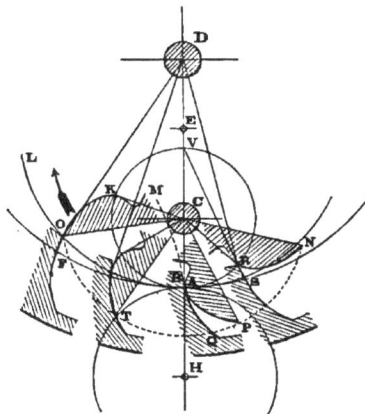

the annular wheel. If these be used as the outlines of teeth, it follows from the mode of their generation that they will transmit rotation with a constant velocity ratio, whatever the form of the describing curve or of the locus of contact: if the pinion drive, the action begins at A, and ends at O.

Now, if it be required that these faces shall be epicycloidal, the curve AOL must be, as in the figure, a circle whose centre E lies between C and D. It is true that OK and OF may then be generated in a different manner ; which, however, though an important coincidence, does not alter the fact that the driving contact between these two curves is due solely to the mode of generation here explained. And it will subsequently be seen that annular wheels may be required to work under conditions which can be satisfied only by this method of construction : upon which also depends the determination of the limiting diameters of the describing circles in either method.

205. Limiting Diameter of Intermediate Describing Circle.—In making these determinations, we avail ourselves of the peculiar property above alluded to, viz : that every epicycloid, internal or external, as well as every hypocycloid, is capable of two generations. **(See Appendix.)**

The face OF being now a true hypocycloid, may be generated not only by the intermediate circle whose centre is E, but by another whose radius is ED. In Fig. 124, AE is equal to CD, whence $ED = AC$ the radius of the smaller pitch circle. Consequently on rolling the pinion within the wheel, the hypocycloidal path traced by the point A of the former, coincides with the face AB of the adjacent tooth of the wheel. The radius AE may be increased, in which case this path AM will lie to the right of AB ; but if it be decreased, AM will lie to the left of AB, that is, within the body of the wheel's tooth, which is clearly impracticable. In this construction then we have the limit, that

The radius of the intermediate describing circle can not be less than the line of centres.

206. Drawing through O a circle about centre D, it cuts the inner pitch circle in R ; at which point the action begins, the face KO then having the position RN.

A marking point at R, carried by the inner pitch circle as a describing curve, will, in going to A, obviously generate the face RS similar to OF; against which this point R of the pinion will act during the approach, the locus of contact being the arc RA.

The face of the pinion having now reached the position AP, the receding action begins ; and here a phenomenon peculiar to inside gearing presents itself. For the face AP is not only an internal epicycloid whose generating circle is AOL, but an external one which may be generated by a circle whose radius is AH, equal to EC. It will, therefore, work correctly with a flank AQ traced by the same describing circle. This action terminates at T, the point in which a circle through O about C cuts the circumference of the exterior describing circle, and during its continuance the locus of contact is the arc AT.

This flank may or may not be used ; but if it be, since the action between the faces AP, AB, previously explained, also begins at A, the singular fact appears, that while the pinion is turning through the angle PCT, its face has *two points of driving contact*. This circumstance is of some practical importance, not only on account of the division of the pressure, but also as affecting the resultant obliquity of the line of action.

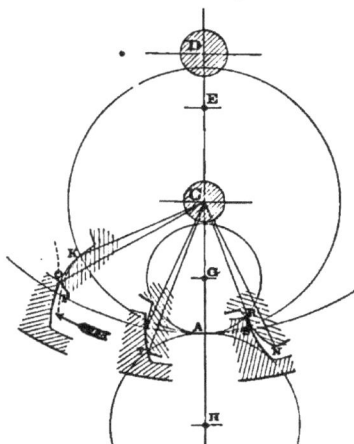

Fig. 125.

207. Limiting Diameters of Exterior and Interior Describing Circles. —In Fig. 125, the radius AE of the intermediate describing circle is greater than CD. The face OF may also be generated by an interior describing circle whose radius AG is equal to ED, and OK by an exterior one of which the radius is AH, equal to EC. In all that relates to the action the only new feature is that the pinion now has a flank of sensible magnitude, traced by the circle whose centre is G ; as will readily be seen, the lettering throughout being similar to that of Fig. 124. Now, either radius, AG or AH, may be diminished without changing the other ; the only result being that the faces OF, OK, as shown in dotted lines, will not be in contact during the recess, and will consequently act only against the corresponding flanks, as in Fig. 123. But if either of these radii be increased without diminishing the other to the same extent, it is apparent that the faces OF, OK, will intersect each other, rendering the construction impracticable : and we have as a limit, that

The sum of the radii of the two describing circles cannot be **greater** *than the line of centres.*

208. This holds true when either radius vanishes, the other then becoming equal to the line of centres. Thus, if in Fig. 125, we make $AE = AD$, we shall have

$$ED = 0 = AH, \quad \text{whence} \quad AG = EC = CD.$$

If, on the other hand, we make $AE = AC$, we shall have

$$EC = 0 = AG, \quad \text{whence} \quad AH = ED = CD.$$

These two limiting cases are illustrated in Figs. 126 and 127 respectively. In the former, it is apparent that when the pinion drives the arc of recess is greater than the arc of approach, but since the action

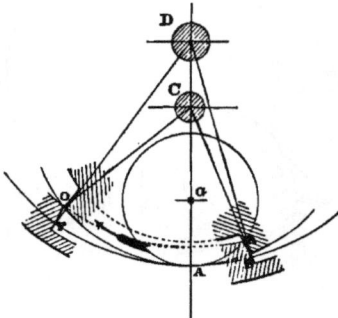

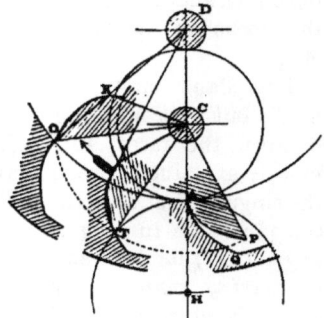

FIG. 126.   FIG. 127.

during the recess is confined to the single point O of the pinion's teeth, the advantage thus gained is more than neutralized. The fact that the pinion can have no face, will be seen from the consideration that if there were one, it must be tangent at O to the radius OC, and would, therefore lie within the body of the adjacent tooth of the annular wheel.

In Fig. 127, on the other hand, the wheel can have no face, for a

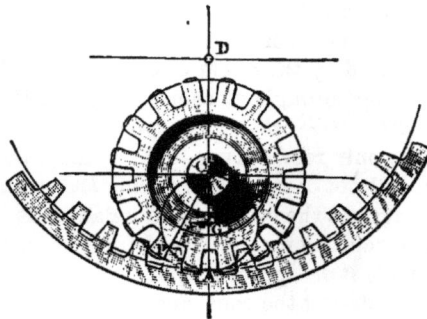

FIG. 128.

similar reason. The pinion consequently having no flank, there is no approaching action, but its face has two points of driving contact during a part of the recess.

209. It is evidently more frequently practicable in inside than in outside gearing, to secure an angle of recess greater than the pitch, and thus to avoid altogether the more injurious friction of approaching action. An instance of this is shown in Fig. 128, where, the wheel driving, an interior describing circle only is employed, whose radius is in this case half that of the inner pitch circle, thus giving the pinion radial flanks. Attention is called to the fact that when the tooth thus has no face, it should, nevertheless, be allowed to project beyond the pitch circle for the sake of strength, the corner being finished by a circular arc tangent to the radius at its extremity.

Fig. 129 shows the appearance of an annular wheel and pinion which differ but little in size ; the teeth are necessarily very short in order that they may escape from engagements and pass each other. Whether they will do so or not, if the height be as-

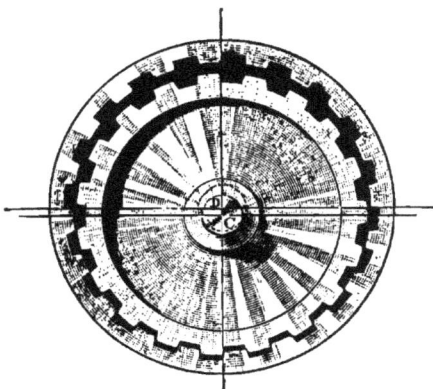

Fɪɢ. 129.

sumed, is readily determined by constructing the epitrochoid traced by the highest point of either tooth when its pitch circle is rolled upon the other one ; this path, obviously, must not intersect the tooth outline of the engaging wheel : and the clearing spaces of both the wheel and the pinion must also be such as not to touch the epitrochoids thus described.

CHAPTER VII.

On the Use of Low-numbered Pinions.

210. In the operation of epicycloidal teeth, the obliquity of the line
of action is continually varying ; diminishing during the approach, it
becomes zero when the point of contact reaches the line of centres, and
again increases during the recess.

If wheels are to do heavy work, it has been found by experience that
the mean obliquity should not in general exceed about 15°, nor the
maximum about 30°. A high maximum is less objectionable when
several pairs of teeth are engaged at once, since the greatest portion of
the pressure will be acting less obliquely. Such distribution of the
pressure, it will readily be seen, is more often to be attained in inside
gearing, and the obliquity is one serious disadvantage when low-numbered pinions act in outside gear. Another is the excessive amount
of sliding, due to the necessarily great length of the faces of the teeth.
And from both combined is deduced the practical rule that, in mill-
work and machinery in general, no pinion of less than twelve teeth
should be used if it be possible to avoid it.

But it is not always possible ; and in lighter mechanism, such as
clock-work, it is often necessary to use much lower numbers. In work
of this description a greater obliquity is often admissible ; and when
it is not so considered, the convexity of the flanks caused by using a
large interior describing circle, is not so objectionable when the work
to be done is light, and strength of form not an imperative necessity.

For the present, then, we will confine ourselves to the following
limits, viz : that the maximum obliquity shall not exceed 36°, and

that the diameter of the describing circle for the flanks shall not exceed ⅙ that of the pitch
circle within which it rolls.

211. In Fig. 130 are shown two equal pitch circles, and two equal describing circles of half the diameter, *TAS* being the common tangent. Draw the line of action with an obliquity of 30°; it cuts the describing circles in *O* and *P*, making the arcs *A O*, *A P*, each equal to 60°. Drawing *CPE* and *DOK*, the arcs *AE*, *AK*, are each equal to 30°, and we thus have two similar wheels which will just work, the angles of approach and of recess being each equal

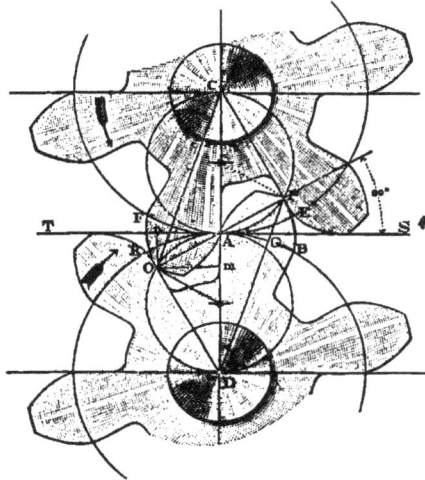

Fig. 130.

to half the pitch, and the mean obliquity exactly 15°. It is seen to be possible, with only a slight increase in the obliquity, to lengthen the faces and secure an arc of action greater than the pitch, as in practice it should be.

In this case the flanks are radial. But if without changing the describing circles, we reduce the pitch circles to ⅚ their present diameter, the arc *AP* of 60° will be equal to an arc of 36° on the new pitch circumference. It is, then, practicable to make two pinions of five teeth each, which, like those in the figure, will just work, with the same obliquity. The flanks will now be convex; still, the diameter of the describing circle is within the assumed limit, being but $\frac{6}{10}$ that of the pitch circle.

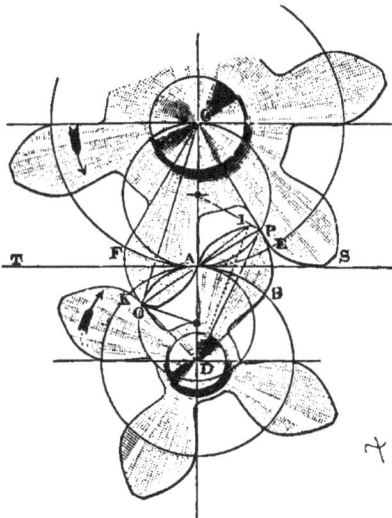

Fig. 131.

212. In Fig. 131 the diameter of the larger pitch circle is 1½ times that of the smaller one; also the diameter of the upper describing circle is equal to CA, so that as in the preceding figure the arc AP of 60° is equal to the arc AE of 30°, and PAS, the maximum obliquity on the right of CD, is also 30°. The diameter of the lower describing circle is $\frac{4}{5}$ that of its pitch circle, therefore the arc AO of 72° is equal to the arc AK of 45°, and OAT, the maximum obliquity on the left of CD, is 36°. Completing the construction, we perceive that the pinion of four teeth will just work with the other of six. The arcs of approach and recess are equal, the mean obliquity being 18° during the one and 15° during the other, giving an average of $16\frac{1}{2}$° for the whole action if the latter be made just equal to the pitch ; and the faces may also be made a little longer than here shown, as in the previous case.

The leaves of the larger pinion still have radial flanks, but proceeding as before, we may reduce the upper pitch circle to $\frac{5}{6}$ of its present diameter, without changing the circle which rolls within it. We shall then have the arc AE equal to 36°, and the five-leaved pinion thus constructed will work with the four-leaved one, the obliquity remaining unchanged, since the points, O, A, P, retain the same positions as in the figure.

213. Now produce OA to I, making $AI = AO$. Then, were another describing circle drawn above TS, equal in diameter to the one below it, the point I would lie upon its circumference ; and since the arc AI would be equal to the arc AB, the epicycloid traced by rolling it upon the lower pitch circle would extend to the point B.

Draw ID ; it will then be apparent that were the angle IDB equal to or less than the angle IDA, two four-leaved pinions would be capable of working together, all the acting curves being traced by a describing circle whose diameter is $\frac{4}{5}$ that of the pitch circles. Now in the triangle ADI, the sides AI and AD, also their included angle, are known ; whence it will be found that the angle IDA is 22° 27′ 42″, whereas it should be 22° 30′ in order that one pinion should drive the other through a total arc of action just equal to the pitch, even if the teeth were pointed.

Five, then, is the least number of teeth that can be used if the pinions are to be alike, and four will work with five or any higher number.

214. The force of the objections urged against the use of small pinions for heavy work, is clearly shown in Fig. 130. Letting OA represent the pressure when O is the point of contact, this may be resolved into the two components Om, On ; of which the latter tends simply to force the journals apart, thus increasing the friction in the bearings. Since only one pair of teeth is in action, and the angle OAT is 30°, this objectionable component is equal to half the total pressure at O, which again is greater than that at the point K on the pitch circumference, in the ratio of DK to DO.

215. Thus far, the angles of approach and recess have been made equal. Should it be desired to make them otherwise, the question whether any assumed conditions can be satisfied or not, is readily settled by the construction of a diagram, as explained in connection with Fig. 115.

If, as usual in such cases, the angle of recess is to be the greater, it will be apparent that with a wheel of a given number of teeth a smaller pinion may in general be used to drive than to follow. For the angle of recess depends upon the length of the face of the driver's tooth ; and the smaller the pinion the smaller will be the describing circle which can be used within it, and, consequently, the less will be the greatest possible length of the face of the wheel-tooth, since the pitch is already assigned by the conditions.

216. An illustration of this is incidentally afforded by Fig. 132, in the construction of the three-leaved pinion. Taking for its flanks a describing circle whose diameter is $\frac{5}{8}$ that of its pitch circle, and making OAT, the maximum obliquity on the left of CD, 36° as before, the angle of action, KDA, on that side of the line of centres is 45°. The pitch arc $K\overset{\frown}{B}$ is 120°, therefore the angle of action ADB, on the right of CD, must be 75°. It will be found that this can be secured, with a very little to spare, by using for its faces a describing circle five times as large as the one used for its flanks.

And keeping to the limits heretofore observed, it follows that the second pitch circle cannot be less than five times as large as that of the pinion ; which, therefore, will just work with a wheel of fifteen teeth; and no less, under the conditions as-

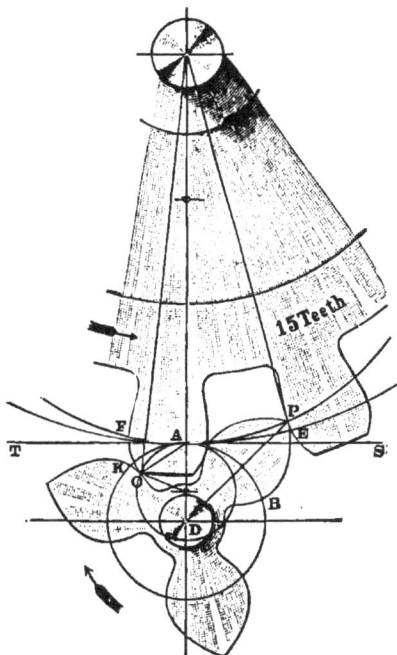

Fig. 132.

signed. In other words, a three-leaved pinion will drive a wheel of

8

fifteen teeth, the angle of approach being $\frac{3}{8}$, and that of recess $\frac{5}{8}$, of the pitch angle. Reasonably well, too, in respect to the obliquity of the line of action; during the approach the maximum is 36° and the mean is 18°, during the recess the maximum is 12° and the mean 6°: but the latter angle is larger than the former in the ratio of 5 to 3, whence we have, as the average obliquity during the whole action,

$$\frac{18° \times 3 + 6° \times 5}{3 + 5} = \frac{84°}{8} = 10\frac{1}{2}°.$$

217. It will be perceived that the above construction involves the solution of the following problem, viz. :

Given the pitch circle, number of teeth, and arc of recess, of the driver, to find the least number of teeth which can be assigned to the follower.

This may be determined graphically, as shown in Fig. 133, where CD is the line of centres, C the centre of the driver, and AF the given arc of recess. Make FL equal to $\frac{1}{4}$ the pitch, that is, half the thickness of the tooth, and draw through L a radial line of indefinite length; then if the tooth be pointed, its point will lie upon the prolongation of CL. Make AG, perpendicular to CD, equal to the arc AF (see Appendix (1)), and set off $AM = \frac{1}{4} AG$. With centre M and radius $MG = \frac{3}{4} AG$, describe an arc cutting CL produced in O. Draw OA, bisect it in N by a perpendicular cutting CD in E; then (see Appendix (1)), the arc OA, whose centre is E, will be equal to AG and therefore to AF.

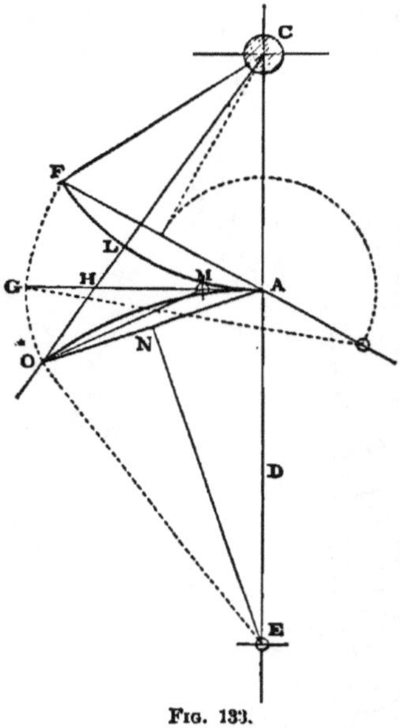

Evidently, then, E is the centre of a describing circle which by rolling on the given pitch circle will trace an epicycloid from O to F, and this will be the face of the driver's tooth. Ob-

Fig. 133.

serving the limits before assigned, the radius of the required pitch circle cannot be less than $\frac{2}{3}$ of AE; but it must be such that the given pitch, viz., four times the arc FL, shall be an aliquot part of the circumference.

It may be noted, that the conditions assigned as above furnish sufficient data to enable us to verify the results by trigonometrical computation, should it be considered necessary.

218. By the above process we ascertain at once the size of the describing circle which will give the driver a pointed tooth, and from this derive the diameter of the follower's pitch circle. Should the latter correspond to a fractional number of teeth, the next higher integer must be used, the pitch circle being increased accordingly. In this case the diameter of the describing circle may be increased or not, at pleasure; though it is better that it should be, since then the driver's tooth may be topped off.

And it may be necessary to increase it for another reason; because the very process of determining the minimum radius AE, Fig. 131, fixes also the maximum obliquity OAG corresponding thereto. Attention must, therefore, be given to this, for it is possible that the obliquity thus fixed may be too great, although the numbers of the teeth be practicable; and in that event the describing circle, and, if necessary, the pitch circle as well, must be increased until the obliquity is reduced to the desired limit.

219. Fig. 134 illustrates the converse case to that of Fig. 132, the conditions being that the wheel of 15 teeth is to drive, and that the angle of recess shall be ⅝ of the pitch angle. Making the construction as explained in (**217**), the least pitch circle of the follower corresponds, as found by computation, to 4.45 teeth, with a maximum obliquity of 40° 36' 2". Five, therefore, is the least practicable number of leaves that the pinion can have, thus making the pitch 72°, and the arc of recess 45°, and by using a describing circle for their flanks, whose diameter is ⅝ that of the pitch circle, the maximum obliquity during the receding action is reduced to 36°, while at the same time the driver's tooth is made of reasonable breadth

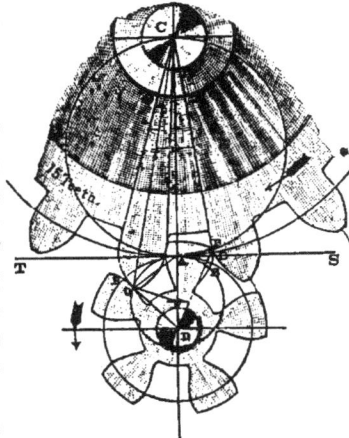

FIG. 134.

at the top. In this diagram, as in Fig. 132, the angle of approach is ⅝ of the pitch angle, but it is obvious that although the driver's flanks are radial, the angle of approach might be very considerably increased in this case, although in that of the 3-leaved driver it could not be.

220. The two-leaved Pinion.—When a spur-wheel is made in the ordinary way, that is, with the teeth in the same plane, a pinion of three leaves is the smallest that can be used either to drive or to follow. But if the alternate teeth are placed in different planes, a two-leaved pinion can be made to drive in a very satisfactory manner ; the arrangement is clearly shown in Fig. 135, the pinion being composed of two heart-shaped cams or teeth, U and W, fixed side by side on the same shaft, and the wheel of the two star-shaped plates similarly fixed upon another shaft, of which the one, M, works with the cam U, and the other, N, with W. The diameter of the describing circle is equal to the radius of the pitch circle of the follower, whose flank is therefore radial. The diameters of the pitch circles are in this case in the

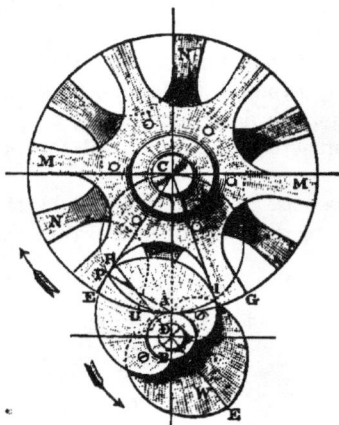

ratio of 6 to 1 ; the arc AP of 60°, is therefore equal to the arc AE of 30°, and to the semi-circumference AB. In the position here shown P is a point of contact, but the epicycloid BP is continued to R, so that the action is not yet ended, although the tooth AIS is just beginning to drive a flank of the other plate N of the wheel.

It is obvious that a half revolution of the pinion will bring the point S to the position R, turning the wheel through the angle ACE of 30° ; G and B then meeting at A, the action will be continuous.

FIG. 135.

The obliquity, it will be observed, is not great, but the amount of sliding is excessive, owing to the great length of the face PB as compared with the flank PE ; still the action is smooth and noiseless, since the driving contact is wholly receding, there being no arc of approach.

The number of teeth in the wheel may be reduced, for the flank need not be radial : by using a describing circle of $\frac{5}{8}$ the diameter of the pitch circle, and making $AP = 72°$, the angle ACE will become 45°, the velocity ratio thus obtained being that of 4 to 1, and the maximum obliquity very little over 36°.

221. Low-numbered Pinions in Inside Gear.—In making investigations similar to the preceding in relation to annular wheels, different cases will be found to arise under varying conditions, the treatment of which will perhaps be best developed by beginning with a specific example.

Let us take in illustration the three-leaved pinion, Fig. 136, the conditions being as follows :

Diam. Inner Des. Circle $= \frac{5}{8}$ Diam. Pitch Circle.

Arc of approach. 45° } Total = 120° = Pitch.
    " " recess,   75°

Required to find the least annular wheel which can be driven.

Evidently a describing circle externally tangent to both pitch circles, as in Fig. 123, must be used to generate the faces of the pinion and the flanks for the wheel; and its least radius may be found as in Fig. 133.

Knowing, then, this radius and that of the pinion's pitch circle, for the reason given in (**208**) the radius of the larger pitch circle must be greater than their sum. Also, it must be such that the pitch which is given shall be an aliquot part of the circumference.

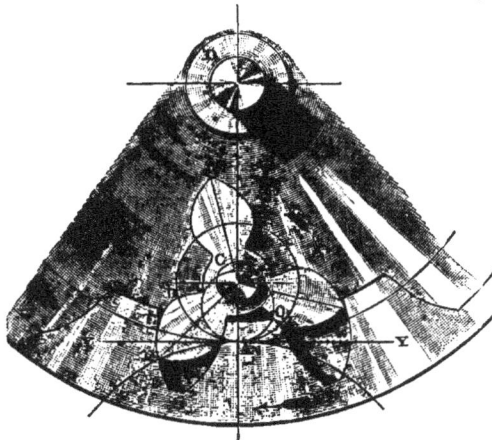

Fig. 136.

Taking, for convenience, 5 as the radius of the inner describing circle, then $AC = 8$, whence by computation we find the least radius of the exterior describing circle to be 24.93.

AD, then, must be greater than $8 + 24.93 = 32.93$; let it equal 33 for instance.

Now the numbers of the teeth are proportional to the radii of the pitch circles ; and letting

$$N = \text{No. teeth of Annular Wheel,}$$

we must have

$$3 : 8 :: N : 33,$$

which gives

$$N = 12\tfrac{3}{8} .$$

But since N must be an integer, the least number that can be used is 13 ; and the corresponding radius of the pitch circle is $34\tfrac{2}{3}$.

222. With this radius accordingly, the wheel shown in Fig. 136 is constructed ; the radius of the outer describing circle is taken as 25, so that the teeth of the pinion are not *exactly* pointed, though their breadth at the top is so small as to be practically inappreciable.

And a glance at the figure is sufficient to show that no matter how many more teeth are given to the wheel, the pinion will work with it equally well. If the outer pitch circle be increased, the chord of the arc *AE* will also increase, and the flank *PE* lying always outside of the face *PB*, will continue to be, as now, a practicable curve; which is self-evidently true in regard to the face *OF* of the wheel-tooth.

This pinion, it will be observed, is precisely the same as the one shown in Fig. 132 ; and is, therefore, capable of working with a rack, or with a wheel of any number of teeth whatever, between the limits of

13 and ∞ in Inside Gear,

∞ and 15 in Outside Gear.

223. From this example, then, we have thus far ascertained that with a given pinion, under certain conditions at least, there is an inferior limit, below which the number of teeth in the annular wheel cannot be reduced, but no superior limit beyond which the number may not be increased.

To these results we call attention, the more particularly because so high an authority as Professor Willis makes, without further comment, this sweeping statement, viz. :

" The case of annular wheels differs from that of spur-wheels in this respect, that, with a given pinion a small-numbered wheel works with a greater angle of action than a large-numbered one, and therefore we have to assign the *greatest* number that will work with each given pinion." It is true that he tacitly recognizes the absence under some conditions of a limit in that direction, by assigning "*any* number" as the greatest, in his tables of limiting cases ; but neither in this fact, nor in the remarks just quoted, is there any recognition of an inferior limit under any conditions. It is still more singular that he gives nowhere any explanation of the circumstances under which there is a maximum number, nor yet of the only mode in which the teeth can possibly be generated when there is one ; neither does any other writer that we know of.

224. Let us now suppose that a pinion of three leaves is to drive, under the conditions that the tooth shall be equal to the space, and the arc of recess just equal to the pitch.

Then in Fig. 137, AO being the diameter of the pitch circle, we shall have on its circumference the space $AL =$ the tooth $LB = 60°$; and if the tooth be pointed, its point must lie upon MC, perpendicular to CD, the line of centres. Evidently, if we draw AL and OB, these lines produced will intersect in P upon MC, making $OP = OA$, and the arc AP, upon the circle whose centre is O and radius OA, will be equal to the arc ALB. That circle may therefore be used as a describing circle, and by rolling upon the pinion's pitch circle it will generate the cardioidal faces PB and PL for its tooth. This describing circle cuts CD in N; drawing PN and LO, they will be parallel to each other and perpendicular to AP. Erect a perpendicular to PL at its middle point S; this will pass through B, and bisect ON in D, whence

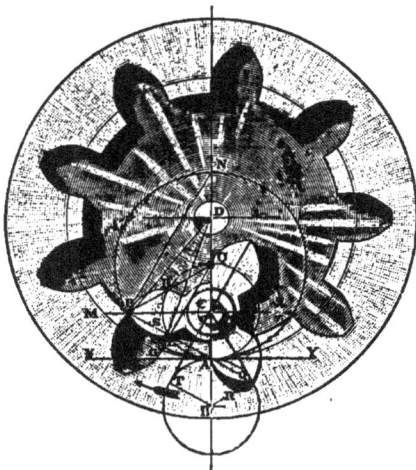

$$DA : OA :: 3 : 2, \; \therefore DA : CA :: 3 : 1;$$

and taking D as the centre of the outer pitch circle, an arc, AE, of $40°$ upon its circumference will be equal to the arc AP of $60°$ and the arc AB of $120°$.

Bisect AE in G, and draw ED, GD; then the angle included between these radii is bisected by SD, since $SDA = 30°$; consequently, L is the highest point on the face LG of one tooth of the annular wheel, just as P is in the face PE of the next one.

225. From this example we draw these conclusions, viz.:

1. When the conditions are such that the highest point in the face of the pinion's tooth, when quitting contact, lies *above* the common tangent YY, the describing circle *must* be an intermediate one, as in Fig. 125.

2. The intermediate describing circle is *largest* when the tooth is pointed.

3. The larger that describing circle, the larger may be the annular

wheel, for (205) AD may be equal to $AO + AC$, but cannot be greater.

Now by the construction above explained we have ascertained the *maximum* diameters of the describing circle and of the outer pitch circle, the latter being three times that of the inner; therefore 9 is the greatest number of teeth that can be given to the annular wheel. The number assigned by Prof. Willis under the same conditions is 12.

226. It will be perceived that when an intermediate describing circle is used, the obliquity varies inversely as its diameter. If, then, a maximum value of the obliquity be assigned, this will determine a minimum diameter of the describing circle, and consequently of the outer pitch circle, which fixes in this case also an inferior limit to the number of teeth in the annular wheel. Thus, in Fig. 137 it is clear that although if we reduce the outer pitch circle we may also reduce the describing circle, we cannot do it without increasing the obliquity, of which the maximum value in the figure is 30°. Now, if it be stipulated that this value shall not be exceeded, we must retain the present describing circle, but we may still diminish the outer pitch circle. If this be done, the face PE will become shorter, and the limit, obviously, will be reached when AD becomes equal to $AO = 2\,AC$. The hypocycloid PE will then have degenerated into the point P, to which the whole action will be confined; all the points of BP, which remains unchanged, coming successively into coincidence with P. We find, then, that under this additional restriction the *least* number of teeth that can be employed is 6. If, however, it should under any circumstances be desirable for the modification of motion to use such a combination, regardless of excessive obliquity, it is proper to note that by making the describing circle smaller the number may be reduced to 4; and in general, a pinion of any given number of teeth may thus be made to work, more or less satisfactorily, with an annular wheel having one more, a fact of considerable importance in the construction of differential trains of wheels.

227. It is also to be observed that in Fig. 137, the point L upon the inner pitch circle is at once the highest point of the face GL and the lowest one of the face PL; whence, as will readily appear if we suppose the motion to be reversed, the arc of approach is equal to half the pitch.

Again, the face AQ will work with a flank generated by the circle whose radius is $AH = OC = AC$. And the point Q lies upon the circumference of this circle; for it lies on PA produced, and $AQ = AL = AC = AH$, whence, if HQ be drawn, the triangles ACL,

AHQ, will be equilateral and equal. But $ACQ = ACT = 30°$; so that the angle QCT, while turning through which the pinion's face has two points of driving contact, is also equal to half the pitch.

In rolling the outer pitch circle upon the inner, the point L traces an epitrochoid, to which AL is normal and LO is tangent at that point (these lines corresponding to AR and RV in Fig. 124). The outline of the clearing space in the pinion must be such that L can move freely within it, but may also be tangent to LO at L. But the face PL is tangent at the same point to the radius LC; consequently the tooth must be formed with a positive, although an obtuse, intersection at that point. Still, were such a wheel and pinion made of the proper material and finish, as for instance of hardened steel finely polished, the combination might be used in light mechanism.

228. The above is a special case, in which the argument is based upon peculiarities of the assigned conditions; and was selected because it seemed most simply and clearly to illustrate the principles involved.

In general, however, a direct geometrical solution like this is not possible, and the required results are reached, as in outside gearing, by the process explained in connection with Fig. 133. In adapting it to the case of inside gearing, the diagram, as shown in Fig. 138, is modified only in this respect, that OA, and consequently the centre E of the intermediate describing circle, will lie above instead of below the tangent line AG. In this case it is necessary to produce CO and AG in order that they may intersect in H; which being done, we have before us all the data for trigonometrical verification.

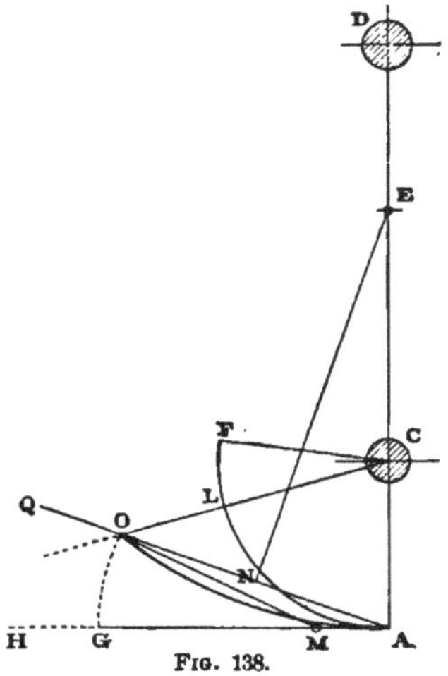

Fig. 138.

If the obliquity be assigned, the construction is made in the following order: Draw the indefinite line AQ, making with AG an angle equal to the given obliquity; then the arc GO cuts AQ in O, the highest point of the tooth. Then as before bisect OA by a perpendicular cutting AD in E; this determines AE, the radius of the *least* intermediate describing circle.

229. The conditions might be such that the point O should fall exactly upon the common tangent. In this event OA, equal to OF, is itself the arc of the describing circle, whose radius AE, whether we choose to regard it as a maximum or a minimum, is infinite. The

pinion, then, can just work with a rack, but if the tooth be pointed
it can drive nothing less; and the face will be an involute of the
pitch circle, precisely as in Fig. 121.   Since, however, the tooth need
not be pointed, the pinion, with the same pitch and arc of recess, can
be made to drive any annular wheel which has at least one more tooth
than itself, by the use of an intermediate describing circle (see **226**).

230. Two-leaved Pinions in Inside Gear.—An annular wheel can be
driven by a pinion of only two leaves, the teeth working in the same
plane, as shown in Fig. 139.   Let AC, AO, AD, be to each other in
the proportion of 1, 2, 3; then the intermediate describing circle is
of maximum diameter.   Let the arc of recess, AB, $= 135°$, then the
angle $AOB = 67\frac{1}{2}°$, and prolonging OB to P, the arcs AB, AP, will

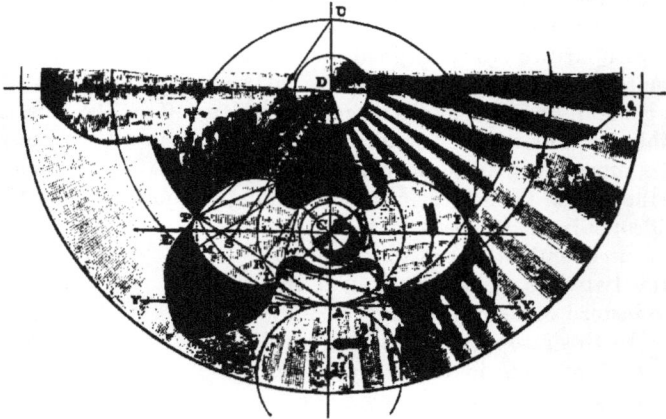

Fig. 139.

be equal; also P will be the highest point of the cardioidal face PB,
and of the hypocycloidal face PE.   Draw AP, cutting the inner
pitch circle in R; then RO and PU will be perpendicular to AP.
Draw DS also perpendicular to AP; then, since AU is bisected at
O, and OU at D, PR is also bisected at S.

Therefore, $DP = DR$, that is to say, the path of P will cut the
inner pitch circle in R.   If then we suppose the motion to be reversed,
the root L of the pinion's face LM must have met the point N of the
wheel's face NG, at R.   In other words, the arc of approach is equal to
AR, which again is equal to RB, because RO bisects the angle AOP.

231. The total arc of action, then, is ample, that of ap-
proach being $\frac{3}{8}$, and that of recess $\frac{3}{4}$, of the pitch; and the
maximum obliquity being $37\frac{1}{4}°$, the combination might be used

in light mechanism. But attention is called to a peculiarity of the action which would require a modification in the finish of the teeth for practical purposes. When contact is just ending at P between the fronts, or acting faces, of one pair of teeth, the *backs* of the next pair are just coming into contact at I. Now, were the teeth of the pinion merely "topped off" in the usual way in the lathe, there will be danger that the points would catch upon each other near the point I, in case of any inaccuracy of workmanship or wear in the bearings. This risk might be obviated by allowing some backlash, but this is highly objectionable in mechanism of the only description for which, if for any, this combination is suitable. A much better expedient in such cases is shown in the figure, the top of the tooth being bounded by a circular arc whose centre is the intersection of the normals PA, MO. The action still ends at P, but the other tooth of the pinion will have been guided into its space without risk of jamming. And it may be added that a similar finish of the wheel's tooth, by lengthening it a little and rounding off the corner, is advisable in order to prevent its catching upon the root of the pinion's tooth, since, as explained in (**227**), there will be a blunt angle at L.

The tooth of the pinion here shown has a considerable breadth, PM, at the top, and it will be seen that the action might have been extended ; also, that a larger pitch circle might have been used for the wheel ; but as will subsequently appear, the maximum number of teeth which can be used with this pinion is seven.

232. A pinion with two leaves in different planes may also be made to drive an annular wheel, as shown in Fig. 140. The velocity ratio being as four to one, let $AC = 1$, $AD = 4$; then the

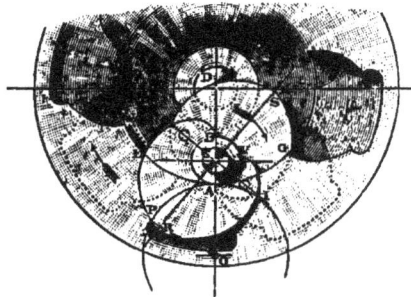

FIG. 140.

limiting value of $AH = 3$, and the pinion is constructed exactly as in Fig. 135, being in fact identical with the one there shown, which is, therefore, capable of driving wheels of any number of teeth between the limits of

6 and ∞ in Outside Gear.

4 and ∞ in Inside Gear.

The tooth of the wheel should be finished as shown in dotted line at
EW, by which the injurious action of a sharp corner may be avoided ;
and the movement is very smooth and noiseless, while the amount of
sliding is much less than in outside gear.

233. In the cases thus far considered the pinion has been the driver ;
let it now be required to determine the least number of teeth that can
be given to a pinion which is to be driven by an annular wheel whose
diameter, pitch and arc of recess are assigned. In Fig. 141 let *D* be
the centre of this wheel, *AF* the arc of recess, *FL* the half thickness
of the tooth, whose face, evidently, must be generated by an interior
describing circle. On the tangent at *A*, set off *AG* = arc *AF*, also

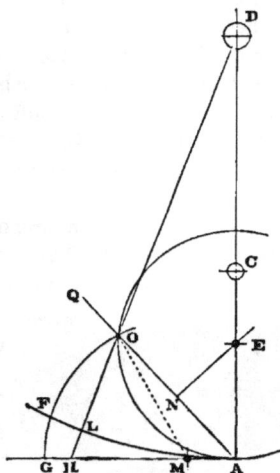

FIG. 141.

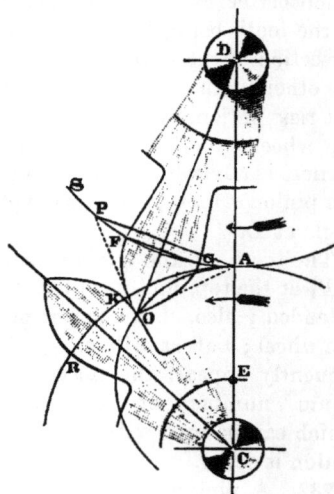

FIG. 142.

$AM = \frac{1}{4} AG$; with centre *M* and radius *MG* describe an arc cutting *DL* in *O* ; draw *OA*, and bisect it by a perpendicular cutting *AD*
in *E*. Then *E* is the centre and *EA* the radius of the *least* interior
describing circle, and if the pinion's flanks are to be radial the minimum radius of its pitch circle will be $AC = 2\,AE$; but if we adopt
the limit mentioned in **(210)** it will be $AC = \frac{8}{5}\,AE$. Since,
however, *AE* cannot exceed *CD*, it follows that if its value as above
determined prove to be greater than $\frac{1}{3}\,AD$, in the one case, or $\frac{5}{13}\,AD$
in the other, the assigned conditions cannot be satisfied.

If a maximum obliquity be assigned, the teeth may or may not be
pointed, and the diagram is constructed by first drawing *AQ*, making
with *AG* an angle equal to the given obliquity, and then describing

the arc about M, to cut AQ in O, which will be the highest point of the wheel's tooth-face, whether the tooth itself be blunted or not.

234. Low-numbered Pinions. Follower Given.—A new phase of the question presents itself when the assigned conditions relate to the follower, and the least number of teeth for the driver is to be determined. This requires a different mode of operation, which, as before, will be best explained by first considering a case in outside gear.

In Fig. 142, let C be the centre of the follower; then AK, the assigned arc of recess, is equal to the arc AO of a describing circle whose radius AE is known, and O will be the highest point of the

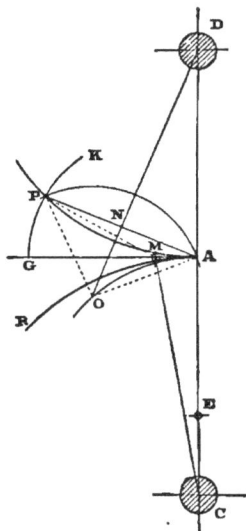

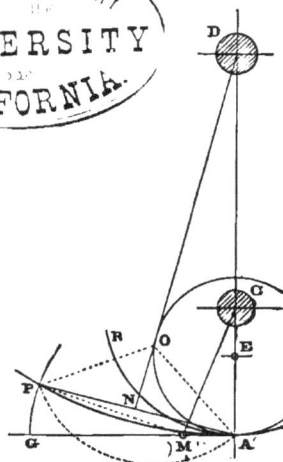

Fig. 143.   Fig. 144.

driver's tooth. Let D be the centre of the driver; then AD will be a minimum when the tooth is pointed as in the figure, in which case OD bisects and is perpendicular to the chord FG subtending the tooth. Draw a parallel to FG through A, cutting the driver's pitch circle in P; then the chord AP is also bisected by OD, and AO, PO, are equal. We have also $AF = PG$, and

$$AP = AF + PG - FG, = 2\,AF - FG,$$

or

$$AP = 2 \text{ (arc of recess)} - \text{(thickness of tooth)}.$$

Hence the construction, Fig. 143, in which AO is, as before, the assigned arc of recess laid off on the describing circle of the follower's flank. Draw a tangent to this arc at A, on which set off

$$AG = 2 \text{ (arc } AO) - \text{ (thickness of tooth)},$$

also

$$AM = \tfrac{1}{4} AG.$$

About centre M with radius MG describe the indefinite arc GK; about centre O with radius OA describe another arc cutting GK in P. Draw AP, and bisect it by a perpendicular which will pass through O and cut CA produced in D, then AD will be the *minimum* radius of the driver.

235. Should the point P fall upon the tangent line, the assigned conditions can just be satisfied by a rack with pointed teeth, but by nothing less in outside gear.

It may, however, fall upon the same side of the tangent with the centres C and E, as in Fig. 144. In that case the driver will be annular, and its radius AD will be a *maximum* when the tooth is pointed. But AE cannot in any event exceed CD; consequently, if AD prove to be less than $\tfrac{1}{8}$ AC (or $\tfrac{3}{4}$ AC if the follower have radial flanks) the assigned conditions cannot be satisfied.

If it prove to be greater, the maximum value may be used or not, at pleasure, for the tooth need not be pointed, and by topping it off,

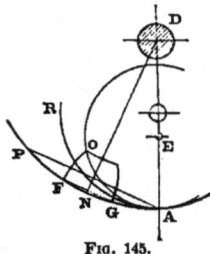

Fig. 145.

as shown in Fig. 145, the radius of the outer pitch circle may be reduced, but not below the limit just named.

It appears, then, that with a given pinion there may or may not be a superior limit to the number of teeth for the annular driver, but in either case there is an inferior one if there be any receding action; the wheel cannot have less than one and a half times as many as the pinion, when the latter has radial flanks.

236. One case remains to be considered, viz., when the given follower is annular; in regard to which it has already been pointed out (**226**) that in general the wheel may be driven by a pinion having one tooth less than itself, which may be called a natural maximum.

By reference to Fig. 129, it will be seen that the larger the pinion the broader will the teeth be at the top, and that its diameter will be a minimum when its teeth are pointed. Now in all the previous cases the assigned conditions have enabled us to fix the position of the highest point of the tooth at the instant of quitting contact. This, however, is not so in the present case, and it is therefore impossible to determine that minimum by direct means.

But it may be found indirectly ; for if we ascertain the limiting numbers of teeth for the annular wheels driven by various given pinions, we shall know from mere inspection of these results, the least number for the pinion which can be used to drive, under like conditions, in inside gear, a wheel of any assigned number of teeth.

CHAPTER VIII.

Limiting Numbers of Teeth for Various Arcs of Action. Details of Trigonometrical Process of Determination. The Nomodont, or Curve of Limiting Values.

Computation of Tables.

237. It can always be determined by construction, whether a proposed pair of wheels will work under given conditions as to pitch, arc of action, etc. But in order to avoid wasting time by attempting impossible cases, it is well that the limiting numbers of teeth, within a reasonable range of varying conditions, should be ascertained and tabulated for reference.

Prof. Willis, in his "Principles of Mechanism," * gives a diagram and deduces from it an expression, confessedly "so involved as to make the direct solution of the equation impossible, although approximations may be obtained." He adds : "However, on account of the practical importance of the question, I have arranged in the following Tables the exact required results, which I derived organically from the diagram by constructing it on a large scale with movable rulers."

This rather obscure expression evidently indicates the use of an adjustable mechanical device of some kind ; but since this, whatever it was, must have been capable of being drawn in any phase of its movements, the inference is a fair one that his process was equivalent to that of deducing results graphically by trial and error.

238. Prof. Willis may perhaps have overestimated the practical importance of the question, since limiting cases are in general to be avoided when possible. Still, the discussion has developed some hitherto overlooked points of considerable abstract interest ; and again, the practical value, whatever it may be, of such tables, depends

---

* 2d. Ed. 1870, pp. 106, 107.

entirely upon their correctness. And by the application of Prof. Rankine's singularly elegant processes relating to circular arcs, as explained in the preceding paragraphs, we have been enabled to determine these limiting numbers *accurately,* we believe for the first time. For the numbers in the tables above mentioned are those of teeth in complete circular wheels, capable of transmitting rotation continuously in the same direction; and though the processes of Prof. Rankine are only approximations, they are such close ones as to preclude the possibility of any error in the computations arising from this fact, of such magnitude as to affect the integers in the final results.

Prof. Willis calls particular attention to the statement that his tables are "geometrically exact." * It has already been demonstrated (**224, 225**) that in at least one instance this is very far from being the case. And the results obtained by the above methods differ from his in so many other instances, and so widely, that in view of his high standing as an authority, we feel called upon, before presenting our own Tables of Limiting Numbers, to illustrate in detail the manner in which the values there assigned were computed.

239. As the first example we select the following case.

$$\text{Given,} \begin{cases} \text{Driving Pinion of 5 leaves, Radial Flanks.} \\ \text{Arc of Recess} = \text{Pitch. Tooth} = \text{Space.} \end{cases}$$

The numbers of teeth being directly proportional to the radii, let rad. pinion = 5; constructing the diagram, we find the point O to lie in relation to AG as in Fig. 138, showing that the pinion cannot under these conditions drive a wheel in outside gear, and in that figure we have

$$
\begin{aligned}
ACF &= 72°, & \text{arc } AF &= \tfrac{1}{5} \times 5 \times 2 \times 3.1416 = 6.2832 \\
ACH &= 54°, & AG &= \text{arc } AF = 6.2832 \\
AHC &= 36°, & AM &= \tfrac{1}{4} \ AG = 1.5708 \\
AC &= 5, & OM &= \tfrac{3}{4} \ AG = 4.7124
\end{aligned}
$$

$$
\text{Triangle } ACH. \begin{cases}
\text{Rad.} & 10. \\
: \tan ACH = 54° & 10.138739 \\
:: AC = 5. & 0.698970 \\
: AH = 6.8819 & \overline{0.837709} \\
AM = 1.5708 \\
HM = \overline{5.3111}
\end{cases}
$$

---

* Principles of Mechanism, p. 110.

Triangle $\left\{ \begin{array}{l} OM = 4.7124 \\ : HM = 5.3111 \\ :: \sin OHM = \quad 36° \\ : \sin HOM = 138° \ 30' \ 44'' \end{array} \right.$

$\begin{array}{ll} \text{ar. com.} & 9.326758 \\ & 0.725183 \\ & 9.769219 \\ & \overline{9.821160} \end{array}$

$\begin{array}{l} OMA = \overline{174° \ 30' \ 44''} \\ \quad \quad \ 180° \end{array}$

$OAM + AOM = \overline{\quad 5° \ 29' \ 16''} \ ; \ \div \ 2 = 2° \ 44' \ 38''$

Triangle $\left\{ \begin{array}{l} OM + AM = 6.2832 \\ : OM - AM = 3.1416 \\ :: \tan. \tfrac{1}{2} \ (A + O) = 2° \ 44' \ 38'' \\ : \tan \tfrac{1}{2} \ (A - O) = 1° \ 22' \ 22'' \end{array} \right.$

$\begin{array}{ll} \text{ar. com.} & 9.201819 \\ & 0.497151 \\ & 8.680580 \\ & \overline{8.379550} \end{array}$

$OAM \quad = \overline{4° \quad 7'} = \text{Max. Obliq.} = AEN.$

Triangle $\left\{ \begin{array}{l} \sin OAM = \quad 4° \ \ 7' \\ : \sin OMA = 174° \ 30' \ 44'' \\ :: OM = 4.7124 \\ : OA \\ \ 2. \end{array} \right.$

$\begin{array}{ll} \text{ar. com.} & 1.143951 \\ & 8.980609 \\ & 0.673242 \\ & \overline{0.797802} \\ & 0.301030 \end{array}$

$AN = \tfrac{1}{2} \ OA \qquad \overline{0.496772}$

Triangle $\left\{ \begin{array}{l} \sin AEN = 4° \ 7' \\ : AN \\ :: \text{Rad.} \\ : AE = 43.7343 = \text{Rad. Interm. Des. Circle.} \end{array} \right.$

$\begin{array}{ll} \text{ar. com.} & 1.143951 \\ & 0.496772 \\ & 10. \\ & \overline{1.640723} \end{array}$

$\begin{array}{l} AC = \quad 5. \quad | \ \ (85 = \text{Max. No. given by Prof. Willis.}) \\ \overline{48.7343} \ \therefore \ \mathbf{48} = \text{Maximum No. Teeth.} \end{array}$

240. We will next take a case in outside gear, thus :

Given, $\left\{ \begin{array}{l} \text{Follower of 10 Teeth.} \quad \text{Radial Flanks.} \\ \text{Arc of Recess} = \text{Pitch.} \quad \text{Tooth} = \text{Space.} \end{array} \right.$

Then in Fig. 143 we shall have

$\begin{array}{ll} AG = \tfrac{1}{2} \times 10 \times \tfrac{1}{10} \times 2 \times 3.1416 = 9.4248. & \\ AM = \tfrac{1}{4} \ AG = 2.3562 & \quad 180° \\ PM = \tfrac{3}{4} \ AG = 7.0686 \quad | \ OAM = & \quad 36° \\ \quad AC = 10 = \text{Rad. Follower.} \quad | \ AOM + AMO = & \overline{144°} \\ ACM = OAM = 36° = \text{Max. Obliquity.} \ | \tfrac{1}{2} \ (M + O) = & \quad 72° \end{array}$

Triangle CAM.
$$\begin{cases} \text{Rad.} & 10. \\ : \sin OAM = 36° & 9.769219 \\ :: AC = 10 & 1. \\ : OA = OP \quad = 5.8778 & \overline{0.769219} \\ \quad AM \quad = 2.3562 \end{cases}$$

Triangle OAM.
$$\begin{cases} OA + AM \quad = 8.2340 & \text{ar. com.} \quad 9.084389 \\ : OA - AM \quad = 3.5216 & 0.546740 \\ :: \tan \tfrac{1}{2}(M + O) = 72° & 10.488224 \\ : \tan \tfrac{1}{2}(M - O) = 52°\ 46'\ 32'' & \overline{10.119353} \\ \qquad\qquad AOM = \overline{19°\ 13'\ 28''} \\ \qquad\qquad \tfrac{1}{2}\ AOM = \quad 9°\ 36'\ 44'' \end{cases}$$

Triangle OAM.
$$\begin{cases} \sin AOM = 19°\ 13'\ 28'' & \text{ar. com.} \quad 0.482449 \\ : \sin OAM = 36° & 9.769219 \\ :: AM = 2.3562 & 0.372212 \\ : OM = 4.2061 & \overline{0.623880} \\ \quad OP = 5.8778 \\ \quad PM = 7.0686 \end{cases}$$

$$2)\ \overline{17.1525} = OM + OP + PM.$$
$$\overline{8.5763} = S.$$
$$1.5077 = S - PM.$$

$$\cos \tfrac{1}{2}\ POM = \sqrt{\frac{Rad^2 \cdot S \cdot (S - PM)}{OP \times OM}}$$

Triangle POM.
$$\begin{cases} Rad.^2 & 20. \\ S \qquad\quad = 8.5763 & 0.933300 \\ S - PM = 1.5077 & 0.178315 \\ OP \quad = 5.8778 \quad 0.769219\ \} & \cdot\ \ \overline{21.111615} \\ OM \quad = 4.2061 \quad 0.623880\ \} & 1.393099 \\ & 2)\ \overline{19.718516} \\ \cos \tfrac{1}{2}\ POM = \quad 43°\ 40'\ 51'' & 9.859258 \\ \qquad \tfrac{1}{2}\ AOM = \quad\ 9°\ 36'\ 44'' \\ \qquad\quad AON = \overline{53°\ 17'\ 35''} \end{cases}$$

$$OAD = 90° + OAM = \quad 126°$$
$$ADN = 180° - (\overline{179°\ 17'\ 35''}) = 0°\ 42'\ 25''$$

Triangle ADN.
$$\begin{cases} \sin ADN = \quad 0°\ 42'\ 25'' & \text{ar. com.} \quad 1.908777 \\ : \sin AON = 53°\ 17'\ 35'' & 9.904014 \\ :: OA = \quad 5.8778 & 0.769219 \\ : AD = 381.9531 & \overline{2.582010} \end{cases}$$

$$\therefore\ \ 382 = \text{Minimum No. for Driver.}$$

241. Prof. Willis correctly observes that when the action begins at the line of centres, no pinion of less than ten leaves can be driven, but in reference to the above case he says (p. 210) *"nothing less than a rack* can drive a pinion of ten ;"* whereas it appears from these figures that it can be driven by any wheel of 382 teeth or more. In corroboration of which we will now assume a *driver* of 382 teeth, the arc of recess being equal to the pitch, and the tooth to the space, as before.

We shall then have in Fig. 133,

$$
\begin{aligned}
ACF &= \quad 0°\ 56'\ 32.67'' \quad & \text{arc } AF &= \tfrac{1}{382} \times 382 \times 2 \times 3.1416 = 6.2832 \\
ACH &= \quad 0°\ 42'\ 24.5'' & AG &= \text{arc } AF = 6.2832 \\
AHC &= \quad 89°\ 17'\ 35.5'' & AM &= \tfrac{1}{4}\ AG = 1.5708 \\
OHM &= \quad 90°\ 42'\ 24.5'' & OM &= \tfrac{3}{4}\ AG = 4.7124 \\
AC &= 382.
\end{aligned}
$$

Triangle ACH.
$$
\begin{aligned}
&\text{Rad.} & 10. \\
&: \tan ACH = 0°\ 42'\ 24.5'' & 8.091169 \\
&:: AC = 382 & 2.582063 \\
&: AH = 4.7123 & \overline{0.673232} \\
&\quad AM = 1.5708 \\
&\quad HM = 3.1415
\end{aligned}
$$

Triangle OHM.
$$
\begin{aligned}
&OM = 4.7124 & \text{ar. com.} \quad 9.326758 \\
&: HM = 3.1415 & 0.497137 \\
&:: \sin OHM = \ \ 90°\ 42'\ 24.5'' & 9.999967 \\
&: \sin HOM = \ \ 41°\ 48'\ 17.5'' & \overline{9.823862} \\
&\quad\quad OMA = \overline{132°\ 30'\ 42''} \\
&\quad\quad\quad\quad 180°
\end{aligned}
$$

$$OAM + AOM = \overline{47°\ 29'\ 18''}\ ; \ \div 2 = 23°\ 44'\ 39''$$

Triangle OMA.
$$
\begin{aligned}
&OM + AM = 6.2832 & \text{ar. com.} \quad 9.201819 \\
&: OM - AM = 3.1416 & 0.497151 \\
&:: \tan \tfrac{1}{2}\,(A + O) = 23°\ 44'\ 39'' & 9.643243 \\
&: \tan \tfrac{1}{2}\,(A - O) = 12°\ 24'\ 5.8'' & \overline{9.342213} \\
&\quad\quad OAM = \overline{36°\ \ 8'\ 45''} = \text{Max. Obliq.} = AEN.
\end{aligned}
$$

Triangle OMA.
$$
\begin{aligned}
&\sin OAM = \ \ 36°\ 8'\ 45'' & \text{ar. com.} \quad 0.229236 \\
&: \sin OMA = 132°\ 30'\ 42'' & 9.867596 \\
&:: OM = 4.7124 & 0.673242 \\
&: OA & \overline{0.770074} \\
&\quad 2 & 0.301030 \\
&\quad AN = \tfrac{1}{2}\ OA & \overline{0.469044}
\end{aligned}
$$

$$\text{Triangle } AEN. \begin{cases} \sin AEN = 36° 8' 45'' & \text{ar. com.} \quad 0.229236 \\ : AN & 0.469044 \\ :: \text{Rad.} & 10. \\ : AE = 4.9921 = \text{Rad. Ext. Des. Circle.} & \overline{0.698280} \end{cases}$$

2 (Flanks to be Radial).

$\overline{9.9842}$ ∴ 10 = Minimum No. for Follower.

242. In the case of a driving pinion of 6 leaves, under the same conditions, we find, by a similar computation,

	$AE =$	71.7083	= Rad. Exterior Des. Circle.
Add	$AC =$	6.	= Rad. Driver.
then		$\overline{77.7083}$	= Rad. Annular Wheel.
also	$2\,AE =$	143.4166	= Rad. Follower in Outside Gear.

These being minimum values, the next higher integers must be taken as the limiting numbers of teeth for complete wheels. Thus the 6-leaved pinion can drive an annular wheel of 78 teeth, or an externally toothed one of 144, but no less numbers can be used when the describing circle thus found is employed.

Prof. Willis assigns 176 instead of 144, and, as previously stated, he gives a *rack*, instead of 382 teeth, as the least that can drive a pinion of 10. In the table for outside gear we have merely marked with an asterisk those of our own values which differ from his, because the above-mentioned discrepancies are the most serious ones. But his method, whatever it was, appears to have been singularly defective in its application to inside gear ; he not only ignores entirely the minimum values, but assigns maximum values which differ so often and so widely from those computed as above, that we have deemed it proper to present his tables for annular wheels side by side with our own, which we think it safe to assert are numerically exact. Had the object been to determine the limiting *radii* with the utmost precision, a correction for the error in Prof. Rankine's approximations would have been necessary in every instance. This, however, we have applied only when the arc to be dealt with was so large as to make it requisite, in order to secure in the computation of converse cases, as in **(240)** and **(241)**, results concordant in respect to the number of teeth for complete wheels.

Limiting Numbers of Teeth.

RECESS = PITCH.		RECESS = ½ PITCH.		RECESS = ⅓ PITCH.	
D.	F.	D.	F.	D.	F.
5.58	∞	3.25	∞	2.58	∞
6	144*	4	34*	3	37*
7	50*	5	19	4	15
8	34*	6	14	5*	11*
9	27	7	12	6	10
10	23	8	11*	7	9
11	21	9*	10	8	8
12	19	11*	9*	11	7
13	18	16	8	21*	6
14	17	33*	7	∞	5.17
15	16	∞	6.35		
17	15				
20	14				
24	13				
32*	12 .				
57*	11				
382*	10				
∞	9.83				

OUTSIDE GEARING.
EPICYCLOIDAL TEETH.   RADIAL FLANKS.
Tooth = Space.
Minimum Values.

* Numbers differing from those given by Prof. Willis.

Limiting Numbers of Teeth.

INSIDE EPICYCLOIDAL GEARING. FLANKS OF PINIONS RADIAL. TOOTH = SPACE.					
RECESS = PITCH.		RECESS = ½ PITCH.		RECESS = ¼ PITCH.	
D.	F.	D.	F.	D.	F.
3	9	2	7	2	11
4	16	3	41	3	∞
5	48	4	∞		
6	∞	*Maximum Followers.*			
F.	D.	F.	D.	F.	D.
7	14	5	12	4	8
8	24	6	65	5	53
9	60	7	∞	6	∞
10	∞	*Maximum Drivers.*			
D.	F.	D.	F.	D.	F.
6	78	4	21	3	22
7	32	5	15	4	12
8	25	6	13	5	11
9	23	7	13	6	11
10	22	8	14	7	12
11	22	9	14	8	12
12	22	10	15	9	13
13	22	11	16	10	14
14	23	12	17	11	15
15	23	*Minimum Followers.*			
16	24				

Limiting Numbers. (Prof. Willis.)

INSIDE EPICYCLOIDAL GEARING.					
FLANKS OF PINIONS RADIAL. TOOTH = SPACE.					
RECESS = PITCH.		RECESS = ½ PITCH.		RECESS = ¾ PITCH.	
D.	F.	D.	F.	D.	F.
2	5	2	10	2	14
3	12	3	77	4	Any No.
4	26	4	Any No.		
5	85				
7	Any No.	*Maximum Followers.*			
F.	D.	F.	D.	F.	D.
7	14	4	5	4	8
8	25	5	12	5	64
9	60	6	77		
10	Rack.	*Maximum Drivers.*			

243. The manner of using these tables is best illustrated by an example or two.   In outside gearing, let the given wheel be a driver of 12 teeth, the arc of recess to equal the pitch.   In this division are two columns, marked *D* and *F*, for drivers and followers respectively : and opposite 12 in column *D*, we find in column *F* the number 19, which is the least that can be driven. If the same number be assigned for a follower, we find in column *D*, opposite 12 in column *F*, the least number, 32, that will drive.

If the given number be not found in the table, the number for the required wheel will be found opposite the next less number.   For instance, the arc of recess being three-fourths of the pitch, let the given wheel have 28 teeth. By reference to the table, it is seen that 16 (or over) will drive 8, but 33 is the least that will drive 7; hence the given wheel will drive 8, but no less. The given number is not found in column F either; but 19 can be driven by 5, or any number greater than five, while a pinion of 4 can drive no less than 34: consequently 5 is the least that can be used to drive the wheel of 28. Thus this table includes all possible limiting numbers for the three values of the arc of recess, in outside gearing.

In the first division of the table for inside gearing, marked Maximum Followers, the columns F contain the *greatest* numbers of teeth for the annular wheels which can be driven by the corresponding numbers in the columns D; for instance, when recess = pitch, a pinion of 5 can drive a wheel of 48, but not more. Since the *least* number that can be driven is always one more than that on the pinion, it is not given in the table; it is necessary only to remark that the lowest numbers which can be thus used are 3 for the pinion and 4 for the wheel, as a two-leaved pinion will not drive on account of the excessive obliquity when the wheel has but three teeth. In all the combinations in this division, the receding action can be secured only by the use of an intermediate describing circle.

In the second division, the columns D contain the *greatest* numbers for the annular wheels which can drive the corresponding numbers in columns F: thus when recess = ¾ pitch, a pinion of 6 can be driven by a wheel of 65 teeth, but not by a larger one. Since the flanks of
· the pinion are to be radial, the *least* number for the driver (**235**) will be one and a half times that of the given pinion, whatever the arc of recess, and therefore is not set down in the table.

In the third division, the columns D contain the numbers for given pinions whose teeth are pointed, the faces being, therefore, generated in each case by the least possible describing circle. The columns F contain the *least* numbers for the annular wheels which can be driven by these pinions, whose faces act against flanks traced by the same describing circles, which are exterior ones; and when these least numbers are used, the annular wheels can have no faces. There being, then, no approaching action, it would at first sight appear that the action could only be continuously maintained when the arc of recess is equal to the pitch. But the *assigned* amount of receding action in these cases is secured by the use of the exterior describing circle only: and the actual amount is greater, owing to the peculiarity of the action explained in (**206**), there being in fact two points of driving contact up to the limit assigned in the table, so that the rotation will be properly maintained in every case.

244. It remains now to illustrate by an example the use of this table in finding, as mentioned in (236), the limiting number of teeth for a pinion which is to drive a given annular follower.

Suppose that the recess is to be equal to the pitch : then, recollecting that, as just mentioned, a pinion of 3 can drive a wheel of 4, we deduce at once, in regard to the *least* number, the following by inspection of the table :

No. on Given Wheel.		Least No. for Pinion.	Describing Circle.
4 – 9 inclusive	....	3	
10 – 16 "	....	4	
17 – 48 "	....	5	.... Intermediate.
49 – 77 "	....	6	
78 – ∞	....	6	.... Exterior.

The *greatest* number for the pinion, if an intermediate describing circle be used, is always one less than the given number on the annular wheel.

If it be stipulated that the assigned arc of recess shall be secured by the use of an exterior describing circle, the limiting numbers may also be found by simple inspection if within the range covered by the table. But in order to include the superior limit for even moderately large wheels, the table would require to be inconveniently extended. For instance, let the given number be 42 ; this is found by computation to be the least that can be driven by a spur-wheel of 36 teeth, which is therefore the *largest* possible driver. A pinion of 6 can drive no less number than 78, but one of 7 can drive any number above 32, and is, consequently, the *least* which can be used under the assigned conditions.

245. If an interchangeable set of spur wheels be constructed by the use of a constant describing circle, the limiting numbers for the annular wheels which will gear with them may be readily found without reference to a table. In the system formerly employed by Brown & Sharpe, of Providence, R. I., the diameter of the describing circle was equal to the radius of a pinion of twelve teeth. If then the teeth of the annular wheels are to have both faces and flanks, the distance between centres of the interior and exterior describing circles may be represented by twelve, since the radii are proportional to the numbers of teeth ; and at the limit this distance is equal to that between the centres of the pitch circles. In these circumstances, therefore, any spur wheel of the set will gear with an annular wheel having twelve more teeth than itself, but no less than that will answer. Either may be used as driver or as follower indifferently, but the relative and the actual amounts of approaching and receding action will depend

not only upon which does drive, but also upon the actual numbers of
teeth used in any given case. Since, however, in the use of this system the smallest pinion ever used was one of twelve teeth, the total
angle of action would clearly be always ample.

By cutting down the pinion to the pitch circle we may use a smaller
wheel (as a driver only), and by similarly treating the wheel we may
use a smaller pinion for the same purpose ; the least number for the
annular wheel being found in either case by adding *six*, instead of
twelve, to the number of teeth upon the given internal wheel.

In the system first adopted by Pratt & Whitney, of Hartford, Conn.,
and subsequently by Brown & Sharpe, the diameter of the describing circle is equal to half that of the pinion of fifteen teeth. These
limiting numbers are, therefore, found by adding to the number of
teeth on the given spur-wheel, fifteen if both wheels are to have faces
as well as flanks, and eight, if either is to be cut down to the pitch
line ; in the latter case, the *exact* number to be added is seven and a
half, which giving a fractional result, the next higher integer must be
taken.

246. The Nomodont.—If the radius of the given wheel be gradually
increased, other things remaining unchanged, it is obvious that the
radius of the greatest or the least wheel, as the case may be, which will
work with it either as a driver or a follower, will vary according to
some regular law. As in many other cases, this law can be best illustrated graphically, by a curve whose abscissas are proportional to the
given, and the ordinates to the required, radii of the wheels ; to which
we have given the descriptive name of the Nomodont, from the law
expressed by it in relation to the numbers of teeth. These limiting
numbers, as given in the tables, are integers, as they must be if complete wheels are to be used ; this would not be the case were sectors
only to be employed, as they often are, and in constructing these
curves the fractional values as computed have been given to the
ordinates.

It is hardly necessary to point out that the regularity of the curves
affords a quite rigid test of the correctness of the results which they
embody. And in order to define them the more perfectly, especially
between the origin and the vertices, in the region of the asymptotical
ordinates, a great number of intermediate fractional radii were assumed for the given wheels, the conditions for all the curves represented being, that the externally toothed wheels shall have radial
flanks, that the arc of recess shall be equal to the pitch, and the tooth
equal to the space.

247. In Fig. 146, the abscissas represent the radii of given exter-

nally toothed wheels which are to drive, the ordinates of the curves
A and B are proportional to the corres-
ponding minimum radii of the followers
in outside and inside gear respectively,
determined as in the case of the 6-leaved
pinion, in **(243)**.

This, as appears from the table, is the
least complete pinion which can drive
at all in outside gear; the exact limiting radius of one which can just drive a

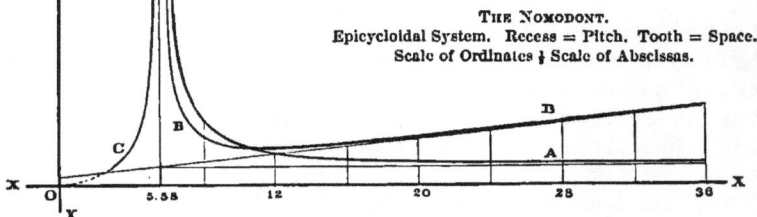

THE NOMODONT.
Epicycloidal System. Recess = Pitch. Tooth = Space.
Scale of Ordinates ⅓ Scale of Abscissas.

RADII OF GIVEN DRIVERS. FIG. 146.

rack, is 5.58, at which point the ordinate is therefore infinite, and asymptotic to the curve A. As the radius of the given driver increases, that
of the follower diminishes, at first very rapidly, then more and more
slowly, until the driver becomes a rack, when the value of the ordinate is 9.83, which agrees with the tabular record that the least whole
number which can be driven is 10 ; consequently this curve has also
a horizontal asymptote at a distance of 9.83 above the axis of abscissas.

Since the driven rack may be regarded as an infinitely large wheel
in inside gear, as well as in outside, the ordinate at 5.58 is also an
asymptote to the curves B and C; the ordinates of the latter being
proportional to the maximum radii of annular followers, whose action
requires the use of an intermediate describing circle, determined as in
(239).

The ordinates of B also diminish rapidly at first, but reaching a
minimum when the radius of the driver is between 10 and 12, subsequently increase. The rate of increase, however, is not constant; the
value of the ordinate is **(239)** $AC + AE$, in which, while AC increases uniformly, AE diminishes more and more slowly, reaching the
limit of 4.92 when $AC = \infty$. If, then, any two ordinates be drawn,
each measuring, by the scale of ordinates, a distance equal to its abscissa

+ 4.92, the right line passing through their extremities will be a second asymptote to the curve B.

Both A and B are, it will be observed, remarkably symmetrical curves, being in fact very nearly true hyperbolas. The ordinates of C, on the other hand, diminish very rapidly as the radius of the driver is reduced, their values, as shown in **(239)** and **(225)**, being 48.73 when that radius is 5, and exactly 9 when the radius is 3. As appears from the table, only three complete pinions, having respectively 3, 4, and 5 teeth, are included in the class here represented. Prof. Willis gives a pinion of *two* leaves, as capable of driving a wheel of five teeth ; but this will be found wholly impracticable on account of the excessive obliquity.

We have, therefore, not considered it worth while to compute an ordinate for a less driving radius than 2.5, at which limit the value is 6.76 ; but the curve is extended in a dotted line to the zero point, on the assumption that driver and follower would vanish simultaneously.

248. In Fig. 147 the abscissas represent the radii of given externally toothed followers. The ordinates of the curve D being proportional to the corresponding minimum radii of drivers in outside gear, it is at once apparent that this curve will have a vertical asymptote at 9.83, and a horizontal one at a distance of 5.58 above the axis of abscissas.

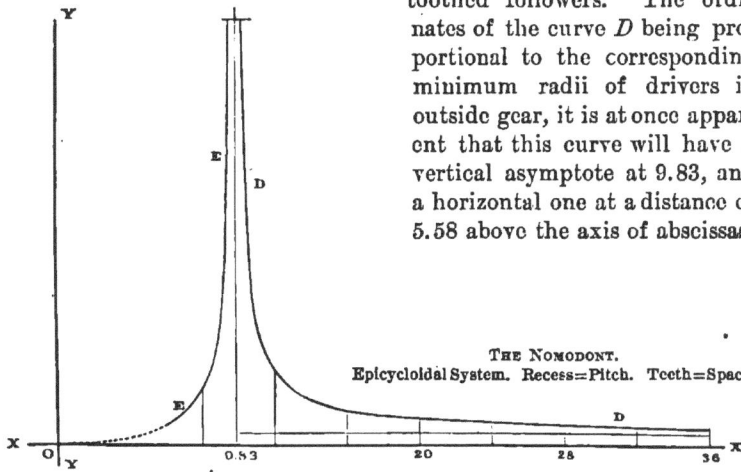

THE NOMODONT.
Epicycloidal System. Recess=Pitch. Teeth=Space.

RADII OF GIVEN FOLLOWERS. FIG. 147.

The curves A and D are very similar, and will become in fact identical if the axes be transposed and the same scale used for both ordinates and abscissas ; for it will readily be seen, by comparing Fig. 133 with Figs. 142 and 143, that assuming the processes of construction and computation to be exact instead of approximative, the radii in those converse cases must under similar conditions be precisely the same. As here shown, however, the curve D has been constructed with ordinates computed as in (240).

When the given radius is less than 9.83, the driver must be annular, and its maximum radius forms an ordinate to the curve E, which is somewhat similar to C, and like it includes the radii of but three complete pinions; which have respectively 7, 8 and 9 leaves.

The maximum radius of the driver for a pinion of six leaves, is 7.45, which is impracticable, as we have seen that it must be at least one and a half times as great as that of the follower. Therefore the computations were carried no further, although this curve, like C, and upon the same assumption, is continued in a dotted line to the zero point.

CHAPTER IX.

SPUR GEARING, CONTINUED—INVOLUTE TEETH.

Involute Generated by Rolling of Right Line on Base Circles. Peculiar Properties. Original Pitch Circle. Rack and Pinion. Annular Wheels. Lownumbered Pinions. Involute Tooth with Epicycloidal Extension. Limiting Numbers for Given Arcs of Recess. Comparison of the Involute and Epicycloidal Systems. Involute Generated by Rolling of Logarithmic Spiral on Pitch Circles.

Involute Teeth.

249. Next to the epicycloid, the curve most extensively used as the tooth-outline in spur gearing, is the involute of the circle.

It will subsequently be shown that this curve can be generated in a manner which conforms to the general law enunciated in (**181**); but its fitness for this purpose is proved much more clearly and simply by deriving it in another way.

In Fig. 148, let *LM, RS,* be portions of two pitch circles in contact at *A* ; *C* and *D* their centres, and *TT* their common tangent. Draw *NN* inclined to *TT,* and upon it let fall the perpendiculars *CE, DF*; with which, as radii, describe the circles *EHY, FGQ.* Suppose these circles to be disks upon which is wound an inextensible thread *EF,* carrying a marking point at *F.* Now let

Fig. 148.

the upper disk turn as shown by the arrow : it will wind the thread upon itself, unwinding it from the lower one, which will thus be

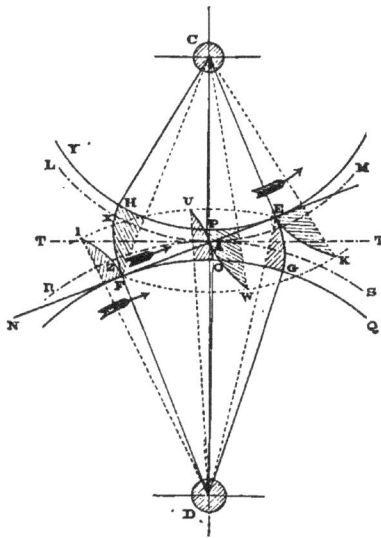

turned in the opposite direction, but with the same perimetral velocity.   Let the arcs *EH*, *FG*, be equal to each other and to the right
line *EF*; then while the pencil is drawn from *F* to *E*, it will trace
upon the plane of the upper disk the curve *KE*, and upon the plane
of the lower one the curve *GE*.

250. These are involutes, not of the pitch circles, but of the *base
circles EHY, FGQ*, the ratio of whose radii *CE, DF*, by reason of
the similar triangles *ACE, ADF*, is the same as that of *AC, AD*, the
radii of the pitch circles.   Being simultaneously generated during the
rotation by a point which lies always in the common tangent to the
base circles, that is to say, in the common normal to the involutes,
these curves will be tangent to each other throughout the generation :
and since this common normal always cuts *CD* at the same point *A*,
they may be used as outlines of teeth for the wheels to which they
respectively belong.   Thus *FI*, similar to *GE*, will drive *FH*, which
is similar to *KE*, in the direction indicated by the arrow, with a constant velocity ratio ; the locus of contact being the line *FE*.   The
two curves are shown in the diagram in one intermediate position, so
that the action will be readily traced without farther explanation ;
and it is evident that this action is precisely equivalent to the rolling
together of the pitch circles *LM, RS*.

251. Now, since the involute does not extend within its base circle, the curves shown in Fig. 148 are of the greatest possible length ;
supposing the lower wheel to drive, the action begins at *F* and ends
at *E*.   Both curves are continuous, and there is no division into face
and flank ; but it is evident that the points *x* and *z* will meet at *A*,
the involutes then occupying the positions *OAU, PAW*: also that
the ratio of the arc *Az* to the arc *FO*, is the same as that between the
arcs *Ax, HP*.

· Reasoning, in like manner, as to the arcs of recess, and recollecting
that the arcs *FO, HP*, are each equal to *FA*, while *OG, PE*, are
each equal to *AE*, we have

$$\frac{\text{Approach}}{\text{Recess}} = \frac{FA}{AE} = \frac{CE}{DF} = \frac{AC}{AD}.$$

That is to say, if each involute be long enough to extend to the root
of the other, the arcs of approach and recess will be to each other in
the direct ratio of the radii of the base circles, or pitch circles, of the
driver and the follower respectively.

In these circumstances, the smaller of two wheels must drive in
order to secure more receding than approaching action.   But as it is
not necessary to make the curves so long, we can vary the proportion

between the arcs of approach and recess at pleasure, by properly regulating the heights of the teeth, whatever the relative diameters of the
wheels.

252. Thus in Fig. 149, which represents teeth of practical proportions, one pair is shown in contact at A, another pair as just quitting

Fɪɢ. 149.

contact at B, the lower wheel driving as indicated by the arrow. The
receding action, then, continues while the point of contact travels
from A to B, a distance equal to OI measured on the base circle.
Meantime the linear motion of the driver, measured on the pitch circle, will be AG, which is greater than OI in the proportion of the
radius of the pitch circle to that of the base circle; but since those
two arcs measure the same angle, either is readily found if the other
be given.

Supposing, then, that the angle of recess is assigned; we first ascertain the length of the arc OI on the base circle by which it is measured, and set off AB on the line of action, equal to it; then the tops
of the driver's teeth are limited by a circle through B, about that
wheel's centre.

A radius through B cuts the pitch circle in H; if GH be equal to
half the thickness of the tooth as determined by the pitch, the construction is just possible, and the tooth will be pointed. If GH be
greater than that, the conditions are impracticable; but if it be less,
the tooth, as in the figure, will be of sensible breadth at the top.

By a similar process the height of the follower's tooth for any assigned angle of approach is determined, and the pitch circles having
been subdivided as usual into equal parts, the construction is completed by drawing the reverse involutes for the backs of the teeth,
and providing suitable clearing spaces. These may be required to extend some distance within the base circles, and if so they may be

10

bounded as in the figure by radial lines tangent to the involutes at
their roots. .

253. Peculiar Properties of Involute Teeth.—In the operations above
described, the line of action *FE* was drawn at pleasure, and the argument in no wise depends upon its obliquity. Consequently, for a
given pair of pitch circles an infinite number of pairs of base circles
may be assigned, and the converse is evidently true, since the common
tangent of any two given base circles will always cut the line of centres, be the same greater or less, into segments having the same ratio
as their radii. From which follow two important practical deductions, viz.:

1. Any two wheels with involute teeth, of which the pitch arcs on
   the base circles are equal, will gear correctly with each other.
2. The velocity ratio will not be affected by any change in the distance between their centres.

We have seen that wheels with epicycloidal teeth may be made interchangeable ; but this second peculiarity gives the involute tooth an
advantage over every other, of special importance in mechanism requiring the greatest smoothness and uniformity of action. The velocity ratio will be correct, although the wheels be improperly located at
the outset, and will remain so in spite of wear in the bearings ; also,
the backlash may at any time be reduced to a minimum, by bringing the
axes as close together as they can be without causing the teeth to bind.

On the other hand, the obliquity is *constant*, and in general greater
than the mean obliquity for epicycloidal teeth having the same angle of
action. While this may be a serious objection to the involute form for
heavy work, it may also be a positive advantage in light mechanism where
smoothness of action is all-important. For the side pressure will always keep the axes at the greatest possible distance from each other,
which tends to prevent shaking if there be any looseness in the
bearings.

254. Original Pitch Circle.—Of the infinite number of pitch circles
which may be assigned for any given wheel, there is evidently but one
upon which the tooth and space, as in Fig. 149, are measured by equal
arcs. This is appropriately called the *original pitch circle*, being the
one given or assumed in laying out a pair of wheels with teeth and '
spaces equal as above described, and it determines what may be called
the *proper obliquity* for both wheels.

Suppose now that a wheel has been made, the teeth being involutes
of a given base circle, the pitch circle unknown ; and let it be required to construct another which shall gear with it in such wise that
the teeth and spaces on the pitch circles shall be equal as in the fig-

ure. In order to do this, it is necessary first to find the proper obliquity of the given wheel. This may be done graphically as follows : draw two radii, the first bisecting a tooth, the second bisecting an adjacent space, of the given wheel. Then bisect the angle between these lines by another radius ; this will cut the involute outline of the tooth in a point through which the original pitch circle must pass. This third radius being taken as the line of centres, the tangent to the base circle, through the point last mentioned, will be the line of action having the required obliquity.

255. Considerations Affecting the Obliquity.—The obliquity is, as above stated, abstractly arbitrary. But it will presently appear that under some circumstances there are definite relations between the obliquity, the velocity ratio and the arc of recess, such that if either two be assigned, the other, in many cases, can vary only within a quite narrow range. For practical guidance when not trammeled by the considerations alluded to, it may be stated that experience has shown that for ordinary purposes the obliquity should not exceed from 15° to 17°. When it is no greater than this, it is safe to say that even for heavy work these teeth are in every respect equal to the epicycloidal or any others ; and in addition to the advantages previously mentioned, it will be noted that their form is essentially a strong one, spreading out rapidly toward the base.

256. Rack and Wheel with Involute Teeth.—In Fig. 150, LM being the pitch circle, and TT the pitch line of the rack, the base circle is determined as in Fig. 148, by drawing FE at an arbitrary angle through A, the point of tangency, and letting fall upon it the perpendicular CE. Then while a marking point travels from E to F, with a linear velocity equal to that of the circumference EHY, it will trace upon the plane of the wheel the involute FH. The rack meantime moves to the left side with a linear velocity equal to that of the pitch circle,

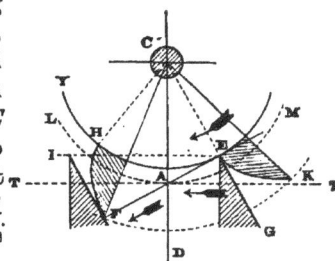

Fig. 150.

which is greater than that of the base circle in the ratio of AC to CE.

Consequently, when the pencil reaches F the point E of the rack will have moved to I, and we shall have EF perpendicular to EC, EI perpendicular to AC, and also

$$\frac{EI}{EF} = \frac{AC}{EC} ;$$

therefore *IF*, traced by the pencil upon plane of the rack, will be a right line perpendicular to *EF*.

Drawing *EK* similar to *HF*, and *EG* similar to *IF*, and supposing the pinion to drive, the action begins at *E*, ending at *F*; and as in Fig. 149, the relative amounts of approaching and receding action

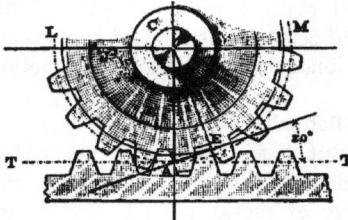

Fig. 151.

may be varied at pleasure by altering the heights of the teeth. In this diagram, as in Fig. 148, practical proportions are disregarded for the sake of perspicuity; the general appearance of the combination under reasonable conditions is shown in Fig. 151, the obliquity being 20°.

The less the obliquity, the shorter will be the acting face of the rack for a given angle of action, and the less will be the amount of approaching action attainable; when the obliquity is *nil*, the rack tooth degenerates into a point, and the wheel-tooth becomes an involute of the pitch circle, as illustrated in Fig. 121. The action is wholly receding, but the sliding is excessive and the wear confined to one point on the rack, thus more than counterbalancing all advantages unless the pressure is very moderate.

257. Annular Wheels with Involute Teeth.—The construction, as shown in Fig. 152, is substantially the same as in the case of outside gearing. And this form of tooth would appear to be extremely well adapted for inside gearing requiring unusual smoothness of action, to which, as above suggested, the constant obliquity is favorable, while, owing to the small difference in the lengths of the acting curves, there is very little sliding. This diagram is so similar to Fig. 150, that no explanation is necessary, beyond calling attention to this fact, viz.: since the action must begin or end at *E*, the root of the pinion's teeth, the greatest possible height of the wheel-tooth is determined by drawing a circle about *D* through that point.

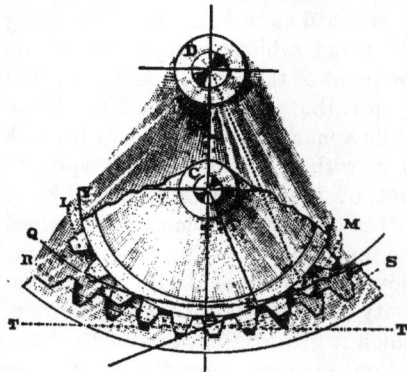

Fig. 152.

258. Low-numbered Pinions with Involute Teeth.—If two wheels be

of equal size, the least number of teeth which can be given to each
depends upon the obliquity.
For if this be assigned, the
points F and E, Fig. 153, are
thereby determined; and the
arcs FG, EH, on the base circles, are each equal to EF.
Taking these arcs, then, as the
measures of the pitch angles,
and making the teeth and
spaces on the pitch circles
equal, the combination, as represented in the figure, is just a
limiting case, and the number
of teeth the least possible.

If the wheels are to be complete ones, a fractional tooth
being impossible, the arc FG
must be contained an exact
number of times in the circumference, or, if there be a

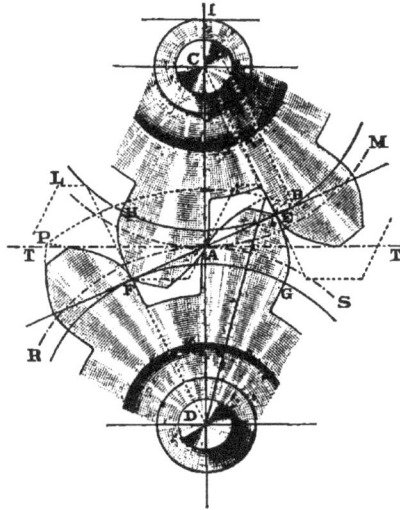

FIG. 153.

remainder, the next higher integer must be taken as the least number
of teeth. In the latter event the arrangement will be practically ser-
viceable, for though the angles of approach and recess will still be
equal, their sum will be greater than the pitch. Each of these angles
is in point of fact a little greater than the obliquity, but the excess, as
can easily be verified by calculation, is so small that the least number
of teeth for practical purposes is at once correctly ascertained, by dividing the circumference by twice the given obliquity : and thus for equal
wheels we have

OBLIQUITY.	LEAST NO. OF TEETH.
10°	18
12°	15
15°	12
18°	10
20°	9
30°	6.

259. Supposing the lower wheel in Fig. 153 to drive, then the
follower may be equal to it, but cannot be less. It may, however, be
made as much greater as we please ;—a moment's study of the diagram
will show that this same wheel can drive even the rack shown in
dotted outlines.

A sufficient increase in the diameter of the follower will permit the tooth GE of the driver to be lengthened to an extent limited by its intersection at P with the reverse involute, forming the back of the tooth. About D, describe a circle through P, cutting the line of action in B; draw BI parallel to EC, cutting the line of centres in I: then if the driver's tooth be pointed, BI will be the least radius of the follower which will secure the greatest possible amount of receding action, viz.: an arc on the base circle equal in length to AB.

260. This leads directly to the consideration that the pitch of the driver may be made greater than in the figure, without changing the obliquity; for by increasing the angle FDP, the point of the leading tooth may be made to fall upon the line of action, the tooth and space, as measured on the pitch circle, still remaining equal to each other.

This is illustrated in Fig. 154, FG being the pitch on the base circle, and LM, the corresponding arc on the pitch circle P, on the line of action, being the point of the tooth, the involute PK must bisect LM in N: then, PD will bisect MN in I, and HD, which bisects FK, will also bisect LN in H. The obliquity being given, the angle FDL, which is equal to MDG, may be calculated;

then letting $FDL = a$,
$$FDH = x,$$

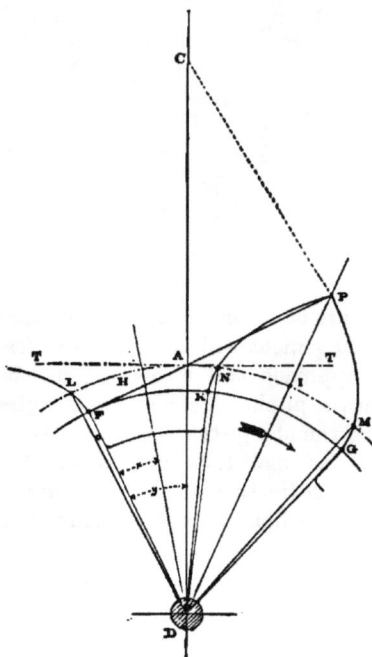

Fig. 154.

we shall have

$$\tfrac{1}{4}\text{ Pitch} = LDH = x + a,$$
$$FDP = 3(x + a) - a = 3x + 2a,$$
$$FDG = LDM = 4(x + a).$$

But FP, the trigonometrical tangent of the arc FI, is equal to the arc FG; that is,

$$\tan (3x + 2a) = \text{arc } (4x + 4a).$$

This equation cannot, we believe, be solved by any process less laborious than the tentative one of assuming a value for x, and if it fail to satisfy the equation, increasing or diminishing it, as the case may be, until it does. Nor does the situation afford any key to a graphic construction, since the distance to be set off on the line of action, and the angle subtended by an arc equal to it on the circumference, are both unknown.

This, however, is of little consequence, as, if the obliquity be assigned, there can be no certainty that the pitch arc when found would be an aliquot part of the circumference : and it is of much more practical interest to ascertain the result of assigning a specific number of teeth, or in other words a definite pitch.

261. When this is done the case becomes more manageable. Considering first the graphic process : The radius FD of the base circle and the pitch angle FDG being given, we have first simply to rectify the arc FG and set off FP equal to it, in a direction perpendicular to DF. Drawing the involutes PG and PK, the tooth is completed as to its acting outlines ; now bisect the angle FDK by the radial line DH, draw PD, and bisect HDP by DN : this latter line will cut the involute PK in the point N, through which is drawn the original pitch circle cutting FP in A, which locates DA the line of centres and determines the proper obliquity.

The pinion thus formed is barely capable of driving, the whole path of contact FP being just equal to FG, the pitch arc on the base circle : and the minimum radius of the follower will be determined by drawing PC perpendicular to the line of action, cutting DA produced in C the centre.

262. Otherwise, by computation : Knowing the radius FD and the length of FP, the angle FDP may be found, whence, the angle FDG being given, the angles GDK, FDK, may also be found, and we shall then have

$$FDL = \tfrac{1}{4} \, (GDK - FDK).$$

Now, FA, the trigonometrical tangent of the angle ADF, is equal to the arc of the base circle, which measures the angle ADL ; therefore, let

$$ADF = y,$$
$$FDL = a,$$

and we have

$$\tan y = \text{arc } (y + a).$$

This equation may be solved by the tentative process above described ; and the angle *ADF* thus determined is equal to the proper obliquity. Knowing this angle and the side *FD*, we can now find *FA*, equal to the arc of approach on the base circle, which taken from *FP* gives *AP* equal to the arc of recess, and the triangles *ADF*, *ACP*, being similar, the radius *PC* is readily ascertained.

263. In Fig. 154 the pitch arc *FG* is 72°, so that the diagram represents the construction of a five-leaved pinion just capable of driving. By comparing this figure with the preceding one, it will be seen that when the point of the tooth thus falls upon the line of action, we have the *maximum pitch* for a given obliquity, or the *minimum obliquity* for a given pitch.

In this case, we find by computation in the manner above explained, the following values, viz. :

Obliquity.............................	28° 8'
Arc of Recess.........................	¼ pitch.
Least No. for Follower................	6.74.

Practically, then, a pinion of five leaves is capable of working with another of seven, but no less without exceeding the obliquity above given. By increasing the obliquity, the number for the follower may be reduced, but it will subsequently be shown that five is not only the least number for equal wheels, but also the least that can either drive or be driven at all.

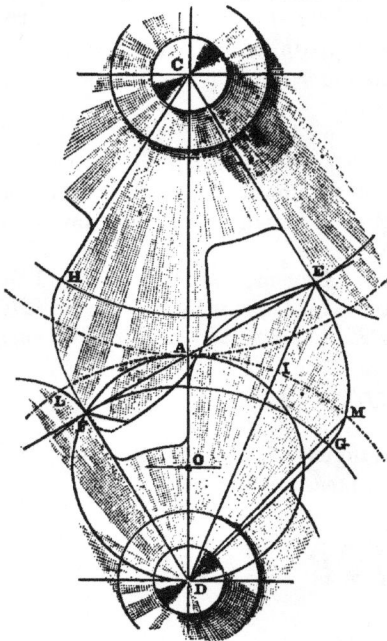

FIG. 155.

264. The greatest possible path of contact is the distance between the feet of the perpendiculars let fall from the centres of the wheels upon the line of action. It has been pointed out that if the obliquity be assumed, there is no certainty that this distance will be an aliquot part of the circumference of the base circle. But if the numbers of teeth be given for a proposed pair of wheels, it may be made so, and the obliquity thus determined.

For example, in Fig. 155, let the numbers be six for the upper wheel and five for the lower. About any centre C, with a radius measuring 6 on any scale of equal parts, describe a circle ; perpendicular to any radius CE, set off EF equal to the arc EH which measures the pitch ; perpendicular to EF, and on the side opposite EC, set off FD measuring 5 parts on the same scale, and draw CD, cutting FE in A. Then taking AC, AD, as the radii of the pitch circles, and CE, DF, as the radii of the base circles, the path of contact, EF, which is equal to one-sixth of the circumference of the upper base circle, will measure one-fifth of that of the lower one.

And this construction fixes the *minimum* obliquity for the given numbers of teeth ; it being evident that a decrease of this obliquity will at once shorten the path of contact and enlarge the base circles.

265. Involute Tooth with Epicycloidal Extension.—The path of contact, as above determined, is exactly equal to the pitch ; now in practice the arc of action must always be greater than the pitch, and yet it may be desirable to use involute teeth in circumstances like those represented in Fig. 155. The involute does not continue within its base circle, but the face HF of the blunted tooth may be extended as shown in dotted lines, the curve beyond F being an epicycloid generated by a circle whose diameter is AD, which will work correctly with the radial flank bounding the clearing space in the lower wheel. A similar construction is applicable, of course, to the other wheel, if its tooth be originally so much blunted as to make it desirable ; and in this way a sufficient angle of action may be secured, and the involute tooth employed in cases where it would otherwise be inadmissible.

Limiting Numbers of Teeth for Given Arcs of Recess.

266. In the employment of the involute system, if a definite pitch and arc of recess be assigned for either wheel of an engaging pair, there will be a limit to the number of teeth which can be given to the other, as was shown to be the case with the epicycloidal tooth.

Practically, a knowledge of these limiting numbers is just as necessary or convenient in using one form of tooth as it is in using another. Next to the epicycloid, the involute is the curve most extensively used, and in the mode of generation, as well as in the peculiarities of the action, there is a marked contrast between the two, and for these reasons a special interest attaches to a comparison of the limiting numbers for these forms under like conditions. We shall therefore proceed to investigate a method of determining these limits for the same values of the arc of recess which were assumed in the tables for epicycloidal teeth, viz.:

1. Recess = Pitch.   2. Recess = $\frac{3}{4}$ Pitch.   3. Recess = $\frac{1}{2}$ Pitch.

267. In Fig. 156, let C, D be the centres of two wheels, VW and HY arcs of the pitch circles tangent at A, FQ the line of action, CQ and DF the radii of the base circles, and let the lower wheel drive in the direction indicated by the arrow.

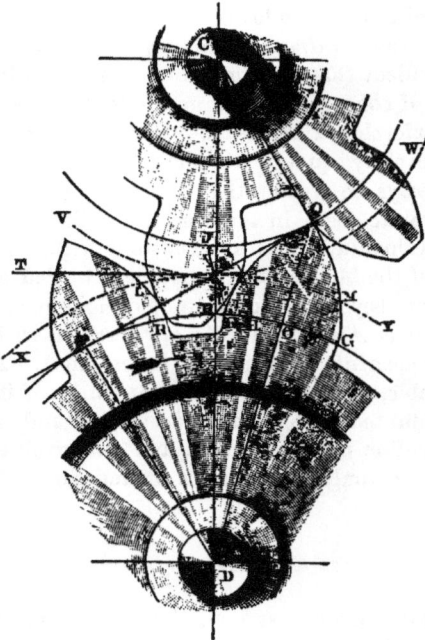

Fig. 156.

The proportions are such that the tooth of the driver is pointed, and that its point, Q, extends just to the root of the other involute; also, the breadth MN of the tooth being equal to the space NL, both measured on the pitch circle, there is no backlash.

This, evidently, is just a limiting case; the pitch ML and the arc of recess AM being assigned, the involutes bounding the tooth GQK must pass through the points M and N, whatever be the radius of the base circle, and the obliquity TAF, which causes their intersection to fall on the line of action, is clearly the greatest possible, which as clearly gives the least possible value for AC.

The diagram, then, represents the problem as already solved, and it remains to deduce from it the means of effecting the solution in any given case.

268. Draw QD, which cuts the lower base circle in O; then the path of contact FQ is equal to the arc FG, and is also the trigonometrical tangent of the arc FO; that is to say, we have, as an equation of condition,

$$\tan FO = \text{arc } FG.$$

The line of centres cuts the lower base circle in B; through L, A, and N, draw LR, AI, NH, involutes of the lower base circle, and through A draw also AJ, an involute of the upper base circle.

Then the arc QJ is equal to QA, the trigonometrical tangent of the angle QCA; the arc FI is equal to FA, the trigonometrical tangent of the angle FDA; and the angles QCA, FDA, are equal to each other and to the obliquity TAF.

We shall now have

$$FG = FR + 2RH \dots\dots\dots\dots\dots\dots \quad (1)$$
$$FO = FG - OG$$
$$OG = \tfrac{1}{2}KG$$

But $\begin{cases} KH = 2BI, \; \therefore \; \dots\dots \quad KG = 2BI + HG \\ MN = NL, \; \therefore \; \dots\dots \quad\quad\quad HG = RH \end{cases}$

$$RG = 2BI + RH$$
$$OG = BI + \frac{RH}{2}$$
$$FO = FR + 2RH - BI - \frac{RH}{2}$$

or
$$FO = FR + \frac{3RH}{2} - BI \dots\dots\dots\dots \quad (2)$$

Also
$$FI = FR + RI = FB + BI \dots\dots\dots\dots \quad (3)$$

269. Now when the definite values mentioned in (266) are assigned for the arc of recess, these equations will be modified as follows:

I. Recess = Pitch.

In this case we have

$$\left. \begin{array}{l} RI = 0, \\ RH = IH, \\ FR = FI, \end{array} \right\} \text{whence}$$

$$\left. \begin{array}{l} FG = FI + 2IH \dots\dots\dots\dots 1. \\ FO = FI + \dfrac{3IH}{2} - BI \dots\dots 2. \\ FI = FB + BI \dots\dots 3. \end{array} \right\} \therefore FO = FB + \frac{3IH}{2}.$$

And for a pinion driving a rack,

$$\left. \begin{array}{l} \cdot FR = FI = 0, \\ FB = 0, \\ BI = 0, \end{array} \right\} \text{whence}$$

$$FG = 2IH\dots\dots 1.$$
$$FO = \frac{3IH}{2}\dots\dots 2.$$

$$\therefore \ FO = \tfrac{3}{4}\,FG.$$

II. Recess $= \tfrac{3}{4}$ Pitch.

We now have

$$RI = \frac{RH}{2},\text{ or } RH = 2\,RI,$$

which gives

$$FG = FR + 4RI\dots\dots 1.$$
$$FI = FR + RI\dots\dots 3.$$

$$\therefore \ FG = FI + 3RI,$$

and

$$FO = FR + 3RI - BI\dots 2.$$
$$FR + RI = FB + BI\dots 3.$$

$$\therefore \ FO = FB + 2RI.$$

When the pinion drives a rack,

$$FR = 0,$$
$$RI = FI,$$

$$\therefore \quad \begin{cases} FG = 4FI, \\ FO = FB + 2FI. \end{cases}$$

III. Recess $= \tfrac{2}{3}$ Pitch.

With this value

$$RI = \tfrac{2}{3}RH,\text{ or } RH = \frac{3RI}{2}.$$

And therefore

$$FG = FR + 3RI\dots 1.$$
$$FI = FR + RI\dots 3.$$

$$\therefore \ FG = FI + 2RI.$$

Also

$$FO = FR + \frac{9RI}{4} - BI\dots 2.$$
$$FR + RI = FB + BI\dots 3.$$

$$\therefore \ FO = FB + \frac{5RI}{4}.$$

And in the case of a pinion driving a rack,

$$FR = 0,$$
$$RI = FI,$$

$$\therefore \quad \begin{cases} FG = 3FI, \\ FO = FB + \frac{5FI}{4}. \end{cases}$$

270. Now suppose the pitch and the arc of recess to be assigned for the driver. The arcs IH and HG are then known; and if we assume a value for FI, we can find the corresponding values of BI and FB, which combined with the others will give certain values for FO and FG; and these are finally to be tested to see whether they will satisfy the equation of condition,

$$\tan FO = \text{arc } FG.$$

Having at last found in this way the obliquity which under the assigned conditions makes FQ equal in length to the arc FG, we know also the length of FA; whence AC, the minimum radius of the follower, is readily ascertained.

If, on the other hand, the wheel whose pitch and arc of recess are given be the follower; we then know the values of QA and the arc QJ, which are equal to each other, and can thence find the angle ACQ, equal to ADF and to the obliquity. We then, knowing the arcs FB, FI, assume a value for RI, and proceed as before to see whether the equation of condition is satisfied.

271. When the given driver becomes a rack, the least radius for the follower is most readily ascertained by a special construction, which is given in Fig. 157.

We have here an exact limiting case, the rack tooth being pointed, and the whole path of contact just equal to the pitch arc on the base circle; thus, the rack driving in the direction of the arrow, a marking point in going from Q to F, generates the right line QU and the involute QG, the latter cut-

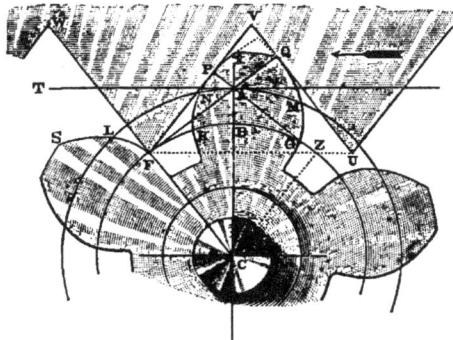

Fig. 157.

ting the pitch circle at M. The action, then, begins at Q, and ends at F, the root of the involute FS, which latter curve cuts the pitch circle in L. Since the tooth and space are to be equal, the side XK of the pinion's tooth must bisect the arc ML at N; and it must also be tangent to the side FV of the rack-tooth, at P its intersection with the other line of action PAZ, to which FV is perpendicular.

It will now be seen that while the marking point is moving from Q to A, it generates the curve QM, the pinion rotating through the arc MA; in the reverse motion, a marking point, in going from P to A, traces the curve PN, while the pinion turns through the arc NA.

But we have

$$MA + AN = MN = \tfrac{1}{2} ML,$$

whence

$$QA + PA = \tfrac{1}{2} QF,$$

or

$$PA = \tfrac{1}{2} QF - QA ;$$

also,

$$QA = QF - AF;$$

therefore, since QF is equal to the pitch arc, and AF to the arc of recess (both measured on the base circle), PA is known when these arcs are given.

We have, then, this simple graphic process : Draw any line FQ, and let it represent the pitch ; set off upon it a distance FA, making

$$\frac{FA}{FQ} = \frac{\text{arc of recess}}{\text{pitch arc}} ;$$

next determine PA as above, and construct the right-angled triangle APF. Then the angle PAF is twice the obliquity, and AT bisecting it is the pitch line of the rack. Draw AC perpendicular to AT, and FC perpendicular to FQ ; these will intersect at C the centre of the pinion, determining CF the radius of the base circle and CA that of the pitch circle.

272. Now introducing the same definite values as before for the arc of recess, we shall have,

I. RECESS = PITCH.

$$\left. \begin{aligned} AQ = \ &0, \\ \therefore FA = \ &FQ, \end{aligned} \right\} \text{ whence } AP = \tfrac{1}{2} AF.$$

II. RECESS = $\tfrac{3}{4}$ PITCH.

$$\left. \begin{aligned} AQ &= \tfrac{1}{4} QF, \\ FA &= \tfrac{3}{4} QF, \end{aligned} \right\} \text{ whence } AP = \tfrac{1}{3} AF.$$

III. RECESS = $\tfrac{2}{3}$ PITCH.

$$\left. \begin{aligned} AQ &= \tfrac{1}{3} QF, \\ FA &= \tfrac{2}{3} QF, \end{aligned} \right\} \text{ whence } AP = \tfrac{1}{4} AF.$$

In these cases, then, no tentative proceeding is involved in determining the least number of teeth for the pinion, which requires merely the solution of the triangles APF, ACF.

273. In Fig. 157 the pinion's tooth has a sensible breadth, QX, at the top, and the involutes might therefore be continued as shown in

dotted lines, and the approaching action thereby prolonged. But the smaller the assigned arc of recess, the less is AP, and the narrower will be the tooth at the top, when as in the figure the total arc of action is made equal to the pitch: there will, then, be some value of the arc of recess which will cause the points Q and X to coincide.

And it is apparent that when the limit is reached, as in Fig. 158, the pinion will have the least possible number of teeth which can be driven by a rack.

Now if in this diagram the arc FI be given, we know

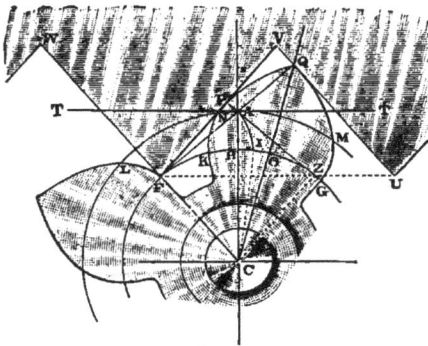

Fig. 158.

FA, the trigonometrical tangent of the angle ACF, which is always one half the angle PAF: whence AP may be found. We have also

$$AQ = AF - 2\,AP,$$
and $$AQ + AF = FQ = \text{arc } FG \quad \ldots \quad \text{(A)}$$

Again, in the equations deduced in (**268**), viz.:

$$FG = FR + 2\,RH,$$
$$FO = FR + \frac{3\,RH}{2} - BI,$$

we have $\quad FR = 0;$

whence $\quad FG = 2\,RH,$

and $\quad FO = \frac{3\,RH}{2} - BI,$

or $\quad FO = \tfrac{3}{4}\,FG - BI. \quad \ldots \quad \text{(B)}$

274. From these data we may now solve the problem in the tentative manner before described.

Taking any convenient value for the radius FC, we first assume the arc FI, and find the corresponding values of FA, of the angle ACF, and of the arc BI. Then solving the triangle APF, we determine AP, whence we find FQ, as in Eq. (A).

The length of FQ being equal to that of the arc FG, we next find the length of the arc FO as in Eq. (B), and knowing the radius, ascertain the value of this arc in degrees.

Finally, we apply the test of determining the length of the trigonometrical tangent of this arc, and comparing it with that of FQ as before found. If these two values are equal, the assumed value of FI is correct; but if they are unequal a new value must be assumed, and the process repeated until concordant results are obtained.

In this case we find that the pitch, at the limit which gives a pointed tooth to the pinion, corresponds to

No. of Teeth.............. $= 4.0906$
Obliquity................. $= 41°\ 20'\ 45''$ } very nearly.
Arc of Recess............ $= \frac{4\ 4}{4\ 5}$ Pitch. }

275. Now referring to Fig. 153, it will be seen that in constructing a pair of equal wheels, making the pitch arc on the base circle equal to the greatest possible path of contact, the teeth will become narrower at the top as the obliquity is increased; evidently, then, there is a limiting value of the obliquity which will make the teeth of both wheels pointed, as shown in Fig. 159.

In this case we have [Eq. (B). **(273)**],

$$FO = \tfrac{3}{4}\ FG - BI\ ;$$
and also $\qquad FG = 2\ FI,\ \therefore\ FQ = 2\ FA.$
Whence $\qquad\quad FO = \tfrac{3}{2}\ FI - BI\ ;$
but $\qquad BI = FI - FB,\ \therefore\ FO = \tfrac{1}{2}\ FI + FB.$

It will now be seen that by assuming a value for FI, all the quantities involved in the above expressions may be determined; and the results, as before, must satisfy the equation of condition,

$$\tan FO = \text{arc } FG.$$

By this process we find

No. of Teeth.................... $= 4.6256$
Obliquity...................... $= 34°\ 10'\ 58.5''$
Arc of Recess (given)........... $= \tfrac{1}{2}$ Pitch.

If complete wheels are to be used, then, five is the least number that can drive or be driven, since, as shown in the preceding section, four and a fraction are required even when the pinion gears with a pointed rack. By adopting the construction shown in Fig. 155, the obliquity just given may be slightly reduced, and we shall have for the smallest possible pair of complete equal wheels,

No. of Teeth.......................... 5
Obliquity.............................. 32° 8′ 31″
Arc of Recess......................... ½ Pitch.

276. The tentative processes above explained are undeniably tedious, yet when systematically conducted they are less so than might be supposed ; and in the preparation of the following tables they have been continued until, with a radius of 10, the linear values of the arc FG and the path of contact FQ agreed up to the sixth decimal place, the approximation being therefore to within the one-millionth part of the radius of the base circle.

It was not to be expected that for a given number of teeth upon either wheel the radius of the other as thus determined would correspond to an exact whole number of teeth upon its periphery. The next higher integer being of course taken as the tabular number, the tooth would not then be absolutely a pointed one ; but we have not made any correction for the slight change in the obliquity due to this circumstance, considering its value as appearing in the calculation, which is given to the nearest second, sufficiently precise for all practical purposes.

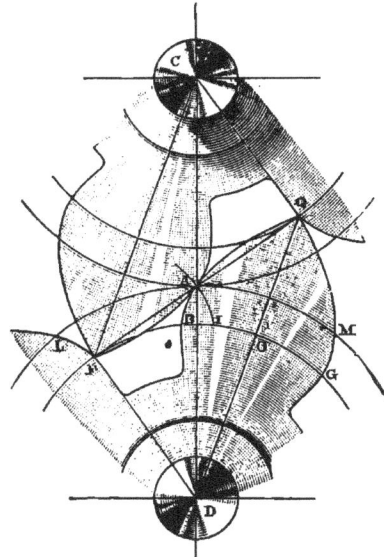

FIG. 159.

It is proper to point out that these tables may be of service, not only as indicating exact limiting cases, which are to be avoided when possible, but upon another account, viz.: although in general terms the obliquity is said to be arbitrary, yet in constructing for special purposes wheels with a given arc of recess, it may be necessary to employ numbers not far removed from limiting values, and in many instances, it will be seen, the possible variation in the obliquity is confined to quite a narrow range.

11

Limiting Numbers of Teeth.

RECESS = PITCH.			TOOTH = SPACE.		
INVOLUTE.			EPICYCLOIDAL.		
CONST. OBLIQUITY.	D.	F.	D.	F.	MAX. OBLIQUITY.
0° 0′ 0″	5.58	∞	5.58	∞	0° 0′ 0″
2° 28′ 42″	6	146	6	144	2° 30′
7° 0′ 47″	7	52	7	50	7° 12′
10° 17′ 4″	8	35	8	34	10° 35′ 18″
12° 45′ 5″	9	28	9	27	13° 20′
14° 40′ 33″	10	24	10	23	15° 39′ 8″
15° 54′ 56″	11	22	11	21	17° 8′ 34″
17° 26′ 26″	12	20	12	19	18° 56′ 50″
18° 17′ 57″	13	19	13	18	20°
19° 2′ 37″	14	18	14	17	21° 10′ 35″
20° 17′ 4″	15	17	15	16	22° 30′
21° 26′ 24″	18	16	17	15	24°
22° 43′ 40″	20	15	20	14	25° 42′ 51″
24° 10′ 12″	25	14	24	13	27° 41′ 32″
25° 47′ 45″	34	13	32	12	30°
27° 38′ 12″	50	12	57	11	32° 43′ 38″
29° 44′ 6″	515	11	382	10	36°
30°	∞	10.88	∞	9.83	36° 37′ 21″

Limiting Numbers of Teeth.

RECESS = $\frac{3}{4}$ PITCH.			TOOTH = SPACE.		
INVOLUTE.			EPICYCLOIDAL.		
CONST. OBLIQUITY.	D.	F.	D.	F.	MAX. OBLIQUITY.
			3.25	∞	0° 0′ 0″
16° 2′ 40″	5.46	∞	4	34	7° 56′ 28″
17° 59′ 6″	6	15	5	19	14° 54′ 4″
20° 43′ 34″	7	13	6	14	19° 17′ 8″
22° 43′ 27″	8	12	7	12	22° 30′
24° 14′ 35″	9	11	8	11	24° 32′ 44″
25° 13′ 56″	10	10	9	10	27°
27° 38′ 12″	13	9	11	9	30°
30° 29′ 58″	20	8	16	8	33° 45′
33° 56′ 54″	70	7	33	7	38° 34′ 10″
35° 46′ 57″	∞	6.54	∞	6.35	42° 31′ 11″

RECESS = $\frac{2}{3}$ PITCH.			TOOTH = SPACE.		
INVOLUTE.			EPICYCLOIDAL.		
CONST. OBLIQUITY.	D.	F.	D.	F.	MAX. OBLIQUITY.
			2.58	∞	0° 0′ 0″
			3	37	6° 29′ 11″
			4	15	16°
21° 32′ 59″	5.30	∞	5	11	21° 49′ 6″
23° 38′ 20″	6	10	6	10	24°
24° 57′ 30″	7	9	7	9	26° 40′
27° 38′ 12″	9	8	8	8	30°
30° 53′ 47″	12	7	11	7	34° 17′ 9″
34° 55′ 12″	28	6	21	6	40°
37° 58′ 55″	∞	5.36	∞	5.17	46° 25′ 18″

277. But the labor of computing these tables was undertaken mainly with the object of making a fuller comparison than has before been made between the two leading systems of spur gearing : and to facilitate this, the limiting numbers and the maximum obliquities for epicycloidal, are here given side by side with those for involute teeth.

One striking point of difference will at once be noticed. In the epicycloidal system, as the arc of recess diminishes, the least number for the driver also diminishes, and the obliquity is always zero at the limit where the follower becomes a rack.

In the involute system, on the other hand, this obliquity rapidly increases, while the number for the least driver remains practically the same, varying but little even at the limit.

The reason is not far to seek. The generating circle for the driving *face* of the epicycloidal tooth, which alone has to do with the receding action, is different from and wholly independent of that for the *flank*, which produces the required amount of approaching action.

But the outline of the involute tooth is one continuous curve, and the rectilinear generatrix, or line of action, is the same for the part without the pitch circle and the part within. The sum of these two segments must always be at least equal to the length of the pitch arc on the base circle. If then the outer one, upon which the arc of recess depends, be decreased, the inner one must be correspondingly increased ; and this obviously involves a greater obliquity.

278. The difference between the two systems is best illustrated by constructing, as in Fig. 160, the nomodont for each under like conditions.

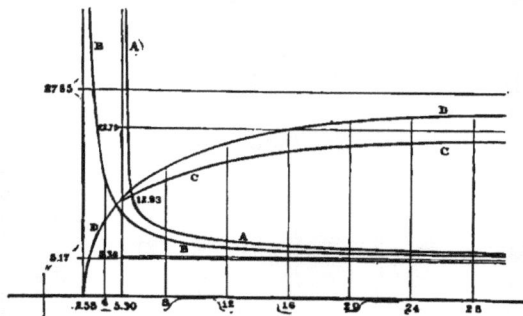

Fig. 160.

The arc of recess is here taken at $\frac{2}{3}$ the pitch, and the scale for the abscissas is twice that for the ordinates : the distances of the vertical and horizontal asymptotes from the axes being respectively 5.30 and 5.36 for AA, which represents the involute system, and 2.58 and 5.17 for BB, the curve for the epicycloidal system.

The law of variation in the obliquities is also most clearly exhibited by the same graphic means : thus, the ordinates of the curves CC and DD are proportional to the tabular values of the obliquity for involute and epicycloidal teeth respectively, the abscissas, as before, representing the numbers of teeth of given drivers.

These ordinates are drawn to the same scale as those of the nomodonts, the numerical value being in each case the one-hundredth part of the obliquity taken to the nearest minute. Thus, the first ordinate for C is the vertical asymptote to A, and its value is

$$\frac{21° \ 33'}{100} = 12\overset{'}{.}93.$$

As the number for the driver approaches infinity, the obliquity approaches the limit 37° 59', as seen by reference to the table, and

$$\frac{37° \ 59'}{100} = 22\overset{'}{.}79,$$

at which distance above the axis of abscissas this curve will have a horizontal asymptote.

In like manner, we find that the curve DD has also a horizontal asymptote, at a distance of 27.85 above the axis of abscissas. But this curve, instead of beginning abruptly like the other, and intersecting the vertical asymptote to BB, is tangent to that line at its foot.

279. The Involute Generated by Rolling Contact with the Pitch Circle.—Before dismissing the involute, it remains to prove that it can be generated in such a manner as to accord with the general principle stated in (**181**).

In Fig. 161, let AT be the common tangent of the pitch circles, and let CQ, DP, the radii of the base circles, be perpendicular to the line of action PQ, drawn at pleasure. Produce DP to intersect AT in E; and let POA be a logarithmic spiral, of which P is the pole and AT the tangent at the extremity of the radiant PA. If now the portion AP of this spiral roll upon AE, the pole P will trace the right line PE, which, therefore, may represent the direction and velocity of the

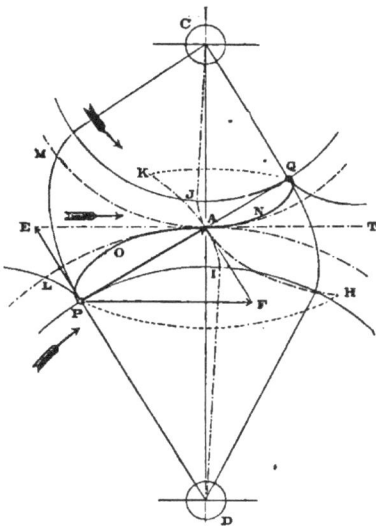

Fig. 161.

motion of P due to this rolling. With this motion let there be compounded a motion similarly represented by PF, equal and parallel to EA, then the resultant is PA ; the effect being the same as if, during the rolling of the spiral upon the tangent, the tangent should move to the right as shown by the horizontal arrow.   Suppose further that the pitch circles meantime move in rolling contact with the tangent ; they will then have turned about the fixed centres C and D, through angles measured by the arcs AM, AL, equal to each other and to AE, and the rectilinear motions PE, PF, being uniform, these rotations will also be uniform.   Also, the pitch circles will, during this rotation, move in rolling contact with the spiral AP, which being taken as the describing line, will carry the pole P in the right line PA with uniform velocity.   But we have already seen that such motion of P will simultaneously trace upon the planes of the two wheels, the involutes AH and AI of the base circles.   And in like manner it may be shown that the continuations AJ and AK of these involutes may be generated by the similar and equal spiral ANQ, rolling in contact with the pitch circles, its pole Q being the tracing point.

CHAPTER X.

Conjugate Teeth. Sang's Theory. Path of Contact. Unsymmetrical Teeth
Approximate Forms. The Odontographs. Diametral Pitch. Manufacture
of Gear Cutters. Determination of Series of Equidistant Cutters.

Conjugate Teeth.

280. It follows from the general principles stated in (**181**), that if
a tooth-outline of any reasonable form be assumed, it may be generated by some describing curve rolling upon the pitch circle; and if
that describing curve be determined, and used in connection with any
other pitch circle, the tooth thus formed will evidently gear correctly
with the assumed one. Teeth thus related are said to be *conjugate* to
each other; since had the second one been given, the application of
the above process would have resulted in the formation of a tooth
identical with the first.

Now it may be required to lay out a wheel to gear with one already
made; and supposing the original drawing to be lost, the tooth of
this given wheel is to all intents and purposes an assumed one: a
ready means of tracing its conjugate is therefore of some practical
importance.

The radius of the original pitch circle, even, may be unknown, but
this does not affect the above reasoning. We must assume a radius:
and if it should not be the same as the one originally used, the form
of the describing curve will be modified accordingly, and the outline
of the conjugate tooth will still be correct.

The form of the describing curve, which by rolling on the assumed
pitch circle will generate the given tooth, can in general be only approximately determined by graphic processes (See Appendix). And
even if it were susceptible of exact construction, the method above
indicated, of ascertaining by means of it the outline of the conjugate
tooth, involves much greater labor, and is withal less satisfactory in

all practical respects, than the simple and direct mechanical expedient illustrated in Fig. 162.

281. Let the form of a tooth, T, of the given wheel, be accurately cut out of a piece of sheet metal, KG, which is properly centred to turn upon a fixed pin at D; and let AD be the radius of the assumed pitch circle.

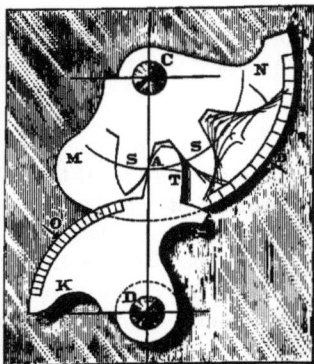

Behind this is another piece of sheet metal MN, being a portion of a disk turning upon a fixed centre C. Since the radius AD and the number of teeth upon the given wheel are known, the radius AC is determined by the velocity ratio required.

The outline of the tooth, T, is now to be traced on the blank disk, MN, with a fine marking point: after which let each piece be rotated through a small angle, as though the pitch circles moved in rolling contact. The proper relative amounts of these angular motions are easily regulated by means of subdivisions on the circular edges, in connection with fixed marks on the board to which this mechanism is secured, as shown at O and L. The outline of T is then to be traced again; and by repeating this process the form of the conjugate tooth may be very accurately mapped out, since each of the marks thus made upon the blank MN, by the very nature of the operation, is tangent to its outline.

Moreover, if both sides of T be traced upon the disk MN in its various positions, thus determining the boundaries of the space between two adjacent conjugate teeth S, S', there will be no backlash; for by construction the backs as well as fronts of the teeth are in contact at all times. It is true that the tooth and the space of the given wheel may not be measured by equal arcs upon the pitch circle, which was drawn at pleasure. Nor is this essential, since the spaces and the teeth of the derived wheel will be correspondingly unequal; the total pitch arc, or sum of a tooth and a space, being however necessarily the same for both wheels.

Sang's Theory of the Teeth of Wheels.

282. By giving different values to AC in Fig. 162, we may construct a whole series of wheels, having various numbers of teeth, all working properly with the given wheel.

Next, let the tooth of any one of these be taken as the assumed tooth in the above process, and used in like manner with various pitch circles, thus forming a second series of wheels.

From the mode of generation it is evident that any wheel belonging to either series will gear correctly with any one of those constituting the other series. But the wheel of a given number of teeth in the first may not be like that one of the second which has the same number. If the outlines of conjugate teeth upon any two wheels of the same number be made alike by any means, it is clear that the two series will be identical, and all the wheels of both will be interchangeable.

Now in the derivation of either series, the assumed* tooth always occupies and maps out the space between two adjacent conjugate teeth; and this holds true when the latter becomes infinite. From this it follows that the racks of the two series are in any case exactly converse; their contours are identical, the teeth of the one being precisely like the spaces of the other, and *vice versa*, as shown in Fig. 163.

When, therefore, the two series become identical, the two racks will be identical; each will be its own converse, and have its teeth and its spaces similar and equal to each other. This requires that the tooth and space shall be equal as measured on the pitch line, and also that the portions a, b, of the contour, shall be respectively similar and symmetrical to the portions a' b' on the opposite side of the pitch line.

And if, as all along supposed, the teeth are symmetrical to a radius, that is, have their fronts and backs alike, it is further necessary that a and b shall be similar and equal.

283. Now it is evident that in the process of Fig. 162, we may assume the tooth of a rack, and from it derive the conjugate teeth of a series of wheels, by a modification in detail which is too obvious to require explanation.

And from the foregoing, the deduction is that if this original rack-tooth be bounded by four similar and equal lines in alternate reversion, the wheels whose teeth are thus determined will form an interchangeable set. It will readily be seen that if the rack be composed of cycloidal arcs, the result will be the familiar epicycloidal system with a constant describing circle; while if it be bounded by oblique right lines, the involute system is at once reproduced.

The contour of the rack being arbitrary, with the limitation just

mentioned, any number of systems may be thus derived, each differing from the others in the peculiarities of its action, more particularly in respect to the variation in the obliquity.

Professor Edward Sang, in a most elaborate treatise, has discussed this method of constructing the teeth of wheels, which consists substantially in reversing the usual order of proceeding. His treatment of the subject is analytical and abstruse, the style obscure and pedantic to the last degree; and although the theory of gearing is fully developed, the equally important practical operations of laying it out are very much neglected.

284. Determination of the Path of Contact.—The theory is, however, more comprehensively seen from this point of view than from any other, since the important part played by the *path of contact* is brought more prominently to our notice : and Professor Sang's investigations relate more particularly to the form of that path, as dependent upon the contour of the assumed tooth.

It is not necessary to know the describing curve in order to determine the path of contact. In Fig. 164, let AB be the face of a tooth for the wheel whose centre is D; draw a series of normals to this face, cutting the pitch circle at 1, 2, etc. During the action, the common normal of the tooth-outlines must at every instant pass through the common point of the pitch circles ; therefore as the points 1, 2, etc., successively reach the line of centres, the normals $1a$, $2b$, $3c$, will take the positions Aa', Ab', Ac', thus determining the path of contact AB'.

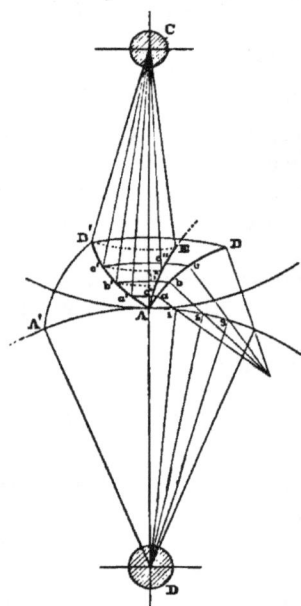

Fig. 164.

Let C be the centre of the engaging wheel ; then the form of the conjugate flank may also be found without making use of the describing curve. For considering the given wheel as the driver, the angular motions of the follower corresponding to the already ascertained angular motions aa', bb', etc., of the driver, may be determined, since the velocity ratio is known. Arcs $a'a''$, $b'b''$, $c'c''$, measuring these angular motions of the follower, then, are to be set off on circles of which C is the centre, and the curve AE thus determined is the acting conjugate flank ; which will, of course, require to be extended as shown in dotted lines for the formation of a clearing space, as usual.

285. But although it is thus possible to determine the path of contact and the conjugate tooth by graphic means, this method is not in general well suited for ordinary practical operations, since it involves the drawing of the normals to a curve whose nature may not be known, and this of itself (see **Appendix,** (7) is a matter of considerable labor; while the whole process above described is not only tedious, but requires most careful manipulation in order to secure satisfactory results. Consequently, if the form of a tooth be given or assumed, the mechanical method of Fig. 162 is for practical purposes by far the best for determining that of its conjugate. It is true that it gives no indication of the path of contact, and therefore does not enable us to ascertain precisely either where the action begins or where it ends. But if neither wheel has less than twelve teeth, and these of proportions in accordance with any of the arbitrary rules given in (201), there can be no question that the total angle of action will be ample.

For when we consider that in the involute system the path of contact is a right line joining the initial and terminal points of action, while in the epicycloidal it is composed of two circular arcs tangent to each other and to both pitch circles at their common point, and also that it must pass through this last-mentioned point in all cases, it will appear more than likely that with any other reasonable form of tooth, this path will be intermediate between these extremes. And if so, the difference, either in respect to the limiting numbers of teeth or the variations in the obliquity of action, is greater between these than between any other two systems of gearing: and either of them, as before stated, will with the above-mentioned numbers and proportions give results practically satisfactory for most uses.

It is beyond the scope of this work to discuss in detail the various forms of teeth which may be constructed by the application of Sang's Theory. In all the processes of "laying out" gearing upon the drawing-board, the simplicity of the paths of contact and of the describing lines, peculiar to the involute and the epicycloidal systems, gives them a great and obvious advantage over others; while years of experience have established the fact that in practical operation these forms have no superiors.

286. Unsymmetrical Teeth.—Thus far, the teeth have been supposed to be symmetrical, that is, to have their fronts and backs alike, as indeed they are made except in rare instances. But this is not at all a matter of necessity, and under some circumstances it may be advantageous to make them otherwise. For example, in Fig. 165, the acting outlines (the lower wheel being supposed to drive to the right)

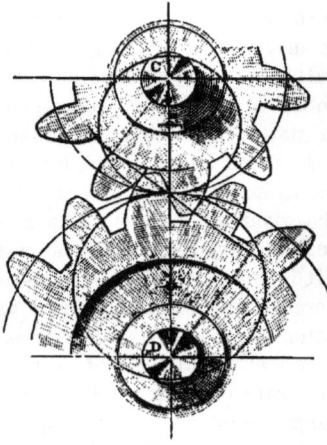

are of the epicycloidal form; the describing circles are large, in order to reduce the obliquity to a minimum, and were the tooth made of the same form on the back, this might render it too weak at the root. ·But if the motion is never to be reversed, this weakness may be avoided by making the back, as shown, an involute of considerable obliquity.

The conjugate racks, derived by the

Fig. 165.   Fig. 166.

process of Fig. 162 from a tooth of this description, would be of the forms shown in Fig. 166; which, it will be observed, are identical, notwithstanding their lack of symmetry.

Approximate Forms of Teeth.

287. For the perfect working of mechanism in which wheels are employed, it is unquestionable that the forms of the teeth should be exactly correct.   And if they are to be cut, there is no reason why

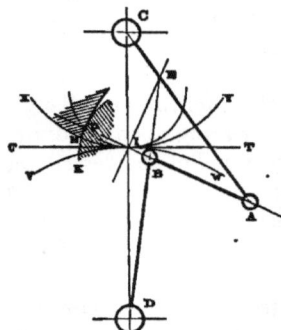

the teeth should not be laid out with the utmost precision, since, the cutter once made, it is as easy to cut one shape as another.

But for ruder machinery, when the wheels are simply to be cast, so great accuracy may be sometimes thought unnecessary; and, especially if time presses, a more ready means of securing a reasonably close approximation may be desirable.

In Fig. 167, draw through I, the common point of the pitch circles XY, VW,

Fig. 167.   a line of action AIP at pleasure, and IE

perpendicular to .it.   Draw CA at pleasure, cutting IE in E, and join ED by a right line cutting AP in B.

Now if CA, DB, be a pair of levers joined by a link AB, then, letting

$$v = \text{ang. vel. about } D,$$
$$v' = \text{ " " " } C,$$

we have the value

$$\frac{v}{v'} = \frac{CI}{DI};$$

and since by the construction E is the instantaneous axis, this velocity ratio will be momentarily constant. (See Figs. 39 and 40.)

If now through any point P on the line of action we describe two circular arcs, one about B, as a tooth-face for the lower wheel, the other one with centre A being the conjugate flank ; the velocity ratio determined by them will at the instant be exactly the same as that of the combination of link and levers, and will not differ appreciably from it during an elementary motion in either direction.

288. The above argument, it will be observed, does not depend either upon the obliquity, or upon the position of the point P, both of which are optional.

But in order to reduce the application of this general principle to a system, let IH, IK, be each one half the pitch arc, and let IP be an arc of the same length upon a describing circle : then PI will be the common normal to an epicycloid KP and a hypocycloid HP, tangent at P, and these are correct tooth-outlines. Let B and A be the centres of curvature for these two curves respectively, at the common point P ; the circular arcs may then be used as approximating closely enough to the exact forms for many practical purposes, provided that the obliquity be not too great nor the teeth too long.

And if the same describing circle, and the same arc IP, be used, the obliquity of the line of action, for the mean position when the curves are in contact at P and the velocity ratio is exactly correct, will be the same for all pitch circles.

289. Willis's Odontograph.—The system above explained was originated by Professor Willis, and upon it is based the invention of his well known Odontograph. Making the arc $IP = 30°$, the obliquity is 15°, and the angle $DIA = 75°$. Then having calculated the distances from I of the centres of curvature at P for the tooth-faces and flanks for various pitches and numbers of teeth, these are tabulated, and the instrument is constructed as in Fig. 168, which also illustrates the method of using it.

It consists merely of two arms of German silver, making an angle of 75° with each other, the edge of one corresponding to the radius

DI, that of the other to the line of action *AP* in Fig. 167; and this latter edge is divided into a scale of equal parts, numbered both ways from the intersection of the two edges, which corresponds to the point *I*. We have thus two scales, the one on the right being the "*Scale of Centres for Faces*," that on the left for flanks; and the table above mentioned has two corresponding divisions.

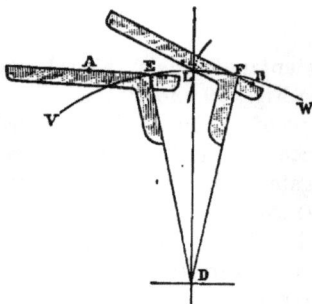

Fig. 168.

Suppose it to be required to lay out a wheel of 20 teeth, the circular pitch being 1½ inches. These data determine the radius of the pitch circle *VW*, Fig. 168; upon which set off the pitch arc *EF*, bisect it at *L*, and draw the radii *DE* and *DF*. Set the odontograph so that the slant edge coincides with *DE*, the point *E* thus coming just at the zero of the scales. Then in the "Table of Centres for Flanks," in the column for 1½-inch pitch, opposite 20 in the column of "Numbers of Teeth," we find the number 37.

The position indicated by this number on the "Scale of Centres for Flanks," fixes the point *A* as the centre of the circular arc through *L*, which gives the approximate form of the required flank. Then setting the odontograph in like manner by the radius *DF*, by a similar process we find on the other scale the location of the centre *B* of the circular arc through *L*, which forms the face of the tooth.

290. It is evident that wheels of the same pitch, laid out by this odontograph, are interchangeable, their teeth being in effect epicycloidal with a constant describing circle. Since half the pitch arc measures 30° upon this circle, the wheel of 12 teeth, the least for which the distinguished inventor designed it to be used, will have radial flanks.

Of course the approximating circular arcs will deviate more and more from the true curves as the tooth is made longer, and will not answer at all for low-numbered pinions. But if the length of the tooth be limited by the arbitrary rules given in (**201**), the instrument may be used with confidence for all ordinary purposes when the wheels are to be cast, and none have less than 12 teeth. In the case of a rack, the radii *DE*, *DF*, of Fig. 168 will become perpendicular to the pitch line; in laying out annular wheels the positions of face and flank are transposed, and due attention must be paid to the limits in regard to the diameters of the pitch circles, which have already been discussed.

It is also to be remarked that the tables which accompany the instrument do not contain radii of curvature for all the consecutive numbers of teeth. The variation in the correct contour, due to the addition of a single tooth, rapidly diminishes as the actual number increases, so that practically the same outline serves for the teeth of several different wheels. If then the number assigned be not found in the table, the nearest number to it is to be taken instead.

Also, if the given pitch be one not inserted in the tables, the radii required may be found by direct proportion from those of other pitches which are tabulated ;—for 4-inch pitch by doubling the radii for 2-inch pitch, halving those given for 1¼ inch pitch in order to find the radii for ⅝-inch pitch, and so on, as occasion may require.

Robinson's Odontograph.

291. This differs entirely from the preceding, both in principle and in the manner of using it. In Fig. 169, let VW be an arc of the pitch circle, to which CT is tangent at A, the middle point of the tooth, AO, being the half thickness. Then the odontograph is to be so set that CT shall not only be tangent to the lower curved edge BFH, but also cut the graduated edge BE, which must pass through O, at

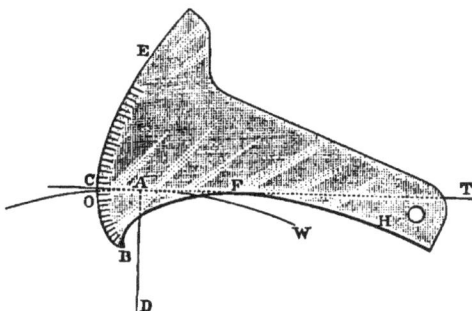

Fig. 169.

a certain division C, determined by the aid of tables and instructions which accompany the instrument. The edge CE is then used as a curved ruler for drawing the face of the teeth, after which the odontograph, being made of thin metal and graduated on both sides, is turned over, and the opposite face of the tooth is drawn in like manner. A different "setting number" is then determined from the tables, which brings a different portion of the curved edge into position for describing the flanks of the teeth. By certain modifications in the determination of the setting numbers and in the placing of the instrument, involute teeth also can be drawn, as well as epicycloidal ones with various describing circles, in either outside or inside gear.

This remarkably ingenious invention, which has a much wider and more flexible range of action than that of Willis, is called the *Tem-*

plet Odontograph, from the fact that the forms of the teeth are drawn by it directly ; and if it be desired to lay out the whole wheel, its use is much facilitated by attaching it temporarily to a wooden rod or radius bar, turning about the centre of the wheel.

292. Its construction, obviously, depends upon the finding of a curve for the graduated edge, of rapidly changing curvature, and of such a nature that its different portions shall closely approximate to the initial parts, or those next the pitch circles, of teeth of all pitches and sizes, of either the involute or epicycloidal forms.   This very peculiar property Prof. Robinson has shown, by a long and elaborate investigation,* to be possessed by a certain logarithmic spiral, of which form, accordingly, the working edge *BCE* is made.   The other curve *BFH*, it may be added, is the evolute of *BCE*, and, therefore, a similar and equal spiral.

Diametral Pitch.

293. In designing and laying out wheels upon any of the systems described, it is necessary to find the *circular pitch,* to which only reference has thus far been made ; not only because it is used in the graphic constructions, but because the strength of the tooth depends upon its thickness.

Were there nothing to the contrary, it would be most convenient to express the pitch in whole numbers or in manageable fractions, as 2-inch pitch, ¾-inch pitch, and so on.   But since for complete wheels the circular pitch must be an aliquot part of the circumference, the diameters of the pitch circles will often contain awkward decimals if this plan be adhered to ; and practically it is much more important to have the diameter a whole number, or a convenient fraction, than that the circular pitch should be either the one or the other.

This consideration has led to the introduction, and the almost exclusive adoption for cut gearing, of what is called the **Diametral Pitch** ; which is merely the quotient found by dividing the *diameter* of the pitch circle, instead of its circumference, into as many equal parts as there are to be teeth upon the wheel under consideration :

whence,

$$\text{Diametral Pitch} = \frac{\text{Diameter}}{\text{No. of Teeth}}\, , \; = \frac{\text{Circular Pitch}}{3.1416}\, ,$$

and

$$\text{Circular Pitch} = \text{Diametral Pitch} \times 3.1416.$$

---

* See Van Nostrand's Eclectic Engineering Magazine, July, 1876.

294. In the use of this system, convenient values of the diametral pitch are selected, each being a fraction with unity for its numerator and an integer for its denominator, as 1, $\frac{1}{2}$, $\frac{1}{4}$, $\frac{1}{8}$, $\frac{1}{10}$, $\frac{1}{12}$, etc.

The denominators of these fractions only are commonly used in giving the diametral pitch; thus an "8-pitch wheel" is one which has eight teeth for each inch of diameter, or whose diametral pitch is $\frac{1}{8}''$. This is, in fact, merely inverting the fraction, and giving the value of $\dfrac{\text{No. of Teeth}}{\text{Diameter}}$; thus, let a wheel of 16 inches diameter have 80 teeth; then, $\frac{16}{80} = \frac{1}{5}''$ = diametral pitch, but $\frac{80}{16} = 5$, and we call it a "5-pitch" wheel. By this system the calculations as to diameter and number of teeth are made very simple; as, for example:

Required, the diameter of a 4-pitch wheel with 37 teeth:

$$\tfrac{37}{4} = 9\tfrac{1}{4} = \text{diameter.}$$

How many teeth of 16-pitch on a wheel of $3\frac{7}{8}$ diameter?

$$3\tfrac{7}{8} \times 16 = 62 = \text{No. teeth.}$$

The tooth may be made to project outside the pitch circle a definite fraction of the diametral pitch, as in the case of the circular pitch; and thus the size of the blank may be readily ascertained. If this projection be made, for instance, $1\frac{1}{8}$ the diametral pitch, the face of the tooth will be nearly as long as that found by the first of the arbitrary rules given in (201) and the diameter of the blank is determined by simply adding to that of the pitch circle $2\frac{1}{4}$ times the diametral pitch.

On the Manufacture of Accurate Gear Cutters.

In cutting a spur wheel, it is obviously essential that the contour of the milling cutter conform precisely to that of the space between two adjacent teeth: in order to this the process most extensively, and until recently exclusively, employed, involves: 1st, the drawing of the required curve; 2d, the filing of a template to that exact form; and 3d, the turning of the cutter to fit the template.

The first improvement upon this, we believe, was the construction by Messrs. Brown & Sharpe, of Providence, R. I., of an ingenious piece of mechanism called an Epicycloidal Engine, by which the curves are traced automatically and with perfect precision upon the template by continuous motion. There still remain the two hand-and-eye operations of filing up the template, and of turning the cutter

12

to fit it when done ; and these operations are so delicate and difficult, that the exact duplication of templates, cutters or wheels in this manner is well nigh impossible.

Yet this duplication, it is very easy to see, is quite as desirable as the accurate formation of a single cutter ; and it has been made a matter of perfect certainty and ready execution by means of two machines recently constructed by Messrs. Pratt & Whitney, of Hartford, Conn.

Of these, which are remarkable for the ingenuity and beauty of their movements, we will now proceed to describe the principles and mode of action, though not the exact details. The whole process under this system is mechanical and nearly automatic ; a template being—not traced, but *milled* out, by one machine, which is subsequently used in the other as a guide by which its motions are so controlled as by another milling operation to finish the contour of the gear cutter, whatever the size of the tooth to be cut, to the precise epicycloidal form—with a minute and practically unimportant exception, as will presently be explained.

295. The Epicycloidal Milling Engine.—In Fig. 170, *A* is a portion of a flat ring, fixed to the framing; this represents a pitch circle. *B,* is a disc, representing the describing circle ; this turns freely upon a tubular stud *E,* fixed in the carrier *C,* which turns about a pivot *D,* fixed to the frame at the centre of *A* ; by means of the clamped socket, capable of sliding upon the rod, the position of *D* may be adjusted to suit the radius of *A*. Thus as *C* moves, the disc can roll upon the edge of *A,* and is compelled to do so by the flexible steel ribbon shown by the heavy line, which is wrapped round and secured to both pieces, due allowance for its thickness being made in adjusting their radii. *E'* is a second tubular stud fixed in the carrier, at the same distance

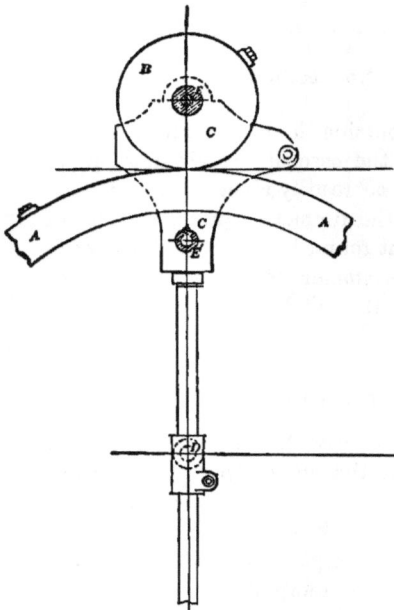

FIG. 170.

from the pitch circle as the other, but on the opposite side ; the centres of the two studs lying on a right line through D. Upon these two studs turn the two worm wheels F, F', shown in Fig. 171 ; these are in a plane above A and B, so that the axis of the worm, G, is vertically over the common tangent of the pitch and describing circles ; the relative positions of these and other parts will be most clearly seen by a study of the vertical section, Fig. 175. The worm G, is supported in bearings secured to the carrier C, and is driven by another small worm turned by the pulley I ; the driving cord, passing through suitable guiding pulleys, is kept at a uniform tension by a weight, not shown, however C moves.

Upon the same studs, in a plane still higher than the worm-wheels,

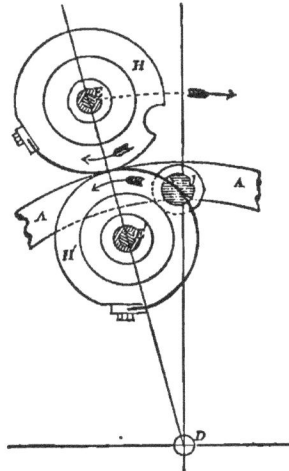

FIG. 171.   FIG. 172.

turn the two discs H, H', Figs. 172, 173, 174. The diameters of these are equal, and precisely the same as those of the describing circles which they represent, with due allowance, again, for the thickness of the steel ribbon by which these also are connected.

It will be understood that each of these discs is secured to the worm wheel below it, and the outer one of these to the disc B ; so that as the worm G turns, H and H' are rotated in opposite directions, the motion of H being identical with that of B ; this last is a rolling upon the edge of A, the carrier C with all its attached mechanism moving around D at the same time. Ultimately, then, the motions of H, H', are those of two equal describing circles rolling in external and internal contact with a fixed pitch circle.

In the edge of each disc a semicircular recess is formed, into which is accurately fitted a cylinder, J, provided with flanges, between which the discs fit so as to prevent end play; this cylinder is perforated for the passage of the steel ribbon, the sides of the opening, as shown in Fig. 172, having the same curvature as the rims of the discs. Thus when these recesses are opposite each other, as in Fig. 173, the cylinder J fills them both, and the tendency of the steel ribbon is to carry it along with H when C moves to one side of this position, as in Fig.

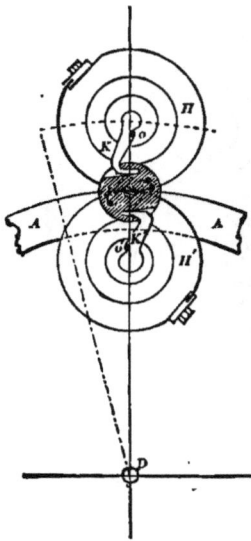
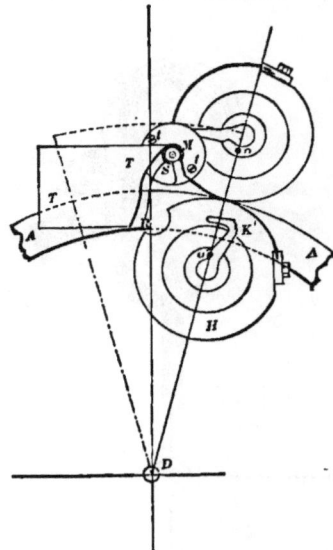

FIG. 173.   FIG. 174.

174, and along with H' when C moves to the other side, as in Fig. 172.

This action is made positively certain by means of the hooks K, K', which catch into recesses formed in the upper flange of J, as seen in Fig. 173. The spindles, with which these hooks turn, extend through the hollow studs, and the coiled springs attached to their lower ends, as seen in Fig. 175, urge the hooks in the directions of their points; their motions being limited by stops o, o', fixed not in the discs H, H', but in projecting collars on the upper ends of the tubular studs. The action will be readily traced by comparing

Fig. 173 with Fig. 174; as C goes to the left, the hook K' is left behind, but the other one, K, cannot escape from its engagement with the flange of J; which accordingly is carried along with H by the combined action of the hook and the steel ribbon.

On the top of the upper flange of J is secured a bracket, carrying the bearings of a vertical spindle L, whose centre line is a prolongation of that of J itself. This spindle is driven by the spur wheel N, keyed on its upper end, through a flexible train of gearing, not shown; at its lower end it carries a small milling cutter, M, which forms the edge of the template, T, firmly clamped to the framing.

When the machine is in operation, a heavy weight, not shown, acts to move C about the pivot D, being attached to the carrier by a cord guided by suitably arranged pulleys; this keeps the cutter M up to

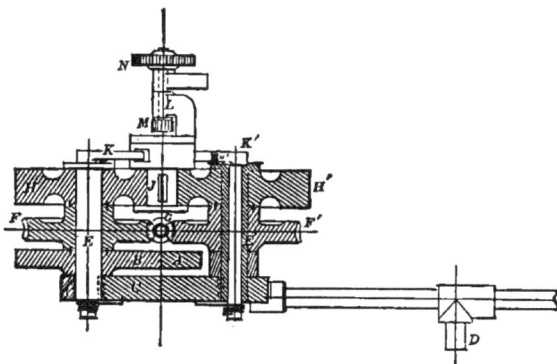

Fig. 175.

its work, while the spindle L is independently driven, and the duty left for the worm G to perform is merely that of controlling the motions of the cutter by the means above described, and regulating their speed.

The centre line of the cutter is thus automatically compelled to travel in the path RS, composed of an epicycloid and a hypocycloid, if A be a segment of a circle as here shown; or of two cycloids, if A be a straight bar. The radius of the cutter being constant, the edge of the template T is cut to an outline also composed of two curves; since the radius of M is small, this outline closely resembles RS; but particular attention is called to the fact that it is *not identical with it, nor yet composed of truly epicycloidal curves of any generation whatever;* the result of which will be subsequently explained.

The diameter of the discs which act as describing circles is $7\frac{1}{2}$

inches, and that of the milling cutter, which shapes the edge of the template, is $\frac{1}{8}$ of an inch.

Now if we make a set of 1-pitch wheels with these describing circles, the one with fifteen teeth will have radial flanks. The curves will be the same whatever the pitch ; but, as shown in Fig. 176, the blank should be adjusted in the epicycloidal engine so that its lower edge shall be $\frac{1}{16}$ of an inch (the radius of the cutter M) above the bottom of the space ; also its relation to the side of the proposed tooth should be as here shown. And, as previously explained, the depth of the space depends upon the pitch. In the system adopted by the Pratt & Whitney Company the whole height of the tooth is $2\frac{1}{5}$ times the diametral pitch, the projection outside the pitch circle being just equal to the pitch, so that diameter of blank = diameter of pitch circle + 2 × diametral pitch.

We have now to show how, from a single set of what may be called 1-pitch templates, complete sets of cutters of the true epicycloidal contour may be made of the same or any less pitch.

The Pantagraphic Cutter Engine.

296. In Fig. 176, the edge TT, is shaped by the cutter M, whose centre travels in the path RS, therefore these two lines are at a constant

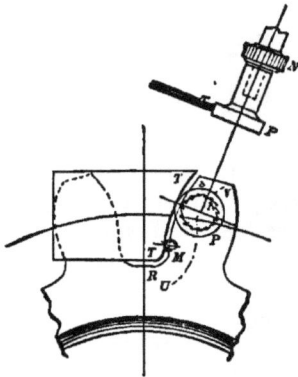

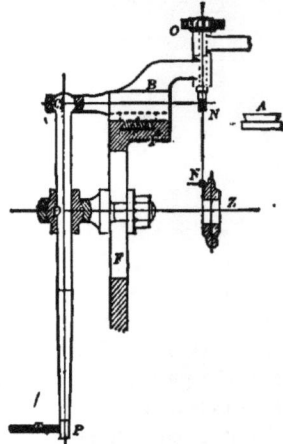

FIG. 176.   FIG. 177.

normal distance from each other. Let a roller P, of any reasonable diameter, be run along TT; its centre will trace the line UV, which is at a constant normal distance from TT, and therefore from RS. Let the normal distance between UV and RS be the radius of another milling cutter, N, having the same axis as the roller P, and carried by it, but in a different plane, as shown in the side view; then whatever N cuts will have RS for its contour if it lie upon the same side of the cutter as the template.

Now if TT be a 1-pitch template above mentioned, it is clear that N will correctly shape a cutting edge of a gear cutter for a 1-pitch wheel. The same figure, reduced to half size, would correctly represent the formation of a cutter for a 2-pitch wheel of the same number of teeth ; if to quarter size, that of a cutter for a 4-pitch wheel, and so on.

But since the actual size and curvature of the contour thus determined depend upon the dimensions and motion of the cutter N, it will be seen that the same result will practically be accomplished if these only be reduced ; the size of the template, the diameter and the path of the roller remaining unchanged.

The nature of the means by which this is effected in the Pantagraphic Cutter Engine is illustrated in Fig. 177. The milling cutter, N, is driven by a flexible train acting upon the wheel, O ; its spindle is carried by the bracket, B, which can slide from right to left upon the piece, A, and this, again, is free to slide in the frame F. These two motions are in horizontal planes, and perpendicular to each other.

The upper end of the long lever, PC, is formed into a ball, working in a socket which is fixed to B. Over the cylindrical upper part of this lever slides an accurately fitted sleeve, D, partly spherical externally, and working in a socket which can be clamped at any height on the frame F. The lower end, P, of this lever being accurately turned, corresponds to the roller, P, in Fig. 176, and is moved along the edge of the template, T, which is fastened in the frame in an invariable position.

By clamping D at various heights, the ratio of the lever arms, TD, DC, may be varied at will, and the axis of N made to travel in a path similar to that of the axis of P, but as many times smaller as we choose ; and the diameter of N is made less than that of P in the same proportion.

The template being on the left of the roller, the cutter to be shaped is placed on the right of N, as shown in the plan view at Z, because the lever reverses the movement.

This arrangement is not mathematically perfect, by reason of the angular motion of the lever. This is, however, very small, owing to the length of the lever ; it might have been compensated for by the introduction of another universal joint, which would practically have introduced an error greater than the one to be obviated, and it has with good judgment been omitted.

The gear cutter is turned nearly to the required form, the notches are cut in it, and the duty of the pantagraphic engine is merely to put the finishing touch to each cutting edge and give it the correct outline. It is obvious that this machine is in no way connected with, or dependent upon, the epicycloidal engine ; but by the use of proper templates it will make cutters for any desired form of tooth ; and by its aid exact duplicates may be made in any numbers with the greatest facility.

Theoretical Defects of the System.

297. It forms no part of our plan to represent as perfect that which is not so. And there are one or two facts which at first thought might seem serious objections to the adoption of the epicycloidal system. These are :

1. It is physically impossible to mill out a *concave* cycloid, by any

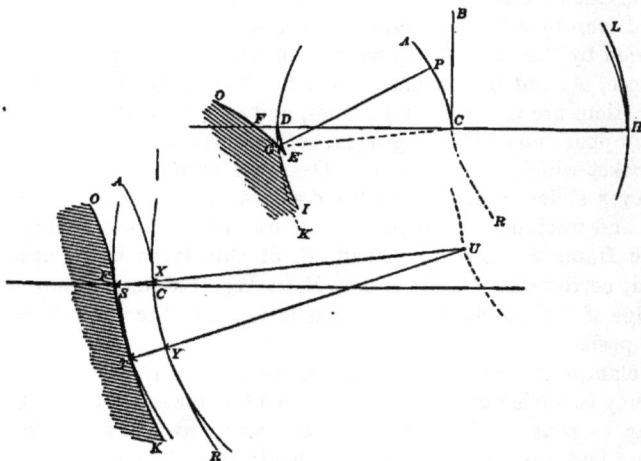

Figs. 178 and 179.

means whatever, because at the pitch line its radius of curvature is zero, and a milling cutter must have a sensible diameter.

2. It is impossible to mill out even a *convex* cycloid or epicycloid, by the means and in the manner above described.

This is on account of a hitherto unnoticed peculiarity of the curve at a constant normal distance from the cycloid. In order to show this clearly, we have, in Fig. 178, enormously exaggerated the radius, *CD*, of the milling cutter (*M* of Figs. 174 and 175). The outer curve, *HL*, evidently could be milled out by the cutter, whose centre travels in the cycloid CA ; it resembles the cycloid somewhat in form, and presents no remarkable features. But the inner one is quite different ; it starts at D, and at first goes down, *inside the circle whose radius is* CD, forms a cusp at E, then begins to rise, crossing this circle at G, and the base line at F. It will be seen then that if the centre of the cutter travel in the cycloid AC, its edge will cut away the part GED, leaving the template of the form OGI. Now if a roller of the same radius CD, be rolled along this edge, its centre will travel in the cycloid from A, to the point P, where a normal from G cuts it ; then the roller will turn upon G as a fulcrum, and its centre will travel from P to C, in a circular arc, whose radius is $GP = CD$.

That is to say, even a roller of the same size as the original milling cutter, will not retrace completely the cycloidal path in which the cutter traveled.

Now in making a rack template, the cutter, after reaching C, travels in the reversed cycloid CR, its left-hand edge, therefore, milling out a curve DK, similar to HL. This curve lies wholly *outside* the circle DI, and therefore cuts OG at a point between F and G, very near to G. This point of intersection is marked S in Fig. 179, where the actual form of the template OSK is shown. The roller which is run along this template is *larger*, as has been explained, than the milling cutter. When the point of contact reaches S (which is so near to G that they practically coincide), this roller cannot now swing about S through an angle so great as PGC of Fig. 12 ; because at the root D, the radius of curvature of DK is only equal to that of the cutter, and G and S are so near the root that the curvature of SK, near the latter point, is greater than that of the roller. Consequently there must be some point U in the path of the centre of the roller, such that, when the centre reaches it, the circumference will pass through S, and be also tangent to SK. Let T be the point of tangency ; draw SU and TU, cutting the cycloidal path AR in X and Y. Then, UY being the radius of the new milling cutter (corresponding to N of Fig. 9), it is clear that in the outline of the gearcutter shaped by it, the circular arc XY will be substituted for the true cycloid.

The System Practically Perfect.

298. The above defects undeniably exist ; now, what do they amount to ? The diagrams, Figs. 178 and 179, are drawn purposely with these sources of error greatly exaggerated in order to make their nature apparent and their existence sensible. The diameters used in practice, as previously stated, are : describing circle, $7\frac{1}{2}$ inches ; cutter for shaping template, $\frac{1}{4}$ of an inch ; roller used against edge of template, $1\frac{1}{2}$ inches ; cutter for shaping a 1-pitch gear-cutter, 1 inch.

With these data the author has found that the *total length* of the arc XY of Fig. 179, which appears instead of the cycloid in the outline of a cutter for a 1-pitch rack, is less than 0.0175 inch ; the real *deviation* from the true form, obviously, must be much less than that. It need hardly be stated that the effect upon the velocity ratio of an error so minute, and in that part of the contour, is so extremely small as to defy detection. And the best proof of the practical perfection of this system of making epicycloidal teeth is found in the smoothness and precision with which the wheels run, in which respects they have proved, after some years of trial, to be unsurpassed. And we repeat, that objection taken, on whatever grounds, to the epicycloidal form of tooth, has no bearing upon the method above described of producing duplicate cutters for teeth of any form, which the pantagraphic engine will make with the same facility and exactness, if furnished with the proper templates.

On the Determination of a Series of Cutters.

299. In making a set of interchangeable wheels, upon any system whatever, every additional tooth changes the diameter of the wheel and the form of the acting curves, so that in order to secure absolute theoretical accuracy, it would be necessary to have as many different cutters as there are wheels. This is clearly out of the question, and fortunately, the proportional increment, and the actual change of form, become less as the wheel becomes larger; and this alteration in the outline soon becomes imperceptible. Going still farther, we can presently add more than one tooth without producing a sensible variation in the contour. That is to say, several wheels can be cut with the same cutter, without introducing a perceptible error. Proceeding in this way we reach the conclusion, that instead of an infinite number of cutters, a quite limited number will suffice to cut all the wheels of such a set, from the smallest up to a rack, with a sufficient degree of accuracy for all ordinary purposes. We have then to decide how many cutters shall be made, and also for what numbers of teeth these should be specially adapted. The answer to the first question evidently depends upon how great a deviation from absolute accuracy is considered admissible. In regard to the second, it appears reasonable, not to say axiomatic, that errors should be uniformly distributed through the series, and that this will be effected by making the greatest difference in form the same between any two consecutive cutters.

300. In relation to this, Professor Willis says :

"—— it appeared worth while to investigate some rule by which the necessary cutters could be determined for a set of wheels, so as to incur the least possible chance of error. To this effect I calculated, by a method sufficiently accurate for the purpose, the following series of what may be termed equidistant values of cutters ; that is, a table of cutters so arranged, that the same difference of form exists between any two consecutive numbers." He gives us no clue to the method by which these numbers were found, nor any explanation of the sign ×, which appears in two places. The teeth are epicycloidal (the describing circle being such as to give radial flanks to the wheel of 12 teeth), and have "the usual addendum"—of which the exact amount is not stated. We here insert the figures as he gives them,* only numbered in the reverse order ; and they appear to have been extensively adopted in practice, and that too under varying conditions, which, as will presently be shown would affect the values of the terms of a truly equidistant series.

301. The only rule or formula for computing such a series that we have met with, is given by Mr. G. B. Grant, as follows :

$$t = \cfrac{an}{n - s + \cfrac{as}{z}} :$$

in which

$a =$ the number of teeth on the least wheel, usually 12 ;

$z =$ "   "   "   "   "   " greatest "   " ∞ ;

$n =$ "   "   " cutters in a series to cut from a to z ;

$s =$ "   "   in the series, of any particular cutter ;

$t =$ "   "   of teeth on the last wheel to be cut by s.

This formula is based upon the following hypothesis : The pitch arc is greater upon the outside of the wheel than on the pitch circle, by an amount which in wheels of equal pitch and constant addendum, depends upon the number of teeth ; the smaller the wheel, the greater is the difference, and *vice versv*, the arcs being equal in the case of a rack. Now selecting any two

1	12
2	×
3	13
4	14
5	. 15
6	×
7	16
8	17
9	19
10	20
11	21
12	23
13	25
14	27
15	30
16	34
17	38
18	43
19	50
20	60
21	76
22	100
23	150
24	300
25	∞

---

* "Principles of Mechanism," p. 141.

wheels as the least and greatest of a series, it is possible to interpolate others whose numbers of teeth are such that this difference, between the pitch arc as measured on the pitch circle and on the extreme perimeter, shall be the same for each : and these numbers are the values of t as found by Mr. Grant's formula, in which it is assumed, although upon what ground is not apparent, that the consecutive wheels of such a series will differ equally throughout in respect to the outline of the tooth.   The proper *location* of each cutter in its own interval, is determined by doubling the number of terms in the series, and taking each alternate value of t as the number of teeth for which the preceding cutter is to be made exactly correct.

It may be admitted that tolerably close approximations, or even

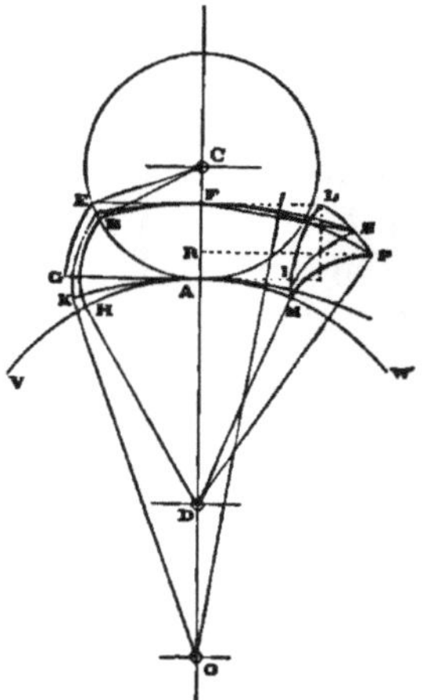

FIG. 180. 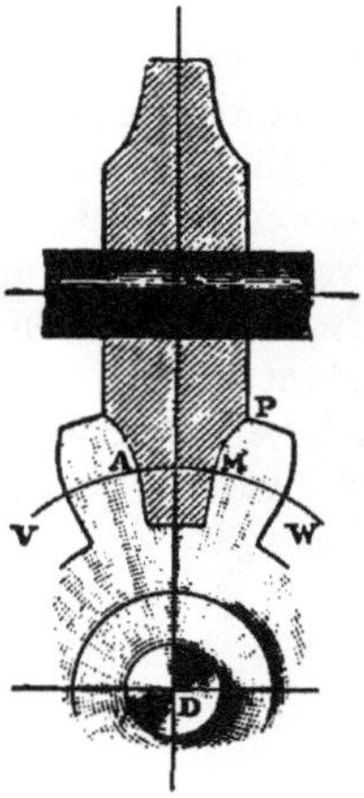   FIG. 181.

correct results, may be obtained by this formula under some conditions.   But it contains no internal evidence as to when or whether this will be the case—it is independent of all the elements of the problem which affect the form or the length of the tooth, and should, therefore, if true at all, be applicable to all systems and under all circumstances.   Which would be very convenient, unquestionably ; but these elements cannot be eliminated without vitiating the results and destroying the " equidistant " characteristic of the series.

302. This is best illustrated by considering the question in relation to the epicycloidal system, which, at least, admits of a direct and exact solution.   In Fig. 180, let C be the centre of the describing circle, VW an arc of the pitch circle of the smallest wheel, and D

its centre ; also let AF be the addendum.   Then a circle through F about centre D determines the arc AB of the describing circle, which by rolling upon an equal arc AH of the pitch circle traces the epicycloidal face BH of the tooth for that wheel.   Again, FE perpendicular to CD cuts off an arc AE, which by rolling upon the tangent at A describes the cycloidal face EG for the tooth of a rack.

Extending the tangent line toward the right, set off upon it $AI =$ arc $AM = \frac{1}{2}$ space, and draw IL, MP, respectively similar and equal to GE, HB.   Then regarding CD as the central plane of a series of cutters, it will be seen, by comparing this figure with the following one, which represents a section of a milling cutter at work, that MP will be the contour of the base of the one for the smallest wheel, and IL the corresponding line of the cutter for the rack.

Proceeding in a similar manner with other pitch circles, it is obvious that the faces for all intermediate wheels will lie between MP and IL, their highest points lying in a curve LP; which is shown on a larger scale in Fig. 182.   Now these epicycloids diverge and differ in form from each other the more the farther they extend from the pitch circles ; if then they be so selected as in this figure that they divide the locus LP into equal parts, the *greatest* variation of each from the one next in order will be constant for the whole series.

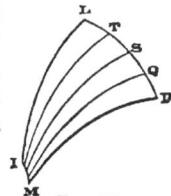
FIG. 182.

And this is sufficient for our purpose, it being self-evident that the errors of which the distribution is to be equalized, should be taken at the maximum ; and again, it is not necessary to take into consideration the flanks of the teeth, to which the like method is equally applicable, because they vary from each other to a much less extent than the faces.

303. We have, then, a key to the perfect solution of the problem. The first step is to determine LP ; this is done by assuming various pitch circles, and constructing the face for a tooth of each, arranging these with reference to the point A as above described, until a sufficient number of points have been located to enable the curve to be properly developed and drawn.   Having decided upon the number of cutters in the proposed series, LP is to be divided into a corresponding number of equal parts.   In regard to the practical execution of this process, it may be remarked that although the curve LP may be of a complicated nature, since it has a sort of transcendental dependence upon the epicycloid, which is itself transcendental, yet in the portion of it with which we have to deal in connection with wheels ranging from twelve teeth up to a rack, its curvature changes so slightly that it may without causing any appreciable error be considered as a circular arc.

We have next to find the radius of the wheel, whose tooth would fall at each point of subdivision ; this is readily done as shown in Fig. 180, thus : let S be one of these points upon LP, we have but to draw FS, and bisect it by a perpendicular, which latter line will cut CD at O, the centre of the required wheel.

This determines the limits between which each cutter is to be used : thus, if in Fig. 182 we suppose a series of four cutters to be required, PM being the face of the smallest wheel and LI that of the largest, then the first cutter must cut from P to Q, the next from Q to S, and so on.   Evidently, each cutter should be made exactly right for a wheel whose tooth would fall at the point bisecting the arc of LP which limits its range of action ; so that the location of the first is found by dividing PQ in half, and determining as above the radius of the wheel corresponding to this point of division.

304. The same general mode of proceeding may be employed with teeth of any form, the operation of determining points in the locus LP differing in detail according to the peculiarities of each system of gearing.   It is apparent that this process requires us to take into account in every case not only the addendum, but the path of contact.   The fact that the latter is in the epicycloidal system identical with the describing circle, renders comparatively simple what in other systems might be very complicated, and enables us by ordinary trigonometrical operations to locate any required number of points in LP with great accuracy.

And pursuing this line of investigation, we find that the curvature, position and magnitude of LP are affected by changes in either the length of the tooth or in the diameter of the describing circle, in such a manner and to such an extent as sensibly to affect in turn at least the higher values of a series of equidistant cutters.

Now it costs no more to make or to use a correct series than an incorrect one ; and in adopting the epicycloidal system, there is no excuse for resting contented with approximations, close or otherwise, since by the process above outlined, results may be obtained which are exact to a single tooth.   On this account, and also because this system has been and is likely to be more extensively employed than any other, we have made the required determinations with great care, in order to reduce the magnitude, and correct the unequal distribution, of errors in the series in use.

305. First, taking the diametral pitch as unity, with a describing circle with diameter $= 7\frac{1}{2}$, and addendum $= 1$, the locus LP was

accurately determined by calculating the positions of a great number
of points. This curve is shown, greatly enlarged, in Fig. 183, LN
in this figure corresponding to the same line in Fig. 180, being the
perpendicular let fall
from L upon PR.
The numbers upon
the horizontal ordi-
nates indicate the
diameters of the pitch
circles assumed for
the computations, or,
what is the same
thing, the numbers of
teeth.

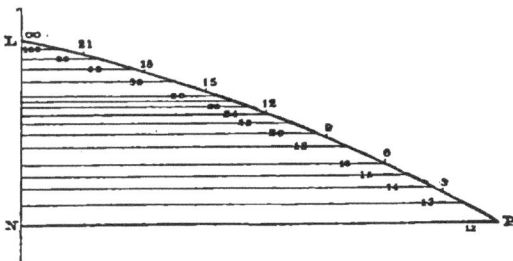

Fig. 183.

Next taking any point thus determined, at pleasure, and assuming
it to lie upon a circle passing through L and P, the radius of that
circle was calculated—which being repeated for various other points,
the close agreement of the radii proved that for the purpose in view
the curve might be regarded as a circular arc. Upon this assumption,
the location of any point of subdivision, with reference to CD and the
point F of Fig. 180, is known, and the number of teeth upon the
wheel from which that point would have been derived, can be calculated trigonometrically ; this converse process is quite different from
that by which a point upon the curve is found, so that the concordance
of the results affords a rigid test of the accuracy both of the work and
of the whole method.

It was not to be expected that the points of exact subdivision of
this curve should correspond precisely to whole numbers of teeth ; on
the contrary, fractional terms were to be looked for as a matter of
course, but since the series must in practice consist of whole numbers
only, the nearest integer was taken in each case.

306. It will be observed that the middle point of LP lies very close
to the ordinate for the wheel of 24 teeth. Therefore, in a set of 24
cutters, a separate one should be made for each wheel from 12 to 23
teeth inclusive, thus confining both the errors and the calculations to
the 12 subdivisions of the upper half of the curve.

A series thus determined is given in the first column of the comparative table following : in regard to which it should be stated that
the 24th cutter is practically used only for very large *wheels*, an extra
one being added for the rack.

When this describing circle of $7\frac{1}{2}$ was first introduced, a series of
cutters was employed, the values for which are given in the second

column ; these figures were copied from those of Prof. Willis, with
the omission of his unintelligible sign ×, and the interpolation of a
cutter for 18 teeth.

In both cases, the cutters used for only one wheel each are correct ;
and the others being "equidistant" in the first series, we have for the
purpose of a graphical comparison set up in Fig. 184, at equal inter-

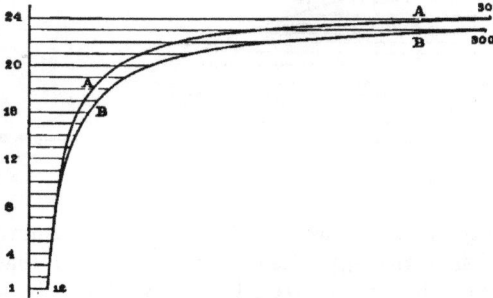

FIG. 184.

vals, an ordinate for
each cutter, of a
length proportional
to the number of
teeth upon the small-
est wheel for which it
is to be used, as given
in the first column.;.
the curve AA thus de-
termined, then, rep-
resents the true series.
The intention was to
make the second series also "equidistant," and with the same difference between consecutive numbers as in the first; accordingly the
curve BB is constructed by setting off on the same ordinates the corresponding numbers in the second column ; and the discrepancy
between these curves exhibits very clearly the inequality in the distribution of the errors in this series, the use of which was promptly
abandoned as soon as this was pointed out.

In the third column we have a series computed by Mr. Grant's
formula, applied to determine the values for 12 cutters to cut from
24 teeth up to a rack, the first 12 being made exact, as in the other
cases. This formula, as already stated, would give the same results
for any describing circle and any addendum ; but changes in these
will in fact sensibly affect the values of the terms in this series. For
example, Prof. Willis made use of a describing circle whose diameter
is half that of the wheel of 12 teeth ; calling this diameter, then, 6,
and the addendum unity as before, we find the curve LP to be different from that corresponding to the larger describing circle before
used. Its middle point, however, still lies so close to the ordinate
for 24 teeth, that the upper half of the curve only need be dealt with
in computation, and the correct series with these data is given in the
fourth column. Again, taking the original describing circle of diameter $= 7\frac{1}{2}$, but reducing the addendum to $\frac{1}{2}$, the values for the
series of 24 cutters appear as given in column fifth. By comparing
this with the first column, it will be seen that the higher values only

arc affected by this reduction of the addendum ; and it is important, since a series once adopted is in the nature of things inflexible, and yet it may at times be necessary to reduce the addendum for the purpose of varying the amount of approaching or receding action, to note that even these values are affected in a far less proportion than that in which the difference between the consecutive cutters is reduced.

This will be clearly seen by the aid of Fig. 185, in which the curves O, B, C, represent the locus LP of Fig. 180, for the first, fourth and fifth columns respectively ; o, b, and c being the corresponding centres of curvature.

Taking the twenty-fourth part of each of these curves as the greatest difference in form between the consecutive cutters of the series to which it belongs, we have the following values ;

Linear Variation. $\begin{cases} \text{I.} & 0.0156645. \\ \text{IV.} & 0.0158606. \quad \text{VI.} \quad 0.0147569. \\ \text{V.} & 0.0046068. \end{cases}$

307. We do not consider it worth while to discuss the determination of any series consisting of a less number of cutters than 24, because the tendency of modern practice is fortunately more and more toward the attainment of the greatest accuracy consistent with a reasonable expenditure ; and this would call for an increase rather than a diminution of the numbers. We therefore, give, finally, in the sixth column of the table, a series of 27 cutters especially constructed for use in connection with the epicycloidal and pantagraphic engines above described. Of these, 14 are exactly right, there being one for each wheel from 12 to 25 teeth inclusive ; the next 12 are computed to cut from 26 teeth up to a rack ; but the 26th is used only for

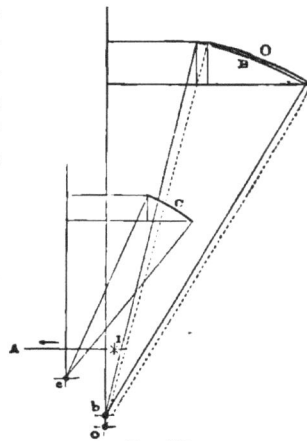

Fig. 185.

wheels having 322 teeth or upward, an extra cutter being added for the rack.

Much smaller series have been employed, some makers contenting themselves with even so few as eight cutters for all pitches ; this, as an examination of Fig. 183 will show without further argument, can hardly be said to give a reasonable approach to accuracy, especially in the case of wheels with comparatively few teeth. Others again

13

NO. OF CUTTER.	NUMBERS OF TEETH CUT BY EACH CUTTER.					
	I.	II.	III.	IV.	V.	VI.
1	12	12	12	12	12	12
2	13	13	13	13	13	13
3	14	14	14	14	14	14
4	15	15	15	15	15	15
5	16	16	16	16	16	16
6	17	17	17	17	17	17
7	18	18	18	18	18	18
8	19	19	19	19	19	19
9	20	20	20	20	20	20
10	21	21–22	21	21	21	21
11	22	23–24	22	22	22	22
12	23	25–26	23	23	23	23
13	24–25	27–29	24–26	24–26	24–25	24
14	26–28	30–33	27–29	27–29	26–28	25
15	29–32	34–37	30–32	30–32	29–32	26–27
16	33–36	38–42	33–36	33–36	33–36	28–30
17	37–41	43–49	37–41	37–41	37–41	31–34
18	42–49	50–59	42–48	42–48	42–48	35–38
19	50–59	60–75	49–58	49–58	49–58	39–44
20	60–74	76–99	59–72	59–73	59–81	45–52
21	75–99	100–149	73–96	74–97	82–109	53–63
22	100–150	150–299	97–144	98–147	110–165	64–79
23	151–302	300–∞	145–288	148–296	166–334	80–106
24	303–∞	∞	289–∞	297–∞	335–∞	107–159
25	∞		∞	∞	∞	160–321
26						322–∞
27						∞

have used eight for small pitches, increasing the number of cutters as the pitch of the tooth increases. This practice seems to be based on the idea that the actual amount of error in the form of the cutter only is to be taken into account; which may be tenable if the formation of the templates and cutters be dependent upon hand-and-eye operations, but not otherwise; since the proportionate error, and the effect upon the velocity ratio, will be the same whatever the size of the tooth.

CHAPTER XI.

Twisted Spur Gearing.

308. Hooke's Stepped Wheels.—Let a pair of ordinary spur wheels, loose upon their shafts, be cut transversely into a number of plates. Let these sections of one of the wheels be first rearranged by rotating them until, as in Fig. 186, the tooth of each overlaps that of the preceding one by the same amount, and then firmly keyed upon the shaft. In passing to the new position, each plate will drive that section of the other wheel with which it is in gear, independently of the others. Thus the second series of plates will necessarily be arranged in a similar manner, and these being now also secured upon their own shaft, we have a pair of the **Stepped Wheels** first introduced by Dr. Hooke.

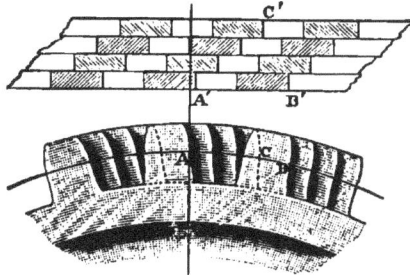

FIG. 186.

By this ingenious device the number of teeth is in effect increased without diminishing their size. Thus, the figure shows a wheel built up of four plates, or thin wheels; supposing each to have say 20 teeth, the resulting action is clearly the same as that of a single wheel of 80, while the fact that the acting faces lie in different planes, enables us to retain the original pitch. The advantages are obvious ; not only is the number of contact points increased, but they cross the line of centres at shorter intervals ; and the action is at its best when this occurs.

The teeth may have any of the forms already described ; and the extent to which they overlap is, abstractly speaking, arbitrary. But clearly the best arrangement is to have the edges of the successive teeth divide the pitch arc AB into equal parts, as shown in the figure. There are in this case four plates, and consequently the arc AC, through which the last plate is rotated from its original position, is equal to $\frac{3}{4} AB$. And in general, letting $n =$ No. of Plates, we shall have

$$AC = AB \left(\frac{n-1}{n} \right).$$

309. In the practical employment of wheels thus constructed there is a limit to the reduction in the thickness of the plates or steps, depending on the material used and the pressure to be transmitted, since if excessively thin they will suffer from abrasion. So long as actual steps of sensible thickness are used, however, the kinematic action differs in no respect from that of any other spur wheels, and the lines of action all lie in the planes of rotation.

If the number of plates, then, be finite, it must be comparatively small ; yet if it be increased to infinity the arrangement again becomes perfectly practical, but the action is modified in a new and peculiar manner.

The steps now disappear entirely, as in Fig. 109 ; and the effect is the same as if the original wheels had been simply twisted, as explained in (**175**). In this process the nature of the acting surfaces is changed. They were in the first place cylindrical, the bases being the involutes or epicycloids forming the outlines of the teeth, and the rectilinear elements being parallel to the axes. Since the twisting is uniform, these elements now become helices, all having the same pitch, but obliquities differing according to the distances from the axes. And the line of contact between two engaging teeth will partake of the helical form, though it will not be a true helix. For it is clear that the transverse sections, by successive planes, will be the original tooth outlines in successive phases, and in each section there will be a point of tangency, which must lie in the projection of the path of contact on a transverse plane.

310. Now, in regard to the common normal at any point of contact. Pass through the point a transverse plane, which cuts out the tooth outlines just mentioned, and draw also the common tangent of the two helices which pass through the point ; these two lines determine the tangent plane, and the normal must be perpendicular to both. The first lies in the transverse plane, but the latter pierces that plane obliquely. Consequently, the line of action *can in no case lie in a plane of rotation,* but will make with it an angle more or less acute. In general, then, the line of action can be resolved into three components, viz :

1. The component of rotation, perpendicular to the plane of the axes.

2. The component of side pressure, parallel to the common perpendicular of the axes.

3. The component of end pressure, parallel to the axes themselves.

When the point of contact lies in the plane of the axes, the second of these components of course vanishes ; of which fact, as will presently appear, advantage may be taken in so forming the teeth that there shall be no sliding between them.

311. When the wheels are thus formed by twisting, instead of with successive steps of sensible thickness, the combination is known as **Hooke's Spiral Gearing** ; and is very commonly described and classified as a modification of *screw* gearing.

That this is an error will be, perhaps, most clearly seen from the considerations, that the direction of the twist does not affect that of the rotation, and its amount does not affect the velocity ratio. Regarding the wheels as built up of exceedingly thin plates or laminæ, each one of those composing the driver turns the corresponding one of the follower precisely as though the thickness were sensible, not only during the formative process of twisting but after. On reaching the limit when the plates become planes and the elements of the tooth surfaces become helices, the action is modified by the deflection of the common normals from the planes of rotation ; but regarding the driver as a screw, its endlong thrust being perpendicular to those planes, may have either direction and any magnitude without affecting the direction or velocity of the transmitted motion. Whereas in screw gearing proper, this endlong thrust either lies in a plane of rotation or has a component which does, and this component is the one which produces the rotation of the follower. In that class of gearing, then, the driver turning in a given direction, the follower will turn one way if the driver be right handed, but the other way if it be left handed ; and the screw pitch of the driver will obviously affect the velocity of the imparted motion.

312. From the very nature of the twisting process, as above explained, it is evident that if the screw pitch of one wheel of a pair be assumed, that of the other is thereby fixed. And it may readily be ascertained, because if the teeth be indefinitely increased in number and diminished in size, any two which are in contact will ultimately become two tangent helices lying on the surface of the original pitch cylinders, and must, therefore, develope into the same right line on the common tangent plane, as in Fig. 32.

Again, it makes no difference in which direction we twist the first wheel ; but it follows at once from the preceding that if the pair be in outside gear, the helical elements of one will be right handed, those of the other left handed. In the case of internal gearing, on the contrary, the helices of both wheels will be either right handed or left handed, as the case may be.

313. Practical Choice of the Screw Pitch.—Although, as has been stated, the amount to which such wheels shall be twisted is abstractly optional, with the limitation mentioned in the preceding paragraph, yet in order to reduce the obliquity of action and the consequent end thrust it is best to make it in practice as small as may be, consistently with securing the advantages due to twisting them at all.

Now when, as thus far supposed, any of the ordinary forms of spur teeth are used, the angle of action is greater than the pitch, so that in any given transverse plane the action of one pair begins before that of the preceding pair ends. Consequently if a wheel of any given depth measured in the direction of the axis *be twisted through an angle equal to the pitch of the teeth*, all the phases of the action will at all times be simultaneously presented, and there will always be a point of contact in the plane of the axes.

This then may be taken as a good practical limit ; and it agrees with the deduction made in **(308)** when the number of steps becomes infinite ;—for the expression there given,

$$AC = AB \left(\frac{n-1}{n} \right),$$

gives

$$n = \frac{AB}{AB - AC},$$

whence if $n = \infty$, we have $AC = AB$.

314. Elimination of Sliding Friction.—If the teeth be so formed that

in any one plane the contact not only begins but ends on the line of
centres, continuing but for a single instant; then the point at which
this contact occurs, which is of course upon the surface of each pitch
cylinder, will be the only point of tangency between the tooth surfaces.
In the next instant this driving point will be found to have shifted
to the consecutive plane, and so on continually, thus moving endlong
in the common element of the pitch
cylinders. Under the circumstances there is no sliding friction, since
the coincident points of the acting
surfaces are at every instant moving
in the same direction and at the
same rate.

In order to comply with the above
condition, the teeth may be formed
as shown in Fig. 187 or as in Fig.
188. In the former, the outlines
are epicycloidal, but the describing
circles for the faces are smaller than
those used for tracing the corresponding flanks. In the second case, sup-

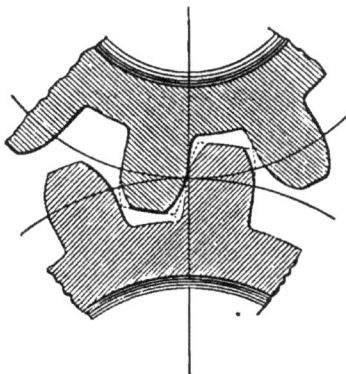

pose the teeth originally to have been involutes, as shown in the dotted
lines, the flank proper, or part within the pitch circle, is still of that
form, but the face, or part without, is of greater curvature, though
it is tangent to the inner part at its intersection with the pitch circle.

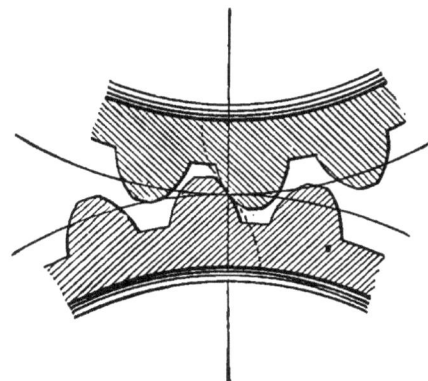

FIG. 188.

In general, then, it will be
easily seen, it is necessary
that when the point of contact is on the line of centres, as in either one of
these figures, all the contour lines which pass through
that point shall be tangent
to each other, but the faces
must lie within the curves
which in ordinary spur gearing would be conjugate to
the flanks determined on.
A very simple construction
on the epicycloidal basis is
shown in Fig. 189, the tops of the teeth being semicircles tangent to
the flanks, which are radial, at their extremities.

The action of plain spur wheels thus formed would, of course, be correct only at the instants here represented ; but by the simple expedient of twisting them, so that the action on ceasing in one plane is continued in the next, these instants become consecutive, the velocity ratio is made constant, and the rotation is transmitted by pure rolling contact.   On this last account, as might be expected, the

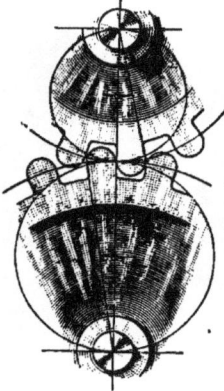

Fig. 180.

action of these wheels is exceedingly smooth and noiseless, almost as much so, if they be well made, as that of belting ; but they are better suited for light work, because the pressure is confined to a single point instead of being distributed along a line. For heavy work it is, therefore, preferable to employ the stepped wheels, or twisted ones in which the teeth, as in Fig. 186, are of the usual forms, although the sliding friction still remains.

315. Neutralization of End Pressure.—But the end pressure due to the screw-like action will exist, even after the forms of the teeth have been modified as above explained.   What now occurs during the rotation, as will be evident from a moment's study, is precisely equivalent to the rolling together of two helices, one on each pitch cylinder, as illustrated in Fig. 32.   And this end pressure, which must be received by a collar on the bearing, causes an amount of friction which it is desirable to avoid.

This disadvantage may in its turn be obviated by the means indicated in Fig. 190.   Each wheel is made in two parts of equal thickness, which are twisted in opposite directions, but to the same extent. The end thrust of one part, then, is exactly counterbalanced by that of the other :

Fig. 190.

and since the action now involves pure rolling contact only, this may be regarded as the very perfection of geared wheel-work.

Pin Gearing.

316. The term **Epicycloidal** is in its ordinary use applied only to teeth all of which have both faces and flanks whose describing curves are circles.

There is, however, a form of gearing, in which the teeth of one wheel of a pair are true epicycloids only at a theoretical limit, upon reaching which those of the other wheel become mathematical points. The latter are in practice actually made cylindrical pins of reasonable diameters, which fact has given rise to the name **Pin Gearing** ; and the former in consequence deviate considerably from the epicycloidal outline whence their working contours are derived.

In Fig. 191, C and D being the centres of the pitch circles, let a marking point be fixed at A in the circumference of the upper one. Then while this point goes to E, in the progress of the rotation, it will trace upon the plane of the lower circle the curve EB ; the arcs AE, AB, being equal. This, evidently, is the familiar epicycloid,

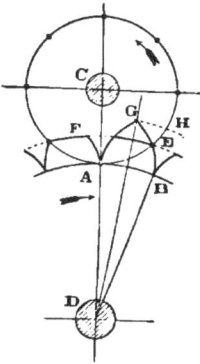

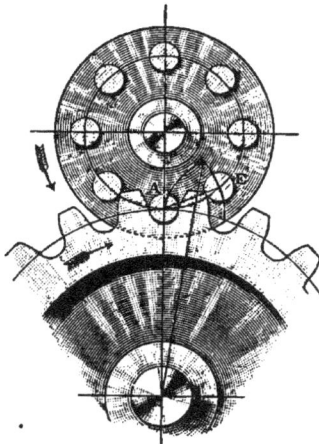

FIG. 191. FIG. 192.

generated by rolling the upper pitch circle upon the lower. Meantime, since the marking point does not change its position in the circumference of the upper circle, it can trace no curve at all upon its plane.

Now let AF be a curve similar to BE, and imagine a pin of no sensible diameter—a rigid material line—to be fixed at A in the upper wheel. Then if the lower one turn to the right, it will drive the pin before it with a constant velocity ratio, the action ending at E, if the driving curve be terminated at F as here shown.

317. Supposing AE to be an aliquot part of the circumference, and assuming it as the pitch arc, we have only to set the pins at equal distances in the upper circle, and after dividing the lower one, to draw through the points of division the reversed curves as shown : this done, the elementary wheels are complete and capable of working in either direction.   From these the practical ones are derived as in Fig. 192 ; the pins being made of sensible magnitude, the outlines of the teeth upon the other wheel are curves parallel to the original epicycloids.   The diameter of the pins is usually about equal to the thickness of the teeth measured on the pitch circle, the radius being, therefore, one fourth of the pitch arc ; this, however, is not imperative, and the pins are sometimes made considerably smaller.   When the radius has been selected, a number of arcs are described with it, having their centres upon the epicycloid, and the envelope of these arcs is the required contour of the working tooth.

The pins are ordinarily supported at each end, two discs being fixed upon the shaft for the purpose, as in Fig. 111 : thus making what from its form is called a *lantern* wheel or pinion.

318. Peculiarities of the Action.—The most striking feature of wheel-work of this kind is, that the action is almost wholly confined to one side of the line of centres.

In the elementary form, this is a direct and obvious consequence of the manner in which the tooth is generated.   Thus, in Fig. 191, if the curve *AF* drive, the action cannot begin until its root reaches the point *A* on *CD*, and is entirely receding ; if on the other hand this curve be driven by the pin, the action will terminate at the same point *A*, and will be entirely approaching.

Consequently, pin-gearing is not well adapted for use in combinations which require the same wheel both to drive and to follow.   But when that is not required, this peculiarity of the action is greatly in its favor ; the pins of course being always given to the follower and the teeth to the driver.   And previously to the introduction of cut gearing, it was very extensively employed, even for heavy work ; the facility of forming the pins, or *staves*, of the lantern pinions, in the lathe, especially adapting it for the construction of wooden wheels. At present its use is almost exclusively confined to clock-work and similar light mechanism, the pins being made of steel wire cut to the proper lengths.

319. It has heretofore been taught, in all the treatises upon this subject which have come to our notice, that when the pins are made of sensible diameter, as in working wheels they always must be, the above-mentioned peculiarity is modified and a certain amount of approaching action introduced in every case.

The fact was too patent to escape notice, that the expansion of the pin from the theoretical point into the practical circle, had the result of shortening the driver's tooth and reducing the amount of receding action. But while this palpable effect of the new condition upon the time when the action shall end was duly recognized, no one seems ever to have inquired whether the same condition might not also affect the time when the action shall begin; and it appears to have been taken for granted that it would not.

Thus in Fig. 192, the pin E is just quitting contact with the tooth which drives it; and the theory as hitherto laid down is based upon the assumption, that correct driving contact with the next pin A is just beginning when it occupies the position here shown; thus giving an arc of approach about equal to the radius of the pin, and making this as it stands an exact limiting case.

It will presently be shown that this assumption is erroneous, and that the next tooth is *not*, in general, tangent to the pin A when in

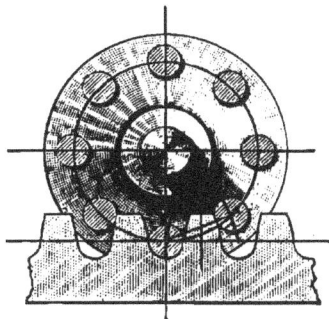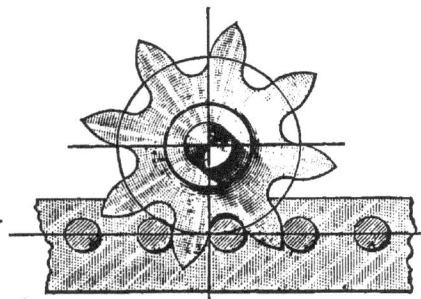

Fig. 193.					Fig. 194.

its present position. The error itself is physically small, which may account for its having remained so long undetected; but its effect is of very perceptible magnitude; not only changing very greatly the amount of approaching action, but in many cases making that action absolutely negative, the first driving contact between each tooth and its pin not occurring until after the latter has bodily passed the line of centres.

320. Pin-Wheel and Rack.—Since the pins are always given to the follower, the construction of a rack will present two cases. If it is to be the driver, as in Fig. 193, the elementary tooth is bounded by cycloids, generated by the pitch circle of the wheel; from which the outlines of the working tooth are derived as above explained.

Pin-Rack and Wheel.—In Fig. 194, the wheel is to drive, and the tooth-outline is the involute of its own pitch circle, the generatrix being the pitch line of the rack. It is usually stated that it is unnecessary to construct the derived curve, since " this process would merely reproduce the same involute in a different position." * Although this is the truth, it is not the whole truth ; and the part which is lacking, will be found to vitiate the results deduced from the part which is given.

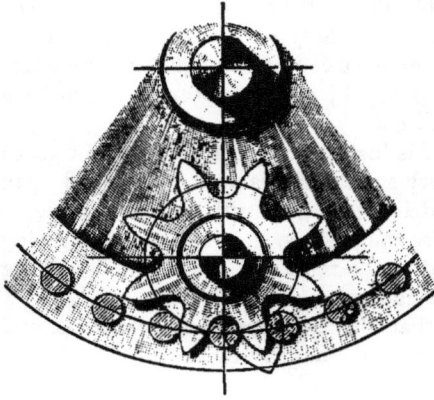

FIG. 195.

321. Inside Pin Gearing.— Here also we have two distinct cases, since the annular wheel may be required either to drive or to follow. In Fig. 195, the pinion drives, and the pins are given to the wheel. The outer pitch circle, then, by rolling upon the inner, generates the internal epicycloid which forms the elementary tooth of the pinion. In Fig. 196, the wheel drives, and its elementary tooth is the hypocycloid traced by rolling the less circle within the greater ; in both cases the curves for the working teeth are derived in the usual manner. If the annular driver be twice as large

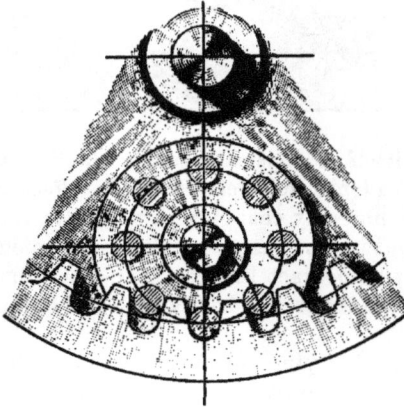

FIG. 196.

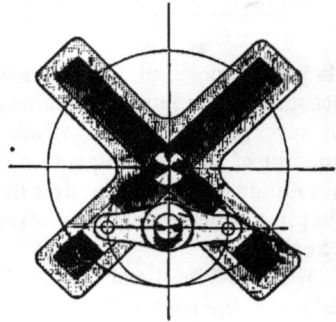

FIG. 197.

as the pinion, the hypocycloid becomes a right line, and the process of derivation gives simply another right line parallel to it. A very practical construction in this case is shown in Fig. 197 ; the pinion has

---

* Willis. (Principles of Mechanism, p. 96.)

but two pins, which turn in blocks sliding in the two slots which, crossing each other at right angles, constitute the disguised annular wheel.

322. Practicability of Assumed Conditions.—It is hardly necessary to state that there are definite relations between the pitch arc and arc of action, the diameter of the pin and the height of the tooth, such that all these cannot be assumed with any certainty that the result will be a practicable arrangement. The preceding figures give merely an idea of the general principles and the general appearance of the various modifications ; and we have now to consider more in detail the processes of construction.

First, in regard to the elementary or ideal form, in which the pin is a point and the tooth-curve a true epicycloid. Referring to Fig. 191, it appears that if we assumed the pitch arc AB, the greatest possible height of the tooth is determined by the intersection of the front and back at G ; and if this height be adopted, the action, beginning at A, will terminate at H, the point on the upper pitch circumference through which G must pass in its rotation about D. Now this point G lies on the radial line which bisects AB, and will, therefore, fall farther to the right, the larger the pitch arc. Should it chance to coincide with H, or in other words should the point

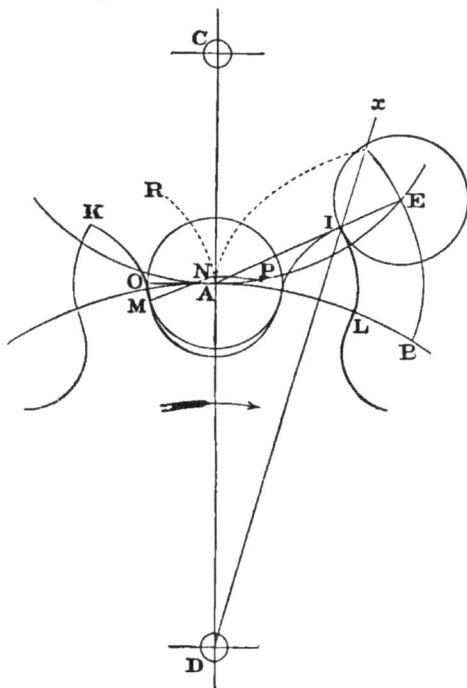

Fig. 198.

of the driver's tooth fall upon the pitch circle of the follower, the proposed case is exactly a limiting one. From an examination of the succeeding figures it will be seen that this is equally true when the driver is a rack or an annular wheel ; and also that in order to render the case practically feasible, the point G should fall within the pitch circle of the follower, except when the latter is annular, in which

case it should fall without. This condition being satisfied, the tooth
may be made pointed, or it may be topped off, at pleasure; and
finally, the arc of action is determined, as in previous constructions, by
the intersection of the path of the highest point of the tooth, with the
path of contact—that is to say with the pitch circle of the pin-wheel.

323. Second, when the pin is of sensible diameter. We give in
Fig. 198, Prof. Willis's
mode of determining the
greatest diameter which
can be used under any
assumed conditions. *C* and
D being the centres of the
follower and driver re-
spectively, suppose the
pitch arc, *AB* or *AE*, to
be assigned. Bisect *AB*
by the indefinite radial
line *Dx*, and draw the
chord *AE*, cutting *Dx* in
I; then *EI* is the greatest
radius for the pin, and
any less one may be used,
as in Fig. 192. Since *I*
is unquestionably the last
point of contact with the
pin *E*, this construction
would be correct, provided
that the next tooth *KM*
were now in driving con-
tact with the pin *A*. Prof.
Willis distinctly states that
it is; * and bases this as-
sertion upon the previous statement † that "by the mode of its description the circle (of the pin) must touch the curve (of the tooth)
when its centre is in any point of the epicycloid." The conclusion,
therefore, is very plausible; but it is not correct, because the curve
derived from the epicycloid *AR* (as was pointed out in Fig. 178),
consists of two branches, one of which, and a part of the other also, is
effaced in the very process of formation. The normal to the epicy-

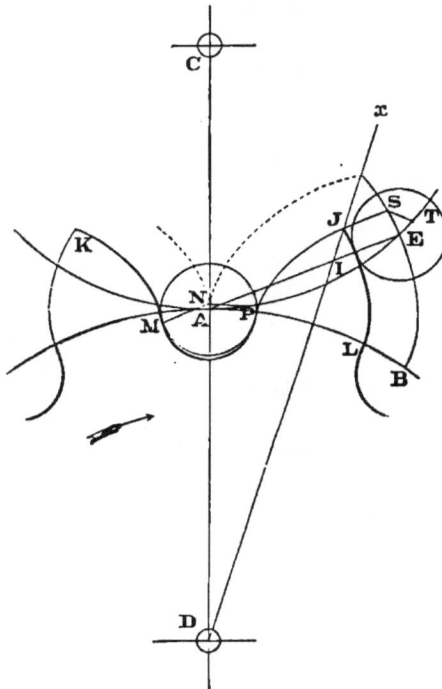
Fig. 199.

cloid at its root, is AO perpendicular to CD; the parallel curve then at first *descends*, having a vertical tangent at O, forms a cusp at a point *within* the pin whose centre is A, and then rising, cuts its circumference at some point M, which is evidently the lowest practicable point in the outline of the working tooth. Draw through M a normal, cutting the original epicycloid at N, and through that point describe an arc about D, cutting the locus of original contact in P. It will now be seen that the point M cannot come into correct driving contact, until N reaches the position P; and the arc AP is the measure of the error in Prof. Willis's construction; that is to say, it is the amount by which the arc of approach as determined, or rather assumed by him, must be changed.

324. Absolute Determinations Impossible.—It appears then that the limiting diameter of the pin as found by the above construction is too great. By reducing it, the tooth IL may be made enough longer to continue in action until the point M comes into driving contact. But the precise amount of reduction which will cause one tooth to quit contact at the instant when the next one begins to drive, it is impossible to determine; because the positions of both the points M and I, which fix the times of the beginning and the ending of the action, depend upon the unknown magnitude, the radius of the pin. We are, therefore, compelled to adopt the tentative process shown in Fig. 199. Having found the point I, as in the preceding diagram, assume a radius for the pin, less than EI, and continue the derived curve to cut Dx in J, which will be the point of the tooth. Through J draw a normal to the epicycloid, cutting it in S; through S describe an arc about D, which will cut the upper pitch circle in T, the position of the pin at the end of the action. Drawing the working outline of the next tooth, we determine, as in Fig. 198, the points M, N, and finally P, the position of the pin when the action begins. Now, if the arc PT prove to be precisely equal to AE, we have an exact limiting case, and the assumed radius of the pin is a maximum: if PT be less than AE, the radius is too great, and must be reduced—but if the contrary, the case is a practicable one.

In the latter event the tooth may be topped off; and it need hardly be added that the above process enables us to determine whether the case be feasible or not, if both the diameter of the pin and height of the tooth are assigned; the normal JS being drawn from the highest point of the given curve.*

---

* *Note.* For details of the graphic processes of drawing these curves and the normals, the reader is referred to the Appendix.

Limiting Numbers of Teeth and Pins.

325. When the Pin is a Mathematical Point, the determination of the limiting numbers of teeth and pins is easily effected by methods precisely like those used in the case of the common epicycloidal teeth. Suppose the number of teeth to be assigned for the driver whose centre is D in Fig. 200, and let AF be the pitch arc. The point, O, of the tooth must lie on the prolongation of DL bisecting AF, in such a position that an arc OA of a circle whose centre lies upon DA produced, shall be equal to the arc AF. On the tangent at A, set off $AG = $ arc AF, and $AM = \frac{1}{4} AG$: with centre M and radius MG describe an arc cutting DL produced, in O; draw OA, and bisect it by a perpendicular NC, which cuts DA or its prolongation in C, the required centre of the follower.

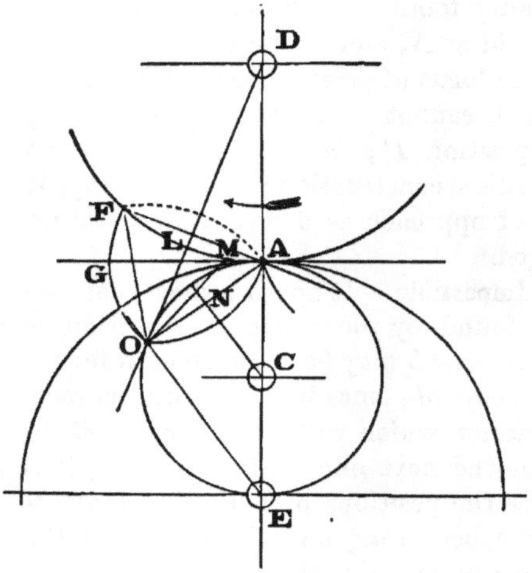

Fig. 200.

In this diagram, the wheels evidently work in outside gear, and AC is a minimum. Should O fall upon the tangent AG, OA will coincide with that line, and the follower becomes a rack: if O falls above AG, the centres C and D will lie on the same side of the point of contact, the follower will be annular, and the radius AC thus found will be a maximum.

If, on the other hand, the number of pins be assigned for the follower; then the position of O is known, and also the arc AO, whence AG and AM are also determined. It is evident that OD when found will bisect the chord AF, making $OA = OF$. Therefore, describe about M an arc with radius MG, and about O another arc with radius OA; these arcs will intersect in F, and a perpendicular to AF through O will cut CA or its prolongation in D, the required centre of the driver, whose radius will be a minimum when, as in this figure, A lies between C and D.

If the point F coincides with G, we have the limiting case when the driver becomes a rack: should F fall below AG, the driver becomes in its turn annular, and its radius as thus determined is a maximum.

326. Wheels with Radial Planes.—Since the face OF of the driver's tooth in Fig. 200 is an epicycloid, it would correctly drive a flank traced by the same describing circle.   And this flank would be radial, were we to use a pitch circle whose radius AE is twice AC.   Hence this driver will work not only with the pin-wheel, but with a wheel of twice the diameter of the latter, furnished with twice as many radial planes as there are pins : the combination presenting the appearance shown in Fig. 201.

In other words, the minimum number of radial planes which can be driven by such a wheel is equal to twice the minimum number of pins.   When stated in this way, it would appear that there should always be an even number of these planes. But it is to be observed, that in computing the least number of pins, we must take the next higher integer in the event of a fractional result ; we may, however, at once double that result in order to ascertain the least number of radial planes, and the next higher integer is not then necessarily an even number ; but on the contrary, is an odd number in the majority of cases, as will be seen by reference to the following tables.

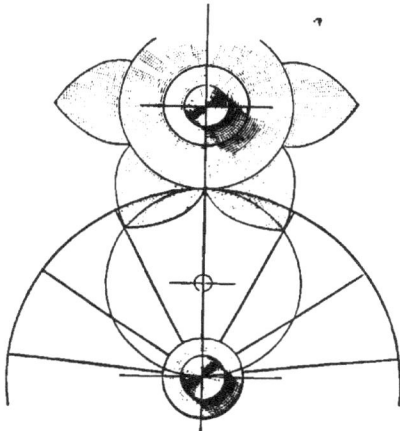

FIG. 201.

327. Limiting Ratio of Pitch Circles in Inside Gear.—In the use of the ordinary epicycloidal teeth, or of those of involute form, the diameters of wheels in inside gear may be made very nearly equal if the teeth are short enough, as in Fig. 129 : and such combinations are frequently met with in differential trains, the annular wheel often having only one tooth more than the pinion.   And at first glance it would appear perfectly feasible to adopt such proportions in pin gearing, whether the annular wheel drive or follow.

But in the former case, the teeth being hypocycloids of which the inner pitch circle is the generatrix, will become radial when the outer one is of twice its diameter ; and if this limit be passed, making the pinion more than half as large as the driver, the teeth of the latter will become *concave:* this, although geometrically satisfying the conditions, is manifestly impracticable.

14

If, however, the internal wheel drive, no such difficulty is met with, and it may have nearly as many teeth as there are pins in the annular follower, if they be duly shortened as the number is increased.

328. When the Pin is of Sensible Diameter, the action, as has been shown, begins at a different time ; and in consequence of the peculiar nature of the derived curve forming the working outline, it becomes in general impossible to determine the limiting numbers with precision. For if the diameter of the pin be assumed, then because one pitch circle is the generatrix and the other is the directrix of the elementary tooth, the diameters of both must be known before the point M of Figs. 198 and 199, which fixes the time when the action shall begin, can be found graphically or otherwise.

In one case, however, this point can be located with exactness, as shown in Fig. 202 ; DD, CC, being arcs of the pitch circles of the driver and the follower respectively, and having the same radius. Let AO be the radius of the pin, which is here enormously exaggerated ; CC cuts its circumference at N and L, and DD cuts it at M and S, making the arcs AN, AM, equal to each other. Rolling CC upon DD, the point A rises in the cardioidal curve AB, and when N reaches M, the chord AN will have the position MB, normal to this curve and also to the derived one.

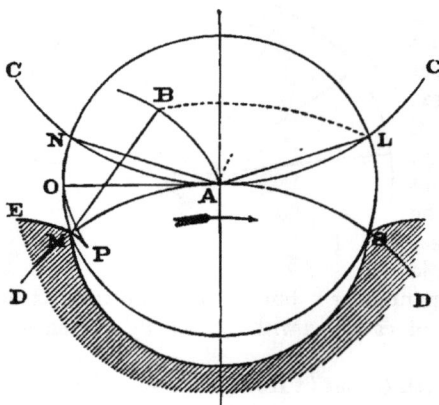

Fig. 202.

The latter is composed of the two branches OP, PE ; the second one, therefore, cuts the base circle and the circumference of the pin, whatever the radius of the latter, at the same point, M. Since $MA = NA$, the points M and N will come together at A as the rotation indicated by the arrow progresses : and moreover, M will then be in correct driving contact, the normal MB at that instant coinciding with the chord AL.

329. We have thus ascertained, that when the driver and the follower are of equal diameters, there is no arc of approach, the action beginning on the line of centres, without regard to the size of the pin. We may now assume a diameter for the latter, and by a slight modification of preceding processes we can determine the least number of pins and teeth that can be used.

Let us suppose the pin to be of such size that the arc NL of Fig. 202 shall be one half the pitch ; then in Fig 203 we shall have $AO =$ $1\frac{1}{4}$ pitch, $AL =$ pitch, and $AF = 1\frac{1}{4}$ pitch ; and the point P, of the working tooth must at the end of the action lie at the intersection of the chord OA with the radial line DP which bisects AF.

Now, were it true that the arc of approach is *nil* in all cases, the limiting numbers could be determined. The graphic process is as follows. First, if the number of teeth be assigned, we make the angle $ADP = \frac{3}{4}$ pitch, set off on the tangent at A, a distance AG equal to the pitch arc AL, make $AM = \frac{1}{4}$ AG, then with centre M and radius MG describe an arc cutting the radial line DP in P. The angle GAP thus determined is evidently equal to ACN, and, therefore, equal to $\frac{5}{8}$ of the pitch angle of the follower, whence the required minimum number of pins is easily found. Second, if the number of pins be assigned, then PA is known, and PF must be equal to it ; also the

Fig. 203.

arc AF when found must be $1\frac{1}{4}$ times the pitch arc ; this is of a known length, and having set off AH equal to it on the tangent, we can find F as in previous cases, thus determining the required least number of teeth, which could drive the given follower on the supposition that there is no approaching action.

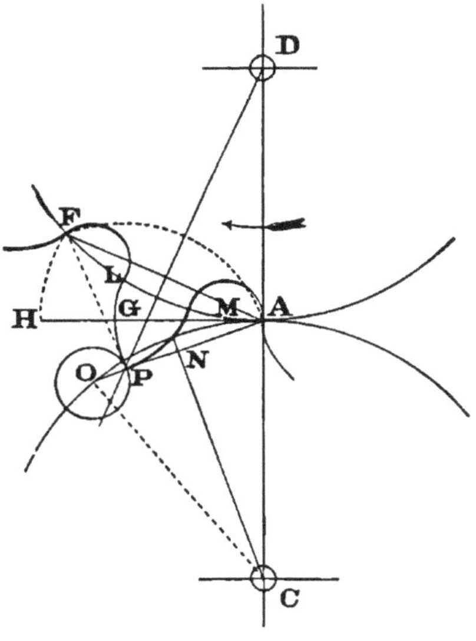

330. We must now proceed tentatively, for this supposition is true only in the one case in which there are just as many teeth as there are pins. If then we assume various numbers of either, the minimum values as thus determined will in general be incorrect ; but the errors will diminish as the number sought approaches equality with the number assumed, and when that equality is reached the result will be exact.

In this way we find, by computation,

Least No. of Pins for 12 Teeth = 11.62.
Least No. of Teeth for 12 Pins = 11.77.

Twelve teeth, then, can be used to drive twelve pins, but no less numbers will answer for equal wheels when the pin is of the size above assumed, which is that most commonly adopted in practice. If the pin be made smaller these numbers may be reduced, the limit being six for each wheel when the pin becomes a point.

331. The above process, with slight and obvious modifications in the diagram, Fig. 203, would be applicable in every case, were the amount of approaching action known for every given ratio of the pitch diameters. But it is not known; and further progress must be made in the face of additional obstacles due to the perverse nature of the derived curve. Consider-

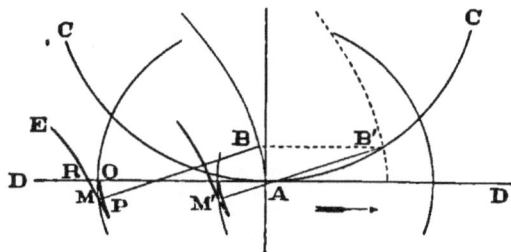

Fig. 204.

ing these now in reference to outside gear, we find that when the pitch circles are unequal, the point M of Fig. 202, in which this curve cuts the circumference of the pin, no longer falls upon the base circle of the epicycloid. If the driver be the larger, the second branch PE cuts the pin first, as in Fig. 204, which represents the limiting case of a driving rack, subsequently intersecting the base line at some point R. The latter will come into driving contact when it has advanced to A, but before that occurs the normal MB will have reached the position $M'AB'$, thus bringing the point M into action, and giving a greater or less arc of approach. If, on the other hand, the follower be the larger, the working branch of the curve will cut the base circle before it does the circumference of the pin. The former intersection is the one which would come into action on the line of centres, but since it is effaced in the process of forming the actual tooth, it does not come into action at all, and the latter intersection cannot, evidently, begin to drive until after passing that line, thus introducing the phenomenon of *negative approach*.

The limiting case in outside gear is that of the driven pin-rack, Fig. 205 ; it is very true that the process of constructing the derived curve "reproduces the same involute in a different position" (PE), but it also introduces a part (PO) of the other branch ; to be sure, this is effaced, but so too is a portion (PM) of what would be the acting tooth ; and M cannot drive until it reaches the position B' in the path of contact CC, the negative approach being, therefore, equal to AB'.

332. Now, the exact mathematical determination of this lowest point M, even when the diameters of both pitch circles and the pin are given, is a matter of extreme complexity, to say the least of it. And were it never so simple, that fact would be of service only in determining the limiting numbers by a series of approximations, since when one pitch circle is given the other is to be found.

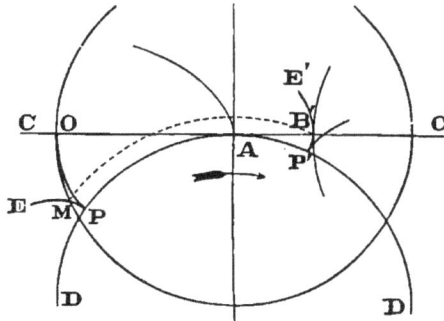

Fig. 205.

Still, these numbers may be ascertained with a degree of precision sufficient for all practical purposes, we think, by the method which we have adopted in computing the subjoined tables. This is based upon the assumption that the amount of approaching action, positive or negative, can be determined as accurately as necessary by graphic processes. Which will appear the more probable, when it is taken into account that in the actual working tooth, the obtuse intersection M will be more or less rounded off, and not sharply defined ; and the effect of this will be to diminish the amount of positive, and to increase that of negative, approach.

Accordingly, it appearing from preliminary trials that about six teeth will be the least that can drive a pin-rack, an accurate diagram upon a large scale (the diameter of the pinion being six feet) gave for a pin of the proportions named in (**329**), a negative approach equal to $\frac{\text{pitch}}{12.41}$. Then, assuming $\frac{1}{12}$ the pitch as the actual amount, we have the equation

$$\tan (\tfrac{5}{6} \text{ pitch}) = \text{arc } (1 + \tfrac{1}{12}) \text{ pitch,}$$

or

$$\tan 10x° = \text{arc } 13x° \text{ ;}$$

which being solved by the tentative process previously described, gives the limiting number, 6.44, for the pinion driving a rack.

Seven teeth, then, will drive in outside gear ; the negative approach would theoretically be a little less—but in view of the consideration above mentioned we have not reduced the amount, but have allowed $\frac{1}{18}$ the pitch in this case also. We have previously found that 12 teeth will drive 12 pins, the approach being exactly zero : what remains, then, is to take for the drivers between 7 and 12, a negative approach diminishing as the numbers of teeth increase, and to test the results of computation by graphic construction.

333. Similarly, we find in the case of a rack driving a pin-wheel, a positive approach of about $\frac{3}{18}$ pitch ; which gives 4.64 as the least number of pins that can be used. As the amount of approaching action here named was determined graphically for a wheel of six pins, it is probably less than it would actually be at the exact limit, and we have allowed the same amount in the case of a wheel of 5 pins, which, on this hypothesis, can be driven by one of 110 teeth. For the pin-wheels between 12 and 5, then, we have a positive approach, gradually increasing from zero to $\frac{3}{18}$ of the pitch.

We repeat, that no pretension to theoretical exactness is here made, the object being to furnish safe rules for practical guidance in using the customary proportions. We have, therefore, in apportioning the positive and negative approach, made the former rather less and the latter rather greater than the precise amount would probably be. It may be observed, that the construction of the nomodont for the whole range from the driving to the driven rack, affords a tolerably reliable check upon the gradation of the approaching action, since any serious error would be manifested by the consequent irregularity of the curve. Besides, the graphic constructions have been made with considerable care and

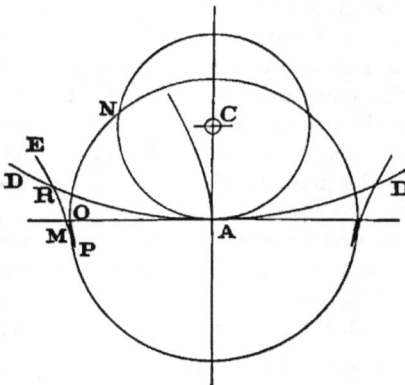

Fig. 206.

upon a large scale ; so that upon the whole we think it safe to say that if the radius of the pin be in each case made equal to half the chord of the pitch arc upon the smaller circle, the numbers here given will work ; and there is a strong probability that these limits cannot be practically transcended.

334. Next, in regard to inside gear. When the annular wheel drives, the positive approach is still greater than with the driving rack, as will be seen by comparing Fig. 204 with Fig. 206 ; the point *R*, which will come into driving contact with *N*, at *A* upon the line of centres, being in the latter case much farther from *M*, the lowest practicable point of the working tooth.

This increase in the approaching action reaches its limit when the diameter of the larger pitch circle is twice that of the smaller; the hypocycloid then becomes a right line, the derived curve becomes another one parallel to it, and whatever the position or the size of the pin, the common perpendicular to these two lines through its centre always passes through *A*. In the case represented in Fig. 197, then, the approaching action continues during half the revolution of the

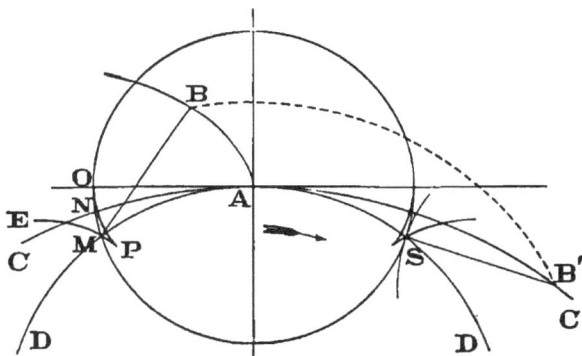

Fig. 207.

follower, and the receding action during the other half, each being, therefore, equal to the whole pitch. A greater number of pins and slots may be used, but the sliding blocks must then be dispensed with, and the angle of action will be reduced, its magnitude depending upon the distance from the driver's centre at which the slots terminate or intersect each other ; that is, in effect, upon the length of the teeth.

335. On the other hand, when the pinion drives, the negative approach becomes greater than in the case of a pin-rack. And here too an exact determination of its amount can be made when the diameter of the outer circle is twice that of the inner, as in Fig. 207; for the original tooth-outline *A B* is in this case also a cardioid, and the lowest point *M* of the working tooth falls upon the pitch circle, *DD*, of the pinion. But the driver must now make two revolutions in order to have the complete cardioid traced upon its plane by a marking point in the circumference of the follower, whereas when two equal wheels work in outside gear but one revolution is necessary : and the same proportion holds true for the generation of any given portion of the curve. Therefore, instead of coming into driving contact at A, as in Fig. 202, the point M must advance twice as far before its action begins; the point B comes into the original locus of contact at B', the normal MB at that instant taking the position SB': thus fixing the arc of negative approach AS, which is readily computed when the size of the pin is given.

Limiting Numbers of Teeth.

D.	F.	Minimum Value, FOR ANNULAR DRIVER, NO. TEETH = 2 (NO. PINS).			
TEETH.	PINS.	INSIDE GEAR.			
6.44	∞				
7	89	D.	F.	D.	F.
8	32				
9	21	TEETH.	PINS.	4	2
10	16	3	4	16	3
11	14	4	8	36	4
12	12	5	14		
14	11	6	35	TEETH.	PINS.
16	10				
18	9	ANNULAR. MAX. VALUES.			
21	8				
27	7				
37	6	PIN GEARING.			
110	5	Tooth = Space.			
∞	4.64				

Limiting Numbers of Teeth.

D. TEETH.	F. PINS.	D. TEETH.	F. PLANES.	D. TEETH.	F. PINS.
2.68	∞	2.68	∞	2	9
3	27	3	54		
4	11	4	21	**F.** PINS.	**D.** TEETH.
5	8	5	15		
6	6	6	12	3	25
8	5	7	11	2	4
16	4	8	10		
∞	3.29	11	9		
		16	8	INSIDE GEAR.	
		53	7	MAX. VALVES.	
RECESS = PITCH.					
PIN = POINT.		∞	6.58		

336. The Path of Contact in Pin Gearing.—In the elementary form, the pin being a mathematical point in the circumference of the follower's pitch circle, that circumference is itself the locus of contact.

When the pin has sensible diameter, its centre yet lies always in that circumference ; the common normals to the pin and its driving tooth are chords of the circle, all passing through the point A, Fig. 208. Supposing the driver to turn to the right, then, we have only to set off on each of these chords a distance equal to the radius of the pin, measured from the circumference toward the left : the line joining the points thus located, which is the curve called the limaçon, is the path of contact.

This might also be constructed from the outline of the working tooth, by the process of Fig. 164 ; but the method above described is more convenient ; and, as will be found more fully set forth in the

Appendix, it involves principles which enable us to determine with greater accuracy than would otherwise be possible, certain critical points not only in the path of contact itself, but in the tooth outline also; since the latter, evidently, may by reversing the construction be derived from the limacon.

Some Practical Considerations.

337. Noise and Vibration.—Although not coming strictly within the scope of this treatise, the practical ill effects of using wheels with incorrectly shaped teeth are so closely connected with the subject as to demand a brief notice.

It is to be observed, then, that the *noise* and the *vibration* which often attend the action of toothed gearing, especially at high speeds, are not necessarily identical in origin. It is true that the causes which produce noise will also produce vibration; but vibration may be produced by other causes, and may at least be imagined to occur without noise.

To explain; suppose two wheels of perfect form and finish to gear with each other, the power and the resistance being absolutely uniform; then, whatever the amount of backlash, there would be neither vibration nor noise. Now in practice, this uniformity of power and resistance seldom or never exists, and the variations in speed cause the fronts, and often the backs, of the teeth to strike together at short intervals. These blows cause a rattling noise, which is worse the higher the speed, and is accompanied by vibrations due to the impact between the teeth.

Fig. 208.

The sole cause of the noise, evidently, is the existence of backlash; but even were the teeth so perfect as to have no backlash at all, these irregularities in the power and the resistance would still give rise to vibrations, more or less injurious according to the suddenness of the changes; they would, however, take place in quiet.

338. But again, vibratory action may result from a totally different cause, namely, incorrect forms of the teeth.

To illustrate this, imagine two engaging wheels, whose teeth are, as before, of perfect finish, but not of proper contour; let the speed of

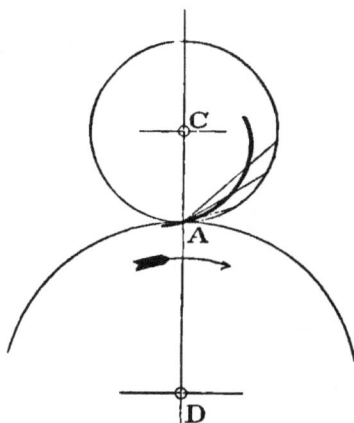

the driver be absolutely uniform, and the resistance such as to keep the acting outlines always in contact, so that there is none of the rattling above mentioned.

The average velocity ratio will be correct ; if the driver has, for instance, 100 teeth and the follower 50, each revolution of the former will cause two revolutions of the latter. And further, $\frac{1}{100}$ of a revolution of the driver will cause $\frac{1}{50}$ of a revolution of the follower ; but during this fractional motion the velocity ratio is not constant, the follower being driven too rapidly during one part of the action, and too slowly during the other part.

Thus the action of each pair of teeth, though correct as a whole, is faulty in detail, being made up of two counterbalancing errors. The speed of the driver is uniform, but that of the follower is fluctuating ; its motion consists of a series of pulsations, not necessarily audible at low speeds, though practically certain to become so at high ones. But even at moderate velocities, this vibration acts injuriously upon the whole mechanism ; and in many cases it is easy to see that the perfection of the work done by the machine may be impaired by the irregularity of its motion, no matter how slowly it runs.

339. Effect of Wear upon Teeth of Incorrect Forms.—It is clearly poor economy to require a machine to finish itself, even were it always able and willing to do so, since in the meantime its proper work must be imperfectly executed. Still it is sometimes asserted that teeth originally of incorrect contour will " wear into shape " and improve by use.

Now wear means abrasion ; so that the only errors which could thus correct themselves, by the removal of superfluous material, must be errors of excess, not of deficiency. But these are the very errors which would hardly be considered admissible by the most zealous defender of bad workmanship, as a little reflection will show. Let us suppose that of two wheels, accurately spaced, the faces only of the driver's teeth are too full, all the other curves being correct. Under these circumstances, it will be seen that during the receding action, the leading tooth of the driver will urge the follower too rapidly ; in order to admit of this, then, considerable backlash is necessary. A little space will be thus left between what should be the acting curves of the following pair ; which cannot come into action until the leading tooth quits contact, allowing the follower for an instant to wait for the driver to overtake it. Therefore, in addition to the disadvantage of having all the work done by one tooth at a time, there will be a constant clattering as the teeth come successively into driving contact.

If, on the other hand, the faces of the follower only are too full; it is then the *last* instead of the leading tooth which will have the work to do, and the follower will be driven beyond the normal speed during the approach.

But it will be driven most rapidly during the first part of the approach, its speed diminishing as the point of contact nears the line of centres : and though there will not be an absolute halt or interruption of the action, as in the previous case, there will still be something like a blow as each tooth of the driver comes into contact, and the wheels will be nearly as noisy as before.

Like reasoning applies if the faces be made correct and the flanks too full ; so that practically the only errors in the forms of the teeth which the worst of workmen would admit in the rudest of mechanism, are errors of deficiency. These may cause less noise, but they will certainly cause the pulsation above described; and when they correct themselves by wear, we shall be called upon to contemplate the phenomenon of growth by abrasion.

340. Wearing to a Bearing.—The fallacy above mentioned, if crushed by theory, is pulverized by practice ; for we believe that the first case has yet to be found, where teeth whose original contours were not full enough, have worn into forms which made the velocity ratio more nearly constant ; but every repair shop can furnish evidence that in this respect such teeth have gone from bad to worse.

Nevertheless, it cannot be disputed that gear wheels, both cast and cut, very frequently run better after being a while in use. In the case of cast wheels, this may be due to the removal by abrasion of rough and irregular spots in the metal ; but in both cases it may be due to other causes. It is likely to happen, especially if the wheels be of considerable thickness, that, owing to some slight error either in the process of cutting or in the keying of the wheel upon the shaft, some of the elements of the teeth are not exactly parallel to the axis. And what is technically called " **wearing to a bearing**," resulting in a general improvement of the action in smoothness if not in constancy of the velocity ratio, consists in point of fact not in the accretion of metal, where a deficiency exists, but in the removal of so much as stands in the way of line contact between the teeth ; and the gain is mainly due to a more equal distribution of the pressure.

341. Effects of Wear in the Bearing.—In order to preserve a constant velocity ratio, the teeth of a pair of wheels must be conjugate to each other ; and if they be of other than involute form, it is also necessary that the distance between centres shall be equal to the sum of the radii of the pitch circles. But unless special provision is made for readjustment, which cannot always be done, this distance is subject to variation on account of wear in the bearings, so that the wheels after a time will either mesh too deeply, or not deeply enough ; and it is easy to see that if the line of centres be thus made too short or too long by a given amount, the effect upon the velocity ratio may be greater with one tooth-system than with another, although both may originally work with perfect precision.

It will subsequently be shown that it is just as feasible to give a practical embodiment to Sang's theory as to the epicycloidal system, and to produce accurate templates for interchangeable wheels, derived from any assumed rack, by automatic machinery. When this shall have been done, the question will naturally arise, which of the various

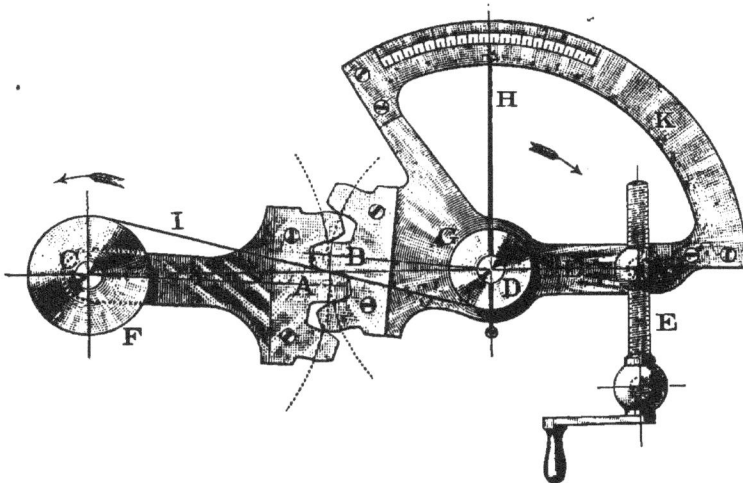

MAC CORD'S ODONTOSCOPE. FIG. 209.

tooth-systems thus placed within equally easy reach will be affected to the least extent by the disturbing causes above pointed out ; for the one which is, will clearly be the best for many purposes.

342. The Odontoscope.—In order to answer this question, it is neces-sary to have some convenient means of comparing the actions of dif-ferent tooth-systems under varying conditions. Neither graphical nor analytical treatment appearing eligible on account of the exceed-ingly intricate nature of the problems involved, the author some years since devised the apparatus of which the principle is illustrated in Fig. 209. This was before the introduction of automatic machinery for the shaping of gear cutters, and the original object was to test the accuracy of such as were made by hand.

For this purpose, two templates, *A* and *B,* are cut out of tolerably thick sheet metal, corresponding to the teeth of a pair of wheels; these are secured to arms turning about the axes *C* and *D,* whose distance from each other is capable of adjustment. The shaft of *D* carries a graduated limb, *K,* and may be slowly rotated by means of the arm, *L,* and tangent screw, *E,* or any equivalent device; motion will thus be communicated to *C,* the tooth *A* being kept in contact with *B,* by a light weight or spring, not shown.

F and *G* are small cylindrical barrels, accurately turned to diameters having the same ratio as those of the pitch circles. *F* is fixed on the axis *C,* while *G,* which carries a pointer, *H,* turns freely on the axis *D,* with which it is connected by a spring, not shown; the tendency being to wind up on *G* the fine flexible wire *I,* secured to both barrels in the manner of a crossed belt. Thus *F* and *G* will always turn in opposite directions, with a velocity ratio perfectly constant from the nature of the connection. The velocity ratio of the motions of *C* and *D,* however, is determined by the templates *A* and *B,* and will not remain constant unless their contours are strictly conjugate.

The pitch circles being scribed on the templates, it will be apparent that when the intersections of these circles with the contours of the teeth are brought into coincidence, the centres of motion are at the correct distance from each other, and the point of contact is also exactly on the line *CD.* By turning the tangent screw in one direction or the other, then, we may examine the action during the arc of approach or that of recess at pleasure; if the templates are correctly shaped, the limb *K* and the pointer *H* will move at the same rate and in the same direction, so that if the latter be set at zero on the graduated arc, it will remain at zero throughout the action; any inaccuracy being manifested by a movement of the pointer over the limb.

343. The sensitiveness of the instrument as designed for actual work is increased by the introduction of multiplying gear between the barrel or pulley *G* and the pointer *H*; thus producing a greater deflection for an error of given magnitude, and causing a very minute one to proclaim its existence. Its exact locality, however, we can not fix; that, indeed, would be equivalent to determining the point of contact of two tangent curves; and if the end in view is the correction of errors in the contours of the templates, the only resource is careful inspection with a magnifying glass.

But for comparative work like that previously alluded to, this instrument is well adapted. Suppose that the templates work with perfect accuracy when the centres are correctly adjusted; by means of a screw, not shown, the bearing of *C* may now be moved either to the right or the left through any given distance : then let the deflection of the pointer be noted for, say every degree or half degree of the angle of action.   If this be repeated after other changes in the distance *CD*, a comparison of these records with similar ones for a different pair of templates, will furnish all needed information as to the effects of wear in the bearings upon the velocity ratio in the two systems of gearing represented.

Teeth of Non-Circular Wheels.

344. When two pitch curves of any form roll together about fixed axes, the point of contact always lies upon the line of centres, and divides it into segments inversely proportional to the angular velocities at the instant.

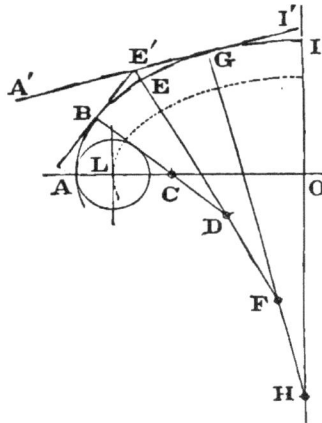

Through this point there must pass the common normal of any two conjugate tooth-curves ; these, consequently, must satisfy the condition of **(181)**, viz.: they can be traced upon the planes of rotation during the action, by a marking point invariably connected with a describing line which moves in rolling contact with both pitch curves.

The forms of non-circular wheels are so numerous and diverse, that it would be an endless task to discuss at length the various possibilities relating to the paths of contact, limiting numbers of teeth, and matters of kindred nature.

Fig. 210.

But the use of such wheels is so limited that the consideration of all these questions has by no means the importance or interest which attach to it in relation to ordinary gearing : and we shall accordingly be content with briefly explaining a purely practical method of constructing the teeth by graphic operations.

345. Laying out the Teeth of Elliptical Wheels.—In Fig. 210, let $A O$ be the semi-major and IO the semi-minor axis of a pitch ellipse. Although no part of this curve is truly circular, yet circular arcs may be found which agree with it so closely that for all practical purposes they may be substituted for it.

Thus by trial and error we find a centre, C, on the major axis, such that an arc described about it will coincide with the ellipse in the region of the vertex A, so nearly that the difference will not be perceptible, even when the curve is drawn with the finest visible line. The deviation will, however, become apparent if the arc be extended, and it should be terminated at some point, *B*, a little before this limit has been reached. If the closest possible approximation is desired, a normal through *B* should pass through *C*; if on trial this prove not to be the case, a minute change in the position of either *B* or *C*, or both if necessary, will cause this condition to be satisfied. On *BC* produced, we now seek for another centre *D*, such that an arc *BE* described about it through *B* shall sensibly coincide with the continuation of the elliptical contour. Proceeding in this way, we shall finally have a centre, *H*, upon the minor axis, for the arc *GI*, which completes the quadrant of the ellipse.

For the purpose of illustration, we have in this diagram represented an ellipse whose eccentricity is much greater than that of any likely ever to be used as the pitch curve of a wheel; yet a very perfect approximation is made by means of the four circular arcs for each quadrant; probably three will suffice for any case to be met with in practice, and in most instances two answer the purpose abundantly well.

346. The next step is to subdivide the perimeter of the ellipse for the location of the teeth and spaces. The outline now being composed of circular arcs, these are rectified by Prof. Rankine's graphic process. Thus in the figure, we lay off upon the tangent at *G*, the lengths *GE'*, *GI* of the arcs *GE*, *GI*; similarly upon the tangent at *B*, the lengths of *BE* and *BA* are laid off, and their sum being added to *I'E'* already found, gives *I'A'* equal in length to the quadrant *IEA*. This right line is next to be properly subdivided according to the number of teeth determined on, and the points of division are then transferred to the ellipse by means of Prof. Rankine's converse process applied to the approximating arcs and their rectifications: this spacing is indicated in the diagram by the alternating fine and heavy lines on both tangents and contour.

A circle is used for generating the tooth-outlines, rolling it first within the ellipse to trace the flanks, and then without to describe the faces; and thus this operation is made identical with that for circular wheels. The radius, *AL*, of this describing circle should not be greater than ½ *AC*, which limiting value will give radial flanks to the teeth whose edges cut the ellipse upon the approximating arc *AB*; the other flanks being concave.

347. The complete wheels in gear are represented in Fig. 211. In regard to the number of teeth, it makes no difference whether it be even or odd; but in either case it is eminently desirable that the two wheels should be alike. In order to make them so, if there be an odd number of teeth, they should be so arranged that one extremity of either the major or the minor axis shall bisect a tooth, the opposite extremity bisecting a space. If the number of teeth be even, then they must be so placed that the fronts of two of them shall cut the pitch ellipse at opposite extremities of one of the axes.

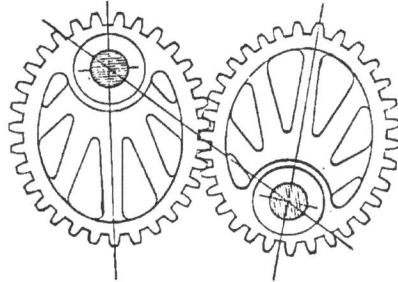

Fig. 211.

Now, the action of the teeth on different parts of the perimeter cannot, under any circumstances, be equalized in all particulars. In order to secure the same amounts of approaching and receding action, it would be necessary to make the faces of varying lengths ; and even then the result is of doubtful utility, since the degree of obliquity is continually changing. Consequently it is held to be sufficient in practice to make the faces of uniform length ; and this should be such as to ensure that in the part of the whole movement in which the obliquity of action is greatest, one pair of teeth shall not quit contact until the next pair has fairly come into engagement. Thus the form of the blank is determined by drawing a curve parallel to the pitch ellipse at a given distance without, the bottoms of the spaces are bounded by another at a given distance within, and the two blanks, secured side by side, may be cut at the same time with milling cutters precisely like those used for circular wheels.

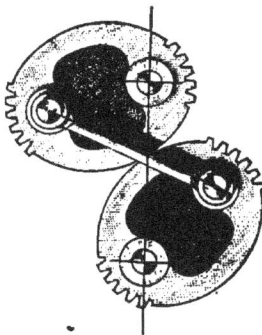

It is not necessary to furnish elliptical wheels with teeth all round their peripheries, when used under circumstances which permit their free foci, as in Fig. 212, to be connected by a link. In this case, a few teeth near the extremities of the major axis, as shown, suffice to carry the wheels past the dead points, which occur when the line of the link coincides with the line of centres. The remaining portions of each wheel must be formed exactly to the contour of

Fig. 212.

the pitch ellipse ; these will then transmit the rotation by direct and purely rolling contact, in either direction, so long as the contact radius of the driver is on the increase, (88). Since while the radius

15

is on the decrease, the work must be done by the link, it is necessary
to use a sufficient number of teeth to prevent excessive obliquity of
action while the link is thus made the sole means of transmission.

348. Elliptical Wheels with Involute Teeth.—It is worthy of notice,
though more as a matter of abstract interest than of practical moment, that the teeth for a pair of pitch ellipses may be made in the
form of involutes of smaller base ellipses having the same foci. In
Fig. 213, let C be the fixed and A the free focus of the ellipse whose
major axis is XX. Let D, at any distance from C, be the fixed focus
of an equal and similar ellipse, the position of the major axis YY
being found as follows : about D describe an arc with radius $= CA$,
and about A an arc with radius $= CD$; these arcs intersect at B,

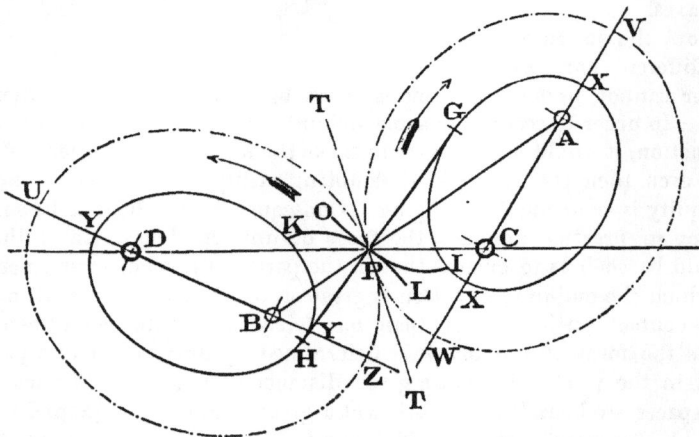

FIG. 213.

which is the free focus of the second ellipse. Draw CD and AB,
intersecting at P. The common tangent to the two ellipses, GH,
will also pass through P (**see Appendix**) ; and if we suppose it to
represent an inextensible band, wrapped around and secured to the
ellipses, it will be wound upon the right hand one if it turn about
C as shown by the arrow, and unwound from the other, which will
thus be caused to revolve about D, with a definite varying velocity
ratio. During this motion, a marking-point fixed at P in this band,
and moving with it, will trace upon the planes of rotation the involutes of the ellipses, KPL, OPI. These curves, therefore, if used as
the outlines of teeth, will maintain the same velocity ratio that was
originally established by the band.

349. But, as will be seen by comparing this diagram with Fig. 55,

P is also the common point, and TPT is the common tangent, of two ellipses confocal with the original ones, whose major axes are WV, UZ, equal to each other and to CD; and these will roll in contact about the fixed foci.

Thus four different elementary combinations are represented in this figure; we may use either

1. The levers CA, DB, connected by the link AB;
2. The smaller ellipses, connected by the band GH;
3. The involutes KL, IO, acting with sliding contact; or
4. The larger ellipses, acting in pure rolling contact. Or, these may be used simultaneously, the velocity ratio being the same in each at any and every instant, and varying between the limits $\dfrac{CW}{UV}$, $\dfrac{CV}{UW}$.

350. The distance CD being arbitrary, it follows that for the same base ellipses, an infinite number of pitch ellipses may be assigned. And on the other hand, for the same pitch ellipses, the teeth may be involutes of any base ellipses having the same foci. From which the deduction is, that so long as the teeth of this form engage at all, they will gear correctly in the sense that the action will be equivalent to the rolling together of a pair of true ellipses of some form: but whereas in the case of circular wheels the velocity ratio was not affected by altering the distance between the centres, it is to be noted that in elliptical gearing the limits between which the velocity ratio varies, *will* be affected by such alteration. For the focal distance CA is invariable, while the major axis WV is always equal to CD; so that a change in the length of the line of centres changes the eccentricity of the pitch ellipse and the value of $\dfrac{CV}{UW}$.

If then it be essential that the limits of variation should be exactly maintained, the involute form of tooth possesses no advantage over that previously described; and it is hardly necessary to add that if the free foci of a pair of elliptical wheels which are provided with teeth, be connected by a link, the distance between the fixed foci must be kept constant, on whatever system the teeth are laid out. If the involute system be adopted, care must be taken lest the base ellipses be so small as to produce excessive obliquity of action. In regard to this, Prof. Rankine, to whom is due the credit of first discussing this form of tooth for elliptical gearing, observes that the common tangent should cut the fronts of at least two teeth between the base and the pitch ellipse; which, although it may not be necessary in all cases in order to ensure the transmission of rotation, is unquestionably a safe practical rule.

351. Construction of Teeth of Lobed Wheels.—As above intimated, the method of operation first described is perfectly general, and may be employed with pitch curves of any form.   The centres of the circular arcs of which the contour is to be made up, may be found by assuming a number of points, at which the curvature begins sensibly to change, and drawing normals through these points (graphic processes for which are described in the Appendix) ; the intersections of the normals being the required centres.

But if the pitch curves be carefully drawn, it will be found that for all practical purposes, these centres can be located with sufficient accuracy by trial and error ; and at best the laying out of teeth for such wheels requires much time, care, and patience.

If the same describing circle be used throughout, its diameter should be such as to give radial flanks to the teeth in that part of the pitch line where the curvature is greatest ; should other parts be very much flatter the teeth may in consequence spread too rapidly at the root or in the flank.   This may be remedied by using different describing circles for the teeth in those parts, care being taken that the same one be always used for the conjugate face and flank.

352. Practical Limit to the Obliquity.—In any form of gearing it would be very desirable to have the line of action perpendicular to

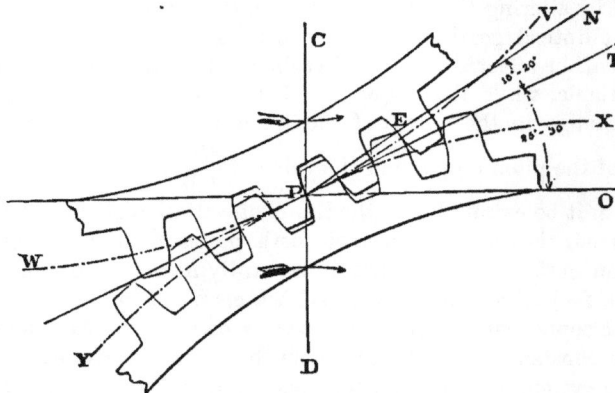

Fig. 214.

the line of centres, since then there would be no component of side pressure.

In speaking of the teeth of circular wheels, the *obliquity* was defined as the angle made by the common normal of the acting curves with the common tangent to the pitch circles.   This, however, was because this tangent is then a perpendicular to the line of centres;

and in general the obliquity is the inclination of the line of action to such a perpendicular, which latter may or may not coincide with the common tangent to the pitch curves. This it seldom does in the class of wheels now under consideration, and in consequence a much greater obliquity than would be admissible in circular gearing is often unavoidable.

Thus in Fig. 214, VW, XY are arcs of the pitch curves, in contact at P, on the line of centres CD, to which PO is perpendicular. The lower wheel driving, as shown by the arrow, let E be the point of tooth-contact, the line of action through which, PEN, has the greatest inclination to PT, the common tangent to the pitch curves.

It will then be apparent that when the angle TPO, which this tangent makes with the perpendicular PO is greatest, the total obliquity NPO will be a maximum.

The results of experience show that the angle TPO

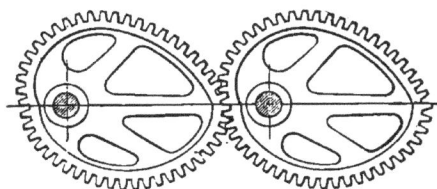

Fig. 215.

should not, if it be practicable to avoid it, exceed from 25° to 30°. In other words, the contours of the non-circular pitch curves should be such that the angle between the tangent and the radius vector shall at no point be less than 60° or 65°. Nor should the angle NPT exceed from 15° to 20°; thus limiting the range of the maximum total obliquity, NPO, to from 40° to 50°; though if the object be rather the modification of motion than the performance of heavy work, higher values may in extreme cases be used.

353. This limit, it will be perceived, has not been regarded in Fig. 215, which shows the appearance of the dissimilar pitch curves of Fig. 75, when furnished with teeth.

In Fig. 216, we have a pair of similar but unsymmetrical unilobes, the pitch lines being elliptical below the horizontal centre line, and of the logarithmic spiral form above, as in Fig. 72.

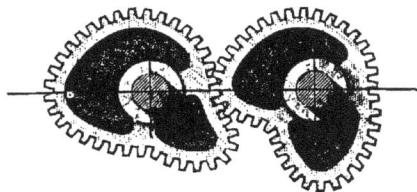

Fig. 216.

The lower halves of the dissimilar pair shown in Fig. 217 are alike, the logarithmic spiral being the pitch curve; while above the line of centres, that spiral is used for a part of the contour, the remainder being elliptical, as in Fig. 74. But the wheels here represented present the peculiarity that the limits of variation in the velocity ratio, which suddenly change when the wheels reach the position given in

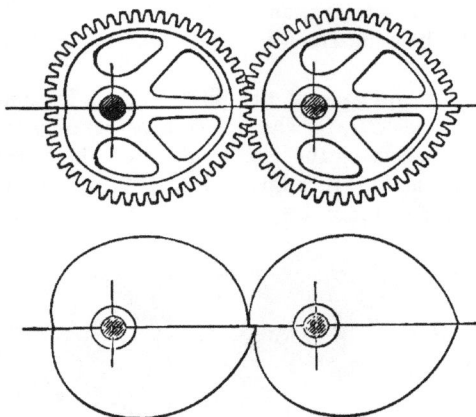

the figure, are not the same for the two halves of the revolution.

A still greater and more abrupt transition is produced by the pair represented in Fig. 218, of which the pitch curves are similar and equal l o g a r i t h m i c spirals. It is tolerably evident that in this case the absolute velocity must be small in order to avoid violent shocks at the instants of the change : yet wheels very similar to these have

FIG. 217.

been used—as for instance in the moulding machine of Gallas and Aufderheide.*

354. Non-Circular Pin Gearing.—The labor of laying out the teeth of a pair of dissimilar wheels may be abridged, and the difficulty of cutting them in part avoided, by making one of them after the manner of a pin-wheel. The principle of the process (which may also be applied to the ellipses or to any similar as well as to dissimilar l o b e d wheels) is precisely the

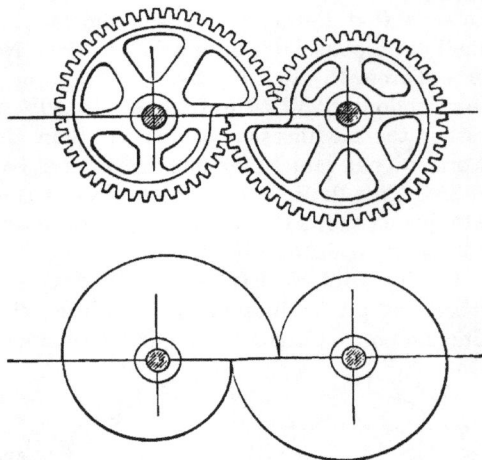

FIG. 218.

same as in the case of circular gearing. That is to say, the pitch curve of the follower is used as the describing line, by rolling which

---

* For description, see "*Mechanics*" of February 18, 1882.

upon the pitch curve of the driver are generated the elementary teeth for the latter; and the working outlines are at a constant normal distance within these, equal to the radius of the pin, as shown in Fig. 219.

In regard to the size of the pins, and their distance from each other, it is sufficient to say that, as the contour of each wheel is regarded as made up of approximating circular arcs, the whole operation is reduced to the familiar one of constructing epicycloids and their parallels; so that the limits in relation to the pitch and the diameter of the pins are readily deduced for each particular case by application of the reasoning given in the discussion of circular gearing of this description. Though it may

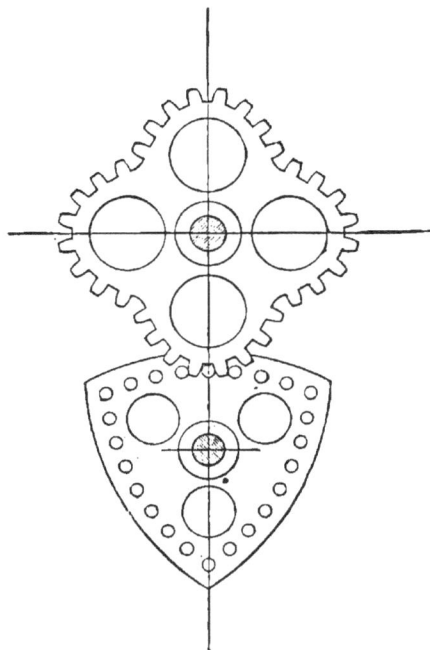

Fig. 219.

be well to repeat that the teeth in all cases should be as small and as numerous as they can be made consistently with securing the necessary strength.

CHAPTER XII.

355. In Treating of Spur Wheels it has been convenient, since all the transverse sections are alike, to consider all the motions as taking place in one plane, and thus to deal with lines instead of surfaces. But we must not lose sight of the fact that the pitch and describing curves, and also the tooth-outlines, are but the bases of surfaces with elements perpendicular to the paper, and acting in right-line contact. Thus in the epicycloidal system, for instance, we must imagine a describing cylinder rolling upon the pitch cylinders, the common element being

the instantaneous axis, about which the describing line is revolving at any given phase of the action. Now the teeth thus generated must touch each other along a right line: and this principle is capable of a more extended application. The very slight modification by which spur wheels are transformed into bevel wheels is at once suggested by regarding the pitch cylinder as the limiting form of the cone, in which the vertex is infinitely remote; and this leads directly to the analogous generation of the tooth surface by the rolling of a describing cone in contact with the pitch cones.

Fig. 220.

This is illustrated in Fig. 220; supposing the arcs AP, AB, to be equal, then while the smaller cone rolls upon the larger, the element CP will generate the surface CPB; to which the plane CPA is normal, since at the phase here represented, CA is the instantaneous axis. This is the essential property, and is independent of the forms of the bases of the cones; but if as here shown these be circular, the extremity P of the generating radius will trace a curve, BP, lying on the surface of a sphere, and called, since it is described by the rolling of one circle upon another (although the two are not in the same plane), a *spherical epicycloid;* which may be considered as the directrix of the conical surface CPB.

In like manner, by the rolling of one cone inside of another, both having circular bases, a *spherical hypocycloid* may be described. And whatever the bases of the cones, it is evident that a surface is generated in this case also, to which the plane determined by the describing line and the element of contact at any instant, will be normal.

It is thus readily seen that, the describing cone being tangent externally to one pitch cone and internally to the other, these two surfaces will be swept up simultaneously as the rotation progresses, and will at every instant have a common normal plane, cutting the plane of the fixed axes of rotation always in the common element of the pitch cones: they are, therefore, correct forms of tooth-surfaces, and will maintain the original velocity ratio, be the same constant or variable.

356. Tredgold's Approximation.—Before considering in detail the

practical operation of laying out the teeth in accordance with the
exact theory as above set forth, it is proper to describe the method
more commonly employed, which involves possibly a little less
labor, and gives results, not
rigidly accurate, but sufficiently so for many purposes
under ordinary conditions.

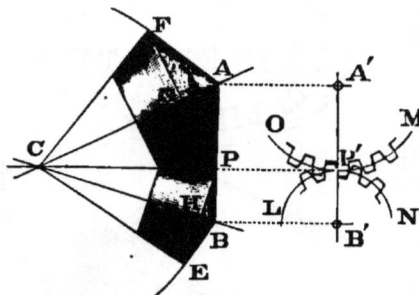

The principle of this process, which, it should be observed, is applicable only in
the case of circular wheels, is
shown in Fig. 221.   Let CA,
CB, in the plane of the paper,
be the axes, and CP the common element, of the pitch

FIG. 221.

cones CPE, CPF; whose bases EP, PF, are small circles of the
sphere shown in dotted outline.   Draw APB perpendicular to CP;
then AP by revolving around CA will generate the cone PAF, and
BP, by revolving around CB, will generate the cone PBE: these
cones are tangent to the sphere, and respectively normal to the pitch
cones.

Now, AB is also the trace of a plane tangent to the sphere at P,
and tangent to both normal cones ; and, in the diagram at the right,
the points A', P', B', are the projections of A, P, and B, upon this
tangent plane, and the arcs $OP'M$, $LP'N$, are the developments of
the bases FP, PE.   These arcs are next to be treated as though they
were the pitch circles of spur wheels, and teeth are to be laid out
upon them according to any of the systems previously explained,
that of pin-gearing of course excepted.

Suppose these teeth to be cut out of a thin sheet of metal, and then
wrapped back upon the normal cones ; their outlines are then to be
traced, and treated as the directrices of the conical tooth-surfaces, all
of whose elements, as before explained, converge to the vertex C of
the pitch cones.

357. The operation of finding the tooth-outlines upon the normal
cone may be performed graphically, as in Figs. 222 and 223.   Since
the tooth projects beyond the pitch circle to R', the normal cone
must be correspondingly extended to R, which determines the extreme diameter of the *blank* CTR; and the bottoms of the spaces
upon its development are limited by the circle whose radius is $A'S'$,
to which AS is, of course, equal.   The points R, P, and S, in revolving about AC, describe circles which in the side view appear as

the right lines RT, PF, SG; these are seen in their true form and size in the end view, Fig. 223, which it is necessary to construct in order to draw the side view. Obviously, the length of the arc which measures the breadth of the tooth at the top, at the bottom, or on the pitch circle, will be the same in the end view as in the development, the chord being a little less in the former. The same holds true in regard to any intermediate circles which may be drawn, and thus the outlines of the teeth in the end view may be determined with any required degree of precision; after which they are projected to the side view, the tops to the line RT, the bottoms to the line SG, etc., in the usual manner.

Since all the elements of the tooth run to the vertex C of the

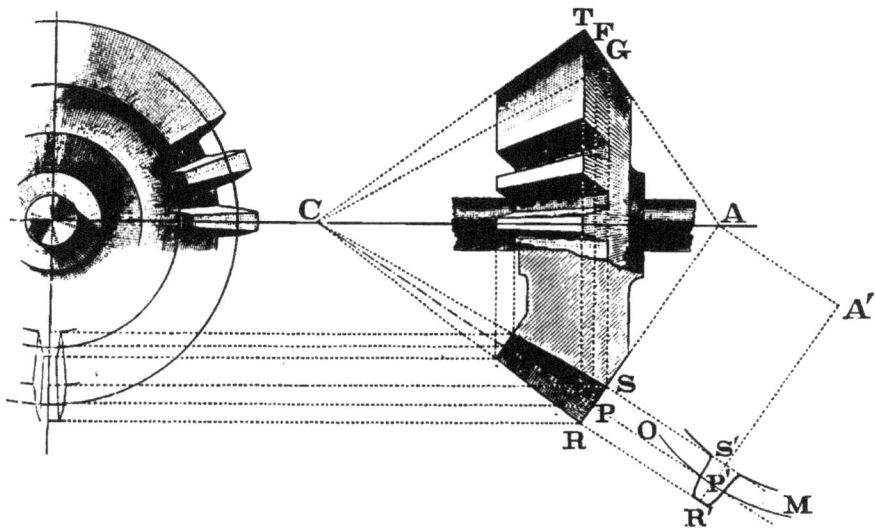

FIG. 223.   FIG. 222.

pitch cone, it follows that if it be limited at the smaller end by another normal cone, as in the figure, its outline there will be of the same form as that of the outer end; and the mode of drawing it is sufficiently indicated without further explanation.

358. The shaping of the blank by the addition of a frustum of the normal cone TAR, is the readiest and best means of giving a presentable finish to the teeth, which would be weak at the extreme points, and project beyond each other in a very unsightly manner, were the larger end of the wheel terminated by a simple transverse plane.

And the assumption upon which this method of laying out the

teeth rests, is that the trace upon this normal cone, of the surface generated by rolling a describing cone upon the pitch cone, will upon development become a true epicycloid, or at least not sensibly differ from it within the limits made use of for the teeth. Let us first, then, see how the exact trace of this surface upon the normal cone may be determined.

359. Construction of the Correct Tooth-Outline.—In Fig. 224, let CPF be the pitch cone, GAH the normal cone, indefinitely extended; and let CL be the axis of a describing cone, ICK, which is tangent to the pitch cone along the element CP, and intersects the normal cone in the curve PON. Making a projection upon a plane perpendicular to CA, the circle FP is seen in its true size as $F'TP'$, the curve of intersection appearing as $P'RN'S$.

Now taking CPI as the describing element, it is clear that if it

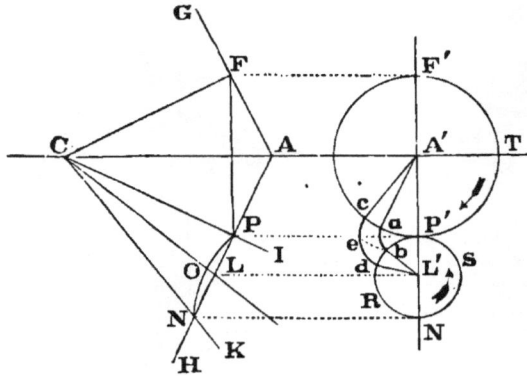

Fig. 224.

revolve about CL, the other cones remaining fixed, it will pierce the normal cone, at every instant, in some point of the curve PON. From this curve, then, the required trace may be found as follows. Suppose the pitch cone to turn as shown by the arrow, through an angle measured by the arc $P'a$. The describing element CP will thus be caused to revolve about CL, through an angle which, the velocity ratio being known, can be readily ascertained; and this will enable us to fix the point b, in which at that time it will pierce the normal cone. And the line ba will be a portion of the required trace. Let the rotation progress until a reaches the position c; then since ba moves with the pitch cone, it will at that instant appear as ec; and the angular motion of the describing element meantime being known, we can find the point h upon the curve $P'RN'$, in which it will then pierce the normal cone, and extend the trace, as *ced* : which operation, repeated at short intervals, will enable us to construct the true form of the tooth-outline with any stipulated degree of accuracy, within the scope of graphic operations.

By proceeding in like manner with the describing cone internally tangent to the pitch cone, we may complete the tooth by finding the trace of the flank-surface.

360. Results of the Two Methods Compared.—In Fig. 225, the full outline represents the tooth of a wheel laid out by the method above explained ; and the dotted curves are the forms of the faces as constructed by Tredgold's process.

The wheel selected for this comparison is one of a pair of *mitre-wheels*, as those are technically called which are of equal diameters, and have their axes perpendicular to each other; the diameter of the base of the pitch cone being 30 inches, the number of teeth 24.

In the application of Tredgold's process, we are at liberty to assume a describing circle which will make the flanks of the developed teeth radial, its diameter being, therefore, equal to the slant height of the normal cone,

Fig. 225.

and this was done in laying out the tooth here shown. On wrapping the developed sheet back upon the cone, then, these radial lines will become elements of that surface ; so that the flanks of the teeth thus constructed will be radial planes.

But a straight line can be generated by the rolling of a circle, only when it rolls within another circle of double its diameter, in its own plane. No circular cone, then, can generate a perfect plane by rolling within another one. However, if the angle at the vertex of the describing cone be half that at the vertex of the pitch cone, the surface swept up will curve so slightly near the first element of contact, that for all practical purposes the flanks may be regarded as truly

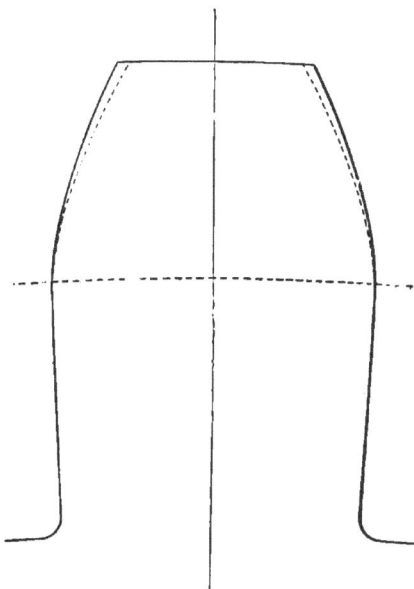

plane surfaces.   But now, the same describing cone must be used for the face-surfaces ; and, as shown in the figure, the discrepancy between the results of the two methods is quite decided, and sufficient to affect materially the constancy of the velocity ratio if the approximate form be adopted.

361. Bevel Wheels in Inside Gear.—In laying out the teeth of a hollow bevel wheel, by Tredgold's process, it is apparent that the size of the describing circles must be fixed with due regard to the limits deduced for the case of annular spur gearing, in order that there may be no interference upon the development of the normal cone. The spherical epicycloids and hypocycloids, it is true, differ from their plane namesakes in this, that they are not capable of two generations ; so that the peculiar feature of double tangency between the face and the flank of a pair of engaging teeth does not exist in inside bevel gearing.   But it is none the less evident that if the outlines of the teeth upon the developed sheet be such as to interfere with each other, the wheels constructed from them by this process will not work.

From this consideration we deduce, as a safe practical rule in selecting the describing cones when laying out the teeth by the exact method, that the diameters of their bases should not be greater than those of the describing circles which can be employed in the approximate method ; the circumferences of these bases being tangent to those of the pitch cones.

A pitch cone rolling in contact with a plane disk (see Fig. 81), presents, upon the application of Tredgold's process, a case analogous to that of a rack and wheel in spur gear, and requires no special notice.   In the use of the exact method, the details of the operation are substantially the same as already explained ; the vertex of the cone normal to the plane disk being infinitely remote, that cone becomes a cylinder, but no additional difficulties result from the introduction of this new feature.

362. Bevel Wheels with Involute Teeth.—When the developed bases of the normal cones are taken as pitch circles, teeth may be laid out upon them by the involute as well as by the epicycloidal system, and the wheels thus made will work as well as the others.   The surfaces of their teeth are obviously approximations to those which would be generated by the rolling of a plane in contact with two base cones. Such a rolling plane must, clearly, cut the plane of the fixed axes always in the same line, and this line will be the common element of two pitch cones whose diameters will have the same ratio as those of the base cones.   And, taking this radius of the rolling plane (whose motion is, during the rotation of the cones, one of revolution about

an axis perpendicular to the plane, passing through the common
vertex), as a describing line, we may follow its movements, and by
finding the points in which at given intervals it pierces the normal
cones, determine the exact tooth-outline with as great facility and
precision as in the preceding case.

This tooth-surface may be also generated in another manner; suppose the pitch cone to be cut from vertex to base along an element,
and the surface to be unrolled into a plane; this plane will, during
the process, be always tangent to the cone, the cut edge will sweep
up the surface under consideration, and, since it is of constant length,
its extremity will trace a curve which may be properly called a *spherical involute*.

It is not necessary to go into a detailed examination of the form
of the surface which by rolling in contact with the pitch cones would
generate teeth of the forms here spoken of, since the method above
given is much more simple. The close analogy between the plane
and the spherical involute, however, is sufficient to make it apparent
that such a conical surface, whose spherical directrix will have a
corresponding analogy to the logarithmic spiral, is capable of thus describing these identical teeth, which consequently form no exception
to the general law.

363. Action of Bevel and Spur Wheels Compared.—It is true of bevel
as well as of spur gearing, that the smaller and more numerous the
teeth, the better will be the action, because there will be less of both
sliding and obliquity.

Now experience shows that, other things being equal, a pair of
bevel wheels will run more smoothly, and in a general way operate
more satisfactorily, than a pair of spur wheels of the same numbers
of teeth. The reason of this will be seen very clearly by reference to
Fig. 221, if we imagine the outer ends of the teeth to be bounded by
the surface of the sphere, and confine our attention to the action
upon each other of the outlines thus formed. The point of contact,
throughout the engagement of any one pair of teeth, will always lie
in or near to the plane AB, perpendicular to CP, upon which are
developed the normal cones, which latter in their turn do not, within
the limits of the height of the teeth, deviate greatly from the spherical surface. The result of this is, that although the velocity ratio is
determined by the lengths of the perpendiculars PG, PH, the dura-
tion of the action is very nearly the same as it would be were the
developed teeth to work together in their own plane about the centres
A', B'. The action of these wheels, then, in so far as it is affected
by the numbers of the teeth, is substantially the same as that of spur

wheels of the same pitch, with PA, PB, for the radii of the pitch
circles. Or, in general, the action of a bevel wheel of a given number of teeth, is in this respect equivalent to that of a spur wheel having a number greater in the proportion of the slant height of the
normal cone to the radius of its base, or of the slant height of the
pitch cone to its altitude.

364. Teeth of Conical Lobed Wheels.—If a cone of any form be intersected by a sphere whose centre is at the vertex, every point of the intersection is equally distant from that centre ; let a describing cone be
rolled upon the first one, as in Fig. 220, then the extremity of any one
of the elements of the rolling cone will trace upon the sphere a curve,
which may be considered as the directrix of the conical surface generated by the element itself ; and that surface will, in accordance with
the general principle, be of the form required for the tooth, the first
or base cone being regarded as a pitch surface ; and, without going
into details, it is apparent that the determination of the spherical
directrices may be effected by graphic means.

Otherwise ; let a plane be passed, say perpendicular to the fixed
axis of rotation, or to the line of symmetry, if there be one, thus
forming a base for the pitch cone. Inasmuch as its contour may be
practically made up of approximating circular arcs, the pitch cone
may be considered as made up of portions of various circular cones ;
and these may be treated as previously explained, by either the exact
or the approximate method. The normal cones, upon which the
traces of the teeth are thus obtained, will not have a common vertex ;
but since each is to be dealt with separately, this is of no consequence.

There is no instance to be cited of the actual employment of these
non-circular conical wheels ; which indeed have not been described
by any previous writer. An objection might be urged on account of
the difficulty of making them : which, however, does not lessen the
abstract interest attaching to them as a new class of kinematic combinations. The above suffices to show that the graphic operations
relating to them are sufficiently simple in detail, although tedious by
reason of their number ; and it may be questioned whether they
would really prove more difficult to make than the skew wheels
which are occasionally met with.

365. Methods of Cutting the Teeth.—Since the teeth taper from end
to end, the outlines of the transverse sections are continually changing ; it is therefore, impossible to cut them with perfect accuracy by
means of a milling cutter of the usual form. Nevertheless, this is
the method almost universally adopted when they are cut at all. In
order to distribute the unavoidable errors as uniformly as may be, it

is customary to make the outline of the cutter agree in form with the cross section of the tooth at the middle of its length ; its travel is directed along the line joining the face and flank of the tooth, that is to say along the element of the pitch cone, first on one side of the space, and then on the other, the thickness of the cutter being less than the breadth of the space.

The effect of this is, that at the larger end of the tooth the flank is too full and the face not full enough, while at the smaller end these errors are reversed ; besides, the tooth surfaces thus cut are not conical, but cylindrical, all the elements being parallel to the line along which the travel of the cutter is directed. These surfaces, consequently, will not work in true line contact, except at the one instant when the junctions of faces and flanks reach the plane of the axes, until they have " worn to a bearing" by use ; but neither then nor thereafter will they work correctly.

366. Corliss's Bevel Gear Cutting Engine.—There are, however, ways and means of accomplishing better results than this; and to the ingenuity and enterprise of Mr. George H. Corliss, of Providence, R. I., is due the production of an engine, capable of doing the work with absolute theoretical precision. In this admirable machine the milling cutter plays no part, but the teeth are *planed* out, element by element.

Without going into details, it will be understood, first, that the cutting point of the tool is made to travel always in a right line passing through the vertex of the pitch cone ; it must, therefore, at each cut plane out an element of a conical surface. Second, that the motions are so controlled with reference to a guide template, that a line drawn from the vertex through the cutting point shall always touch the outline of the template, which is, then, the directrix of this conical surface.

Now, all that is necessary to the production of a perfect wheel, is to provide this machine with a template whose outline accurately conforms to that of a transverse section of a tooth. This, it should also be stated, is practically laid out upon a larger scale than if the outer end of the tooth were to be drawn in the usual manner, because it is placed in the engine considerably farther than the base of the blank from the vertex of the pitch cone ; thus reducing the proportional magnitude of any unavoidable errors in the graphic processes.

367. Twisted Bevel Wheels.—The feasibility of placing upon the same shafts a series of bevel wheels cut from the same cones, but in different phases of their action, is self-evident. The advantages of such stepped wheels would be exactly the same as in the similar ar-

16

rangement of spur gearing; but there are practical drawbacks which, it may be admitted, would prevent their employment. But the process of twisting, by which the spur wheels are transformed into Hooke's spiral wheels, may be equally well applied to conical ones; and to the form of bevel gearing thus produced no reasonable objection can be made, except the difficulty of making it; and this, as will presently appear, is not insuperable, nor so great as might be supposed.

In Fig. 226, it is evident that if each plate of the vertical pitch cone be twisted through the same angle, it will drive the corresponding plate of the inclined cone through an angle depending upon the relative diameters; so that if these plates be indefinitely increased in number, the rectilinear elements AB, EF, will become conical helices, as shown.

The graphical construction presents no difficulty; we have in each case a uniform rotation, combined with a uniform advance along an element. Consequently since if the amount of twist upon one cone be assumed that on the other is known, both curves are readily drawn; they move in perfect rolling contact, and develope upon the common tangent plane into the same Archimedean spiral.

Just as with the spur wheels, any form of tooth which will work correctly before twisting will do so afterward; and if the faces be purposely made not full enough to be conjugate to the engaging flanks, there will be at each instant but a single point of tangency, which will alway lie in the plane of the axes, and the wheels will run in pure rolling contact.

Now, although such teeth as these cannot be made by the use of a milling cutter, it will be apparent that if the blank be placed in the Corliss machine above described, and given a motion of uniform rotation during the advance of the cutting point, they can be planed out as easily and as accurately as any others.

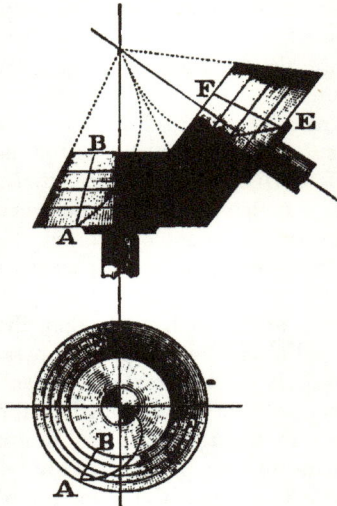

Fig. 226.

The Teeth of Skew Wheels.

368. The Problem of Determining the Forms of Teeth which shall work in right-line contact while transmitting rotation about fixed axes, presents itself in a new form when those axes lie in different planes.

All wheels with teeth which act in this manner must ultimately reduce to pitch surfaces whose class depends upon the relative positions of the axes; these will be tangent to each other along a right line, and move in contact of which the sliding, if there be any, is also along the common element. A third surface of the same class being placed in contact with these along the same line, will move in contact of the same nature with either or both; and if one pitch surface drive the other, the motion of the third is the same whether we regard it as derived from the first or the second. When the velocity ratio is constant, these will be surfaces of revolution—cylinders, cones, or hyperboloids; and in either case all three rotate in contact about fixed axes.

During such rotation, when the axes are parallel, an element of the third cylinder, by its motion relatively to the pitch cylinders, simultaneously generates the conjugate surfaces for the teeth of spur wheels. Now the cylinder is the limiting form assumed by the cone when the vertex is infinitely distant; and accordingly, when the axes intersect, the third cone in like manner describes the surfaces for the teeth of bevel wheels.

Again, the cone is but the special case of the hyperboloid in which the generatrix approaches infinitely near to the axis, and by pursuing the analogy we should reach the deduction that an element of the third hyperboloid, if taken as a describing line, will generate the conjugate tooth-surfaces for skew wheels.

369. This conclusion is plausible, and has received the indorsement of high authorities. Thus, Prof. Willis says : *

" The surfaces adapted for teeth in the case of rolling hyperboloids might be obtained in a manner similar to those of rolling cones, by taking an intermediate describing hyperboloid ; but it does not appear possible to derive from this any rules sufficiently simple for application."

He adds that a sufficiently close approximation may be made by drawing two cones normal to the hyperboloidal frustum selected, developing these, and after laying out teeth, as in Tredgold's method for bevel wheels, wrapping them back in their proper relative positions. The forms and proportions of the teeth traced upon the larger and the smaller cone respectively, are presumably the same ; and the tooth-surface is composed of right lines joining the corresponding points of

* Principles of Mechanism, p. 151.

their contours after the developments are restored to their original conical forms, in due relation to each other as determined by the generatrix of the pitch hyperboloid.

Prof. Rankine,[*] again, remarks: "The surfaces of the teeth of a skew-bevel wheel belong, like its pitch surface, to the hyperboloidal class, and may be conceived to be generated by the motion of a straight line which, in each of its successive positions, coincides with the line of contact of a tooth with the corresponding tooth of another wheel. Those surfaces may also be conceived to be traced by the rolling of a hyperboloidal roller upon the hyperboloidal pitch surface."

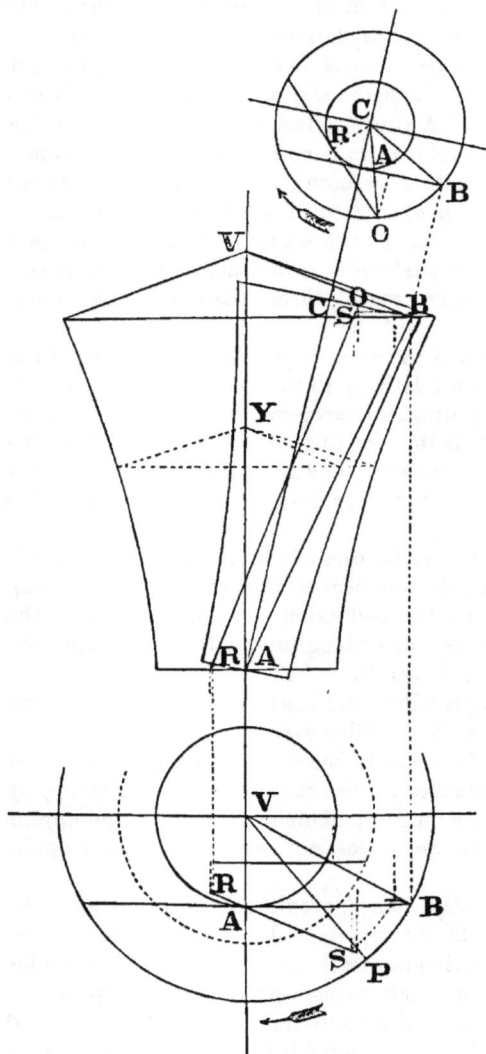

Fig. 227.

And he in turn proceeds to describe at some length a process of constructing the teeth which, although differing in detail from that above explained, gives nearly, and in some cases precisely, the same results; both are, then, to be regarded as approximations more or less close.

370. Direct Process of Construction.—But, assuming that the surfaces generated by the use of an auxiliary hyper-

* Machinery and Mill-Work, p. 146.

boloid are correct, it is nearly, if not quite, as easy to construct them directly as in the exact method for bevel wheels ; supposing the pitch surface and the generating surface to rotate in contact, the axes being fixed, we have simply to follow the movements of the describing line, and find the points in which it pierces the normal cones at various phases of the action.

In Fig. 227, V, Y are the vertices of the cones normal to the vertical pitch surface, and the inclined describing hyperboloid, touching the other internally, is shown in position for generating the flank of the tooth.

For convenience, the velocity ratio of these two hyperboloids is taken as 2 to 1 ; hence a rotation of the vertical surface through the angle BVP, will cause the other to turn through the angle BCO twice as great. The describing line, which was originally at AB, will then have the position RO, piercing the upper normal cone in some point S, which being found, occupies a known position in relation to the element VP ; and a point similarly situated with reference to the element VB, will evidently lie in the trace of the required flank upon that cone. By repeating this process any desired number of points in the curve may be located ; and applying it to the lower normal cone, we determine the outline of the smaller end of the tooth. Or by extending the describing line, and finding where it pierces the plane of the gorge circle, we may in a similar manner map out the trace upon that plane.

371. Next, the describing surface being placed in external contact along the same line, AB, and the normal cone being sufficiently extended, the face of the tooth is constructed in a similar manner, needing no illustration. The method of finding the point in which the describing line pierces the normal cone is shown in Fig. 228. V being the vertex of the cone, AB its base, and MN the line, draw through V a line, VO, either parallel to or intersecting MN; this line pierces the plane of the cone's base in O, and MN pierces it in P. Therefore the plane determined by MN and VO cuts the plane AB in the line OP, which, produced if necessary, cuts the circumference of the base in the point L ; now drawing VL, that element

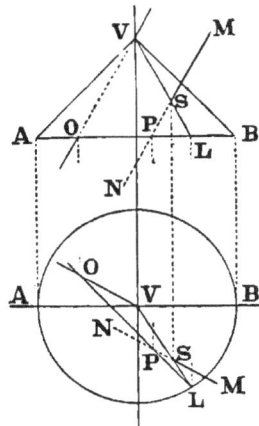

FIG. 228.

of the cone cuts the given line MN in S, which is the required point.

372. The Teeth not Symmetrical.—It will be perceived that the trace of the tooth upon the normal cone is derived from the intersection of that cone with the describing surface, just as it was in constructing the correct teeth for bevel wheels. In that case the axes of these two surfaces were in the same plane, which divided the curve of intersection symmetrically. But this is not so with the hyperboloid, and in Fig. 227 that curve will not be symmetrical with reference to the element VB of the normal cone. The result of this is, that the traces of the two flank surfaces, derived from the parts of the curve on opposite sides of the point B, will not be similar. The same holds true in regard to the face surfaces, so that *the trace of the complete tooth upon the normal cone will not be symmetrical to a radius,* as it is in the constructions of Prof. Willis and Prof. Rankine.

This may be seen from another point of view, thus : The inclinations of the describing line to the elements of the normal cone will not. vary in the same manner nor in the same degree in the opposite directions of its motion from the position AB. The difference of form is not great for the flanks, but is more conspicuous in the faces, as might be expected from their greater length and more rapid changes of curvature.

373. Determination of Height of Tooth.—In Fig. 229, let D be the axis of a wheel, perpendicular to the paper. LBO the base

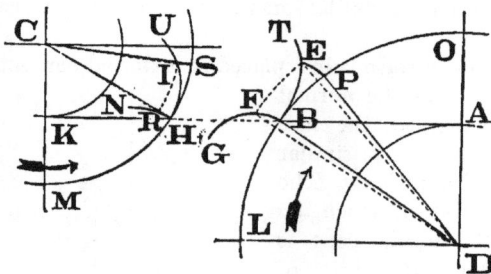

of the normal cone, AB the generatrix of the pitch surface, BT the intersection of the normal cone with the describing hyperboloid, f r o m which is derived the trace BFG of the face of the tooth. The ac-

Fig. 229.

tion of this face begins at B, and BT is the locus of contact ; hence if the angle of recess, BDP, be given, we draw through P a curve similar to BG, cutting BT in E, which will be the highest point of the face. Conversely, if the height BF of the tooth be assumed, an arc through F about the centre D cuts BT in E, and FDE or its equal BDP will be the angle of action between this face and the engaging flank of the other wheel.

Let C be the axis of that second wheel, also perpendicular to the paper ; HK the generatrix (coincident with AB when the wheels are in position); HU the trace, on the second normal cone, of the same describing hyperboloid that was used for the face of the first wheel : and HN the trace of the flank.   Now the turning of the first wheel through the given angle, BDP, causes the describing surface to turn through a known angle ; which enables us to find the point I in HU, corresponding to that angular motion.   Then an arc about centre C through this point I, cuts HN in R, which limits the *acting* depth of the flank. Or, otherwise ; the given rotation of the first wheel will drive the second through a known angle HCS ; then a curve through S, similar to HN, cuts HU in I ; and thus the acting depth is found without reference to the rotation of the describing surface.

The same process is to be repeated with respect to the face of the second wheel and the flank of the first, which completes this operation and determines the whole angle of action for each wheel. But the flank-curves are to be continued beyond the points thus found, because, as in any other form of gearing, clearing spaces must be provided in each wheel, for the passage of the teeth of the other.

374. Determination of the Form of the Blank.—When the height of the tooth has thus been ascertained, or assumed, the position of the describing line which passes through the outer point of its face is definitely fixed.   Now, this element of the tooth-surface, in revolving around the fixed axis of the wheel, generates a new hyperboloid, and the blank should be made in the form of a frustum of that surface, limited by the cones normal to the pitch hyperboloid.   For otherwise the teeth will begin and end their action at a single point, instead of along a line : the obvious result being that the points, especially of the follower's teeth, will suffer from abrasion.

If the wheels are located in the immediate neighborhood of the gorge planes, the true hyperbolic outline must be carefully followed in making the blank, in order to avoid this danger.   But if not, the hyperbola becomes so flat at a short distance from the vertex, that it will ordinarily suffice in practice to use a frustum of a cone tangent to this external hyperboloid, at the middle of the length of the tooth as measured along its outer element.

375. This is illustrated in Fig. 230 ; the dotted curve EF is the outline of the pitch surface, to which the cones whose vertices are V and Y are normal, and ON is the outline of the external hyperboloid, generated by revolving the describing line which passes through N the highest point of the tooth, about the vertical axis VX. NM being the length of the outer hyperbola to be used, let this be bisected at R, at which point draw a tangent to the curve ; this will intersect the axis at X, the vertex of the tangent cone.

The method of drawing the tangent to the hyperbola is also shown. Since the position of the generatrix of the outer surface with respect to the axis is known from the antecedent constructions, it will, when parallel to the paper as in the diagram, take a known direction, AB, asymptotic to the hyperbola ON. The companion generatrix then becomes the other asymptote AL; now draw through R a parallel to AB, cutting AL in F, and on AL set off $PC = AP$, then RC is the required tangent.

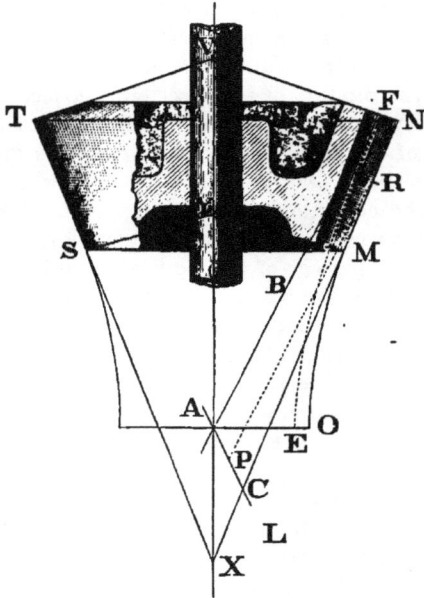

FIG. 230.

376. The forms of the teeth for a pair of wheels, laid out by the processes above described, are shown in Figs. 231 and 232, the wheels being seen from the smaller end; the long tooth on the right, in the upper part of each figure, is continued till cut by the gorge plane, the shorter teeth on the left being limited by an inner normal cone as in Fig. 230. In the lower part of each figure are shown the common element of the pitch surfaces, the trace of one side of the tooth upon the outer normal cone, and the parts of the curves of intersection used in the construction. The lines employed in the determination of the other side of the tooth are omitted in order to prevent confusion; but it is proper to state that these figures are copied from drawings executed with great care upon a large scale, in exact accordance with the methods above set forth. The conditions assumed are, that the projections of the axes upon a plane parallel to both, shall cross each other at an angle of 60°, as in Fig. 97, and that the wheels shall have eighteen and twenty-seven teeth respectively.

It is to be noted that in consequence of the peculiar relation between the pitch and describing surfaces, not only are the fronts and backs unlike, but the traces on the various normal cones are also dissimilar, the whole form of the tooth changing with the distance from the gorge plane.

377. Fallacy of the Preceding Construction.—The correctness of the above method of generating the conjugate tooth-surfaces for skew

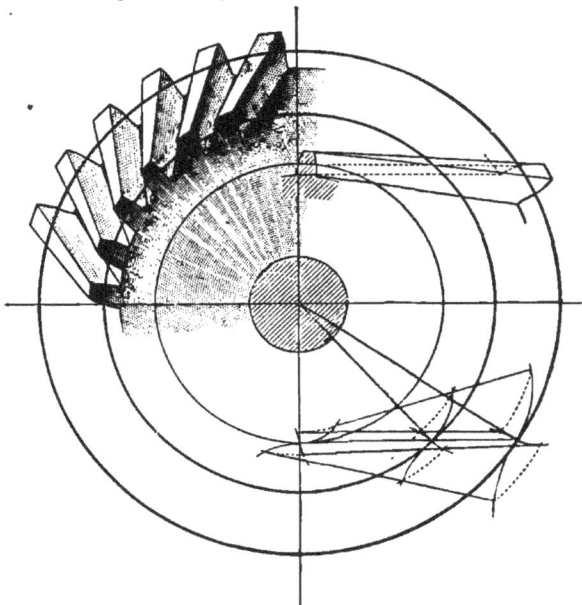

Fig. 231.

wheels has hitherto passed unchallenged; a fact perhaps not to be wondered at, so plausible is the analogical reasoning by which the deduction is reached.

Either one is unquestionably correct in relation to the describing hyperboloid. Let the upper and lower bases of the latter be two discs fixed upon one axis, and in these let a single wire be secured, to represent the describing line. If this axis be properly placed in relation to either pitch surface, it will

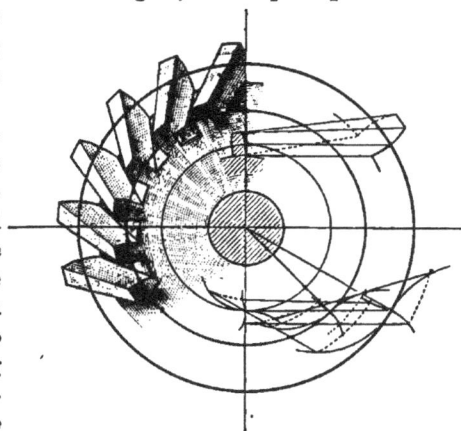

Fig. 232.

be correctly driven by the corresponding tooth-surface acting against

the wire, after the manner of a pin-wheel in spur gearing. And by
hypothesis the engaging surfaces have at every instant in common
this describing line, by which they were simultaneously swept up
during the action.

So far, well ; but one thing more is necessary—this common line
must be a line of tangency ; and it appears to have been assumed that
this is the case, probably because up to this point the analogy is perfect.
Every link in the chain has stood the test ; now let us examine the hook.

378. It is necessary, first, to gain a clear idea of what occurs when
one hyperboloid rolls around another which is stationary ; a question
not previously considered.

If the two surfaces shown
in Fig. 233 turn as shown by
the arrows (the axes being
for the moment regarded as
fixed), let AL, AI be the
linear motions of the points
which fall together at A ;
these will have the same component, AK, perpendicular to
the common element, AB,
and IL represents the amount
of sliding between the two
hyperboloids along that element.

Now suppose the motion
of the inclined surface to be
suddenly arrested, and the
axis of the vertical one to be
at the same instant released
from its bearings, its rotation
continuing. The latter must
then revolve around the axis
of the former in the direction
opposite to that in which the
former was turning, and with
the same angular velocity.

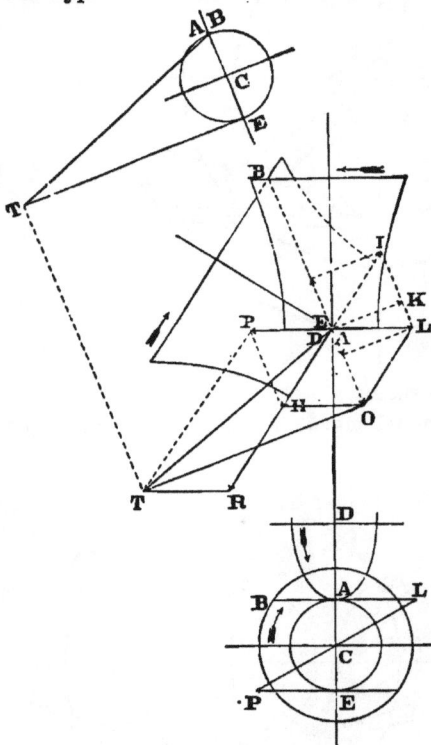

FIG. 233.

The point A of the vertical hyperboloid will then have a component
motion, AL, due to its original rotation, and another, AH, equal and
opposite to AI, due to the revolution.

The resultant of these is AO, in BA produced, and equal to IL.
This is as it should be, evidently, since the relative sliding of the two

surfaces must remain the same; but, moreover, the motion of every point in or connected with the travelling hyperboloid must have a component equal and parallel to AO. The only other motion of which any such point is capable is one of revolution about AB, and since the angular velocity must be the same for every point, it will suffice to find it for one.

The point E has a component motion, EP, equal and opposite to AL, due to the rotation; and another due to the revolution, in the direction of AH, but greater, in the proportion of DE to DA. The resultant is DT, the tangent to the helix which the point E is at the instant describing. Therefore OT, perpendicular to AB, is the linear velocity of the rotation of this point about the common element of the two surfaces, and $\dfrac{OT}{AE}$ is the angular velocity for every point in the rolling hyperboloid.

379. Precisely the same result has already been reached in a different manner, as will be seen by reference to Fig. 25, and the accompanying text. The nature of the action is most explicitly stated by Prof. Rankine, thus * : " If one of those bodies is fixed, and the other made to roll upon it, they continue to touch each other in a straight line, which is the instantaneous axis of the rolling body, and the rotation about that instantaneous axis is accompanied by a sliding motion along the same axis so as to give, as the resultant compound motion, a helical motion about the instantaneous axis."

If, then, we take any element of the rolling hyperboloid as a describing line, it is at any instant in the act of generating a helicoid, which will be tangent all along this line to the tooth-surface generated by a continuation of the rolling. Therefore the plane tangent to the latter surface at any point, is determined by the describing line itself, and the tangent to the helix at that point constructed as above.

380. Now in generating the conjugate tooth-surfaces for a pair of wheels, the describing hyperboloid rolls upon the outside of one pitch surface, and upon the inside of the other, as in Fig. 234. The direction of the rotation about the instantaneous axis is of course the same in both cases, but that of the sliding is not; in the rolling upon the inclined pitch surface the sliding is represented by AO, while in the rolling within the vertical surface it is represented by AP, in the opposite direction.

In Fig. 235 let the vertical line AB represent the common element of the pitch surfaces or instantaneous axis, and OE the describing

---

* Machinery and Millwork, p. 71.

line, the rotation being indicated by the arrows. The tangent to the helix described at the instant by the point P will lie in the plane tangent to the cylinder upon which that helix lies, and in the horizontal projection will have the direction $P'I'$ perpendicular to $P'A'$. Let the motion parallel to AB be upward, then PI will be the vertical projection of this tangent. Let MN be a horizontal plane; it cuts PI at L, and OE at O, giving $O'L'$ as the trace upon MN of the

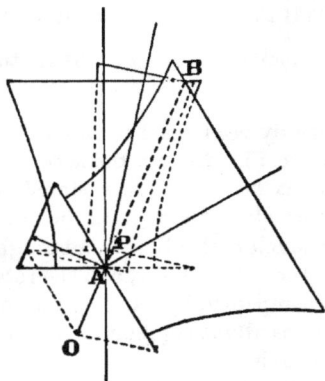
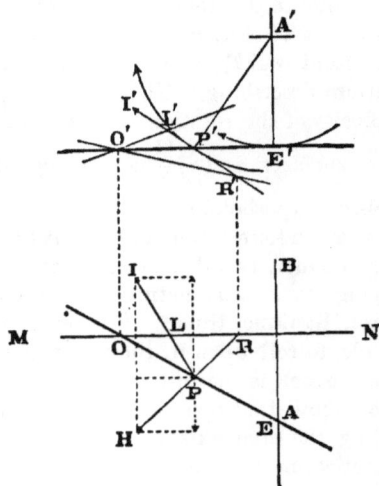

FIG. 234.   FIG. 235.

plane tangent to the helicoid at P. If, on the other hand, P have a downward motion parallel to AB, $P'I'$ will be horizontal and PH the vertical projection of the tangent to the helix, which pierces MN at R, giving $O'R'$ as the trace of this tangent plane, which cuts the first one in the line OE. A vertical plane through the moving line will evidently also contain the tangents to the helices described by the point E of that line, which is at the least distance from AB. At this point, then, and at no other, will the two helicoids have a common tangent plane.

381. Consequently the hook breaks, the chain is useless, and these teeth will not work at all. The common line of the engaging face and flank is a line not of tangency, but of intersection, and these surfaces, simultaneously swept into existence by the auxiliary hyperboloid, are instantaneously swept out again by each other. That this would occur in the special case in which the pitch hyperboloids retain their limiting forms, the one being a cone and the other a plane disc or rather annulus, as in Fig. 101, was pointed out by the author in a paper published some years since ; * but not at the time of that writing, nor until the present, was it perceived that the analogy fails in general, and that this whole method of constructing teeth for skew wheels is radically wrong.

382. A Practicable Method.—A different line of procedure must, therefore, be sought; and a basis of operations which will lead to reasonable results is found in a peculiar property of the involute, first pointed out by M. Théodore Olivier.

In Fig. 236, AB is the common tangent of the two circles whose centres are C' and D, and is, therefore, the locus of contact of the involutes $O'P$, PH, shown in contact at P. Let the first circle be now revolved on the line AB, as on a hinge, until its centre reaches the position C, the circle then appearing as an ellipse, and the tooth taking the position OP. The axes will then lie in different planes, but it is evident that OP will still drive PH exactly as before, AB being the locus of contact, since it is the intersection of the planes of the base circles, whatever their inclination to each other.

Therefore the action of the fronts of the teeth is not affected by the change in the position of the originally parallel axes ; but with regard to

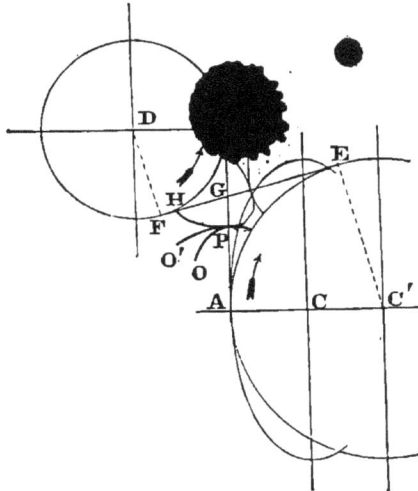

Fig. 236.

the backs the case is quite different.   Their locus of contact, when the circles were in the same plane, was the other common tangent EF, which cuts AB at G, and it will readily be seen that the two points of the backs of the teeth which would meet at G in the first place, will do so still, but no other points of those curves will come into contact at all.   Consequently, regarding the wheels as of no sensible thickness, we perceive that either of them can drive continuously in only one direction.

383. Of this combination Prof. Willis remarks, †

---

* Scientific American Supplement, Nos. 174, 176 and 178.
† Principles of Mechanism, p. 152.

"Involute wheels, therefore, may be employed to communicate a constant velocity ratio between axes that are inclined at any angle to each other, but which do not meet.   But the demonstration supposes the wheels to be very thin, since they coincide with the planes that intersect, and the invariable points of contact are situated in this line of intersection.   The edge of one of the wheels must be in practice rounded so that it may touch the other teeth in a point only."

This, however, is not the only expedient for securing substance or body for the teeth, although it is strictly true if both the curves are involutes.   But, retaining one of them, it is possible so to modify the other, when sensible thickness is given to the wheels, that the locus of contact shall not coincide with the intersection of the two planes, and to form teeth which shall work together in contact along a right line.

384. In Fig. 237, let AB, the common tangent to the bases of the

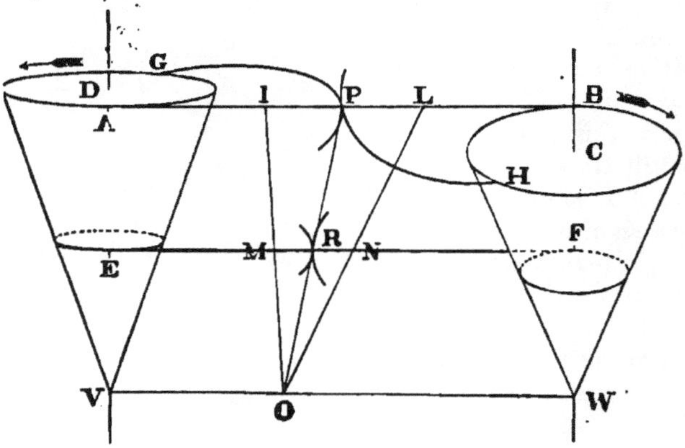

FIG. 237.

two cones, be parallel to the common perpendicular of their axes, whose extremities, V and W, are the vertices.   In the tangent plane thus determined, draw EF parallel to AB; it will be tangent to the two circles cut from the cones by planes parallel to their bases, and dividing proportionally both the axes and the elements.

The involutes PG, PH, of the bases, will work together as in Fig. 236; so too will those of the other sections, shown in contact at R. These curves may be considered as traced, each in its own plane, by marking points at P and R, drawn along by two cords, AB and EF, while the cones turn as indicated by the arrows; in which motion the linear velocities of the marking points are directly proportional to the radii of the bases of either frustum, and these to their distances from the vertex, measured either on the axis or on an element.

Draw PR and produce it to cut VW in O; then if P move to a new position I or L, the corresponding position of R will be M or N, such that

$$\frac{PI}{RM} = \frac{PL}{RN} = \frac{AV}{EV} = \frac{BW}{FW} = \frac{PO}{RO};$$

that is to say, the point O and the two marking points will always lie in one straight line.

385. The Tooth-surface a Single-curved One.—Consequently any number of tracing points may be placed between P and O, each in like manner generating a pair of involutes which will work together as in the preceding figure ; the point O itself describes two circles, which may be regarded as the involutes of their centres V and W. And the straight lines joining the corresponding points of the involutes belonging to either cone constitute a surface ; which may also be generated by sliding a right line upon any three of these involutes as directrices.

In the latter mode of generation, every point of the right line is at any instant moving in the same direction, because the tangents to all the involutes are parallel : but the upper end is moving faster than the lower end. Therefore, the consecutive elements intersect each other two and two, and the surface is not a warped one, but single curved, and the plane tangent to it at any point is tangent to it all along a right line.

But it does not follow, nor is it true, that the two surfaces thus simultaneously generated by the line PO will be tangent to each other. The plane tangent to either at any point is determined by the rectilinear element, and the tangent to the involute, which pass through that point. But the two involutes generated as above lie in different planes, their tangents have different directions, and in consequence the plane tangent to one surface is not tangent to the other. Retaining one of them, however, it is possible to construct another which shall be tangent to it in each of the different positions, by a process analogous to that of deriving a conjugate tooth from a given form.

386. Adaptation to Hyperboloidal Wheels.—The application of this in the case of hyperboloidal wheels will be understood by reference to Fig. 233 ; if through C and D two lines be drawn parallel to the common element AB, these lines, by revolving about the vertical and inclined axes respectively, will generate two cones situated relatively to each other as in Fig. 237, and the tooth-surfaces for the wheels may be formed as above explained.   Those for the first wheel, being composed of true involutes, may be constructed without reference to the second, as shown in Fig. 238, which is drawn without regard to practical conditions or proportions, for the purpose of exhibiting distinctly some peculiarities which must be taken into account in laying out such teeth for actual use.

Let *KIM* be the upper base and *EOF* the gorge circle of the hyper-

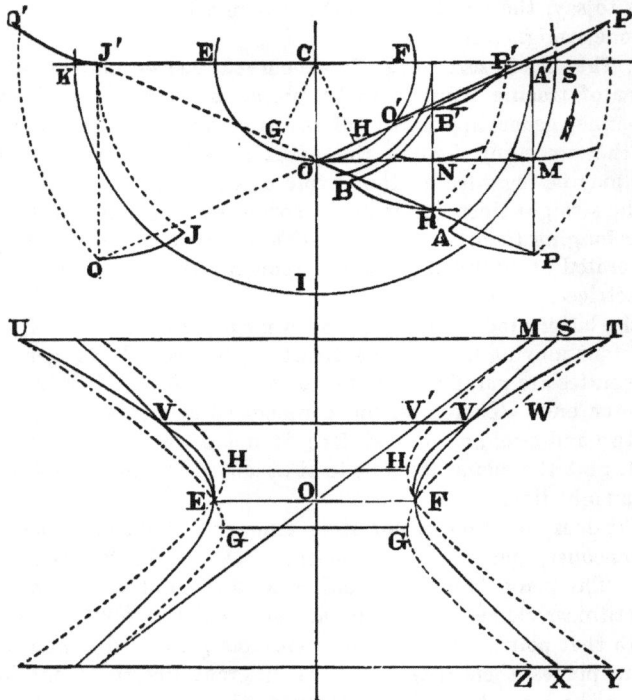

Figs. 238 and 239.

boloid, whose generatrix, *OM*, is parallel to the vertical plane, *KS*, containing the axis *C*.   Through *M* draw a perpendicular to *OM*, piercing this plane at *A'* ; then the circle about *C* through *A'* will be the base of the cone.   On *A'M* produced take any point *P*, and let *PO* be the position of the generatrix of the tooth-surface at the beginning of the action.   The wheel turning as shown by the arrow, let the action continue until a marking point at *P* reaches *A'*, tracing the involute, *PA*, whose position at the end is *P'A'*.   Any other point *R* on $P\,O$ will meantime trace another involute, RB, whose final position is $R'B'$, while O will trace the arc OO' ; and the final position of the generatrix is $OB'A'$. When P has advanced to M, R will have reached N, and OM is, it will be noted, tangent to all the involutes traced by points upon PO.

387. The Teeth Vanish at the Gorge Plane.—In Fig. 239, MO is the generatrix, and $SVFX$ the meridian outline, of the pitch hyperboloid.

Now, considering first the part which lies above the gorge plane EF, we perceive in Fig. 238 that the outer points of the teeth, as P, R, lie in the line PO, whose least distance from the axis is CG, and G is below the plane EF. Similarly, the lowest points, A', B', lie in a line, $A'O$, of which the point, H, nearest to the axis, lies above that plane. These two lines generate two other hyperboloids, whose meridian outlines in Fig. 239 are respectively TWF and MHF.

The teeth then vanish at the gorge circle, but if the generatrix be continued below it, they will reappear ; and their highest and lowest points will also lie upon an exterior and an interior hyperboloid. It is here to be noted, that it is not necessary thus to prolong the same generatrix ; but if it be done, as in these diagrams, the interior hyperbola, FGZ, will be a continuation of the exterior one, TWF, and FY an extension of MHF. In this case it is also necessary to make the angle POM in Fig. 238 equal to the angle MOA, if it be required that the teeth of the upper and lower portions shall be symmetrically disposed, so that for example $Q'J'$ may be equal, similar, and opposite to $P'A'$. Such symmetry, though not essential, is obviously desirable if the wheels are to be used in double pairs (**161**) ; but the angles just mentioned need not be made equal ; and if not, the result will be that in Fig. 229, although TWF and FGZ will be respectively similar to FY and FHM, they will be parts of different hyperbolas.

388. Limit to the Prolongation of the Tooth.—It is to be understood that in Fig. 238, as in previous cases, the vertical plane containing the axis C is parallel to the axis of the engaging wheel, the point O lying upon the common perpendicular. If then the angle MOA' be assumed, the point A' in which the generatrix of the tooth-surface pierces that plane limits the distance of the upper base from the gorge circle. For the transverse plane through that point cuts from the fundamental cone a circle to which PMA' is tangent, so that A' is the root of the involute traced by the marking point in that plane ; and a marking point in any higher plane, in the prolongation of OP, would advance beyond the point of tangency between its path and the corre-

17

sponding section of the cone, and thus begin to trace the reverse branch of the involute of that circle, which is clearly inadmissible. Conversely, if the distance from the gorge be assumed, the point A' is thereby determined, which in turn fixes the greatest possible value of the angle $A'OM$.

We have thus far referred only to the lowest *acting* points of the teeth which lie upon the interior hyperboloid in Fig. 239. But it may be necessary to cut more deeply into the blank in order to provide clearing spaces for the passage of the teeth of the engaging wheel. The involutes are then to be continued as far as required below the last acting points; should the root of the involute have been already reached in any transverse plane, as at A' in Fig. 238, the continuation in that plane will be a radial line.

389. Construction of the Back of the Tooth.—The complete working teeth must be provided with backs as well as fronts; and we have now to consider the results of this operation. In Fig. 240 we have a

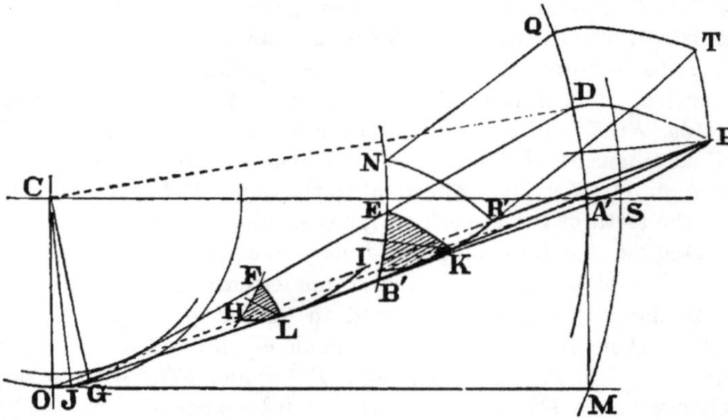

Fig. 240.

single tooth upon a larger scale, the curves $A'P'$, $B'R'$, corresponding to those similarly lettered in Fig. 238. For our present purpose we may disregard the clearing space, and suppose the depths determined by the points A', B', to be sufficient. Considering $A'P'$, then, as the front of the tooth, its back is to be a similar involute in the same plane, and for illustration we will assume the pitch to be such as to make the tooth pointed, as shown; then the radius through P' will bisect the arc $A'D$, which measures the thickness of the tooth on the circle through A'.

Make the angle OCG equal to the angle $A'CD$; then the arc QG is the thickness of the tooth at the gorge circle, and the lowest (acting) points of the backs will lie upon the right line GD. Thus, a circle through B', about C', cuts GD in E, and $B'E$ is the thickness of the tooth in the plane containing the involute $B'R'$. The radius which bisects $B'E$, cuts $B'R'$ at K, through which point passes the reverse involute, EK, or back of the tooth in that plane; in a like manner we determine the section HLF of the complete tooth by a plane still nearer the gorge, and so on indefinitely.

390. The Extension of the Tooth Practically Limited in Both Directions.—It will be observed that the points of these sections become more and more obtuse, and the depths less and less, as we approach the gorge plane. And in that plane, bisecting OG by the radius CJ, the front of the tooth is the arc OJ, and its back is the arc GJ; these being, as previously suggested, the involutes of the centre, C.

If then we make the section of the tooth by the plane farthest from the gorge pointed, we perceive that the fronts of the sections nearer to the gorge cannot be extended to the full heights, as I, R', found as in Fig. 238, since they are intersected by the backs before those heights are reached. And a sufficient number of these intersections, as L, K, being determined, a line, $JLKP'$, is drawn through them, which, by revolving round the axis, C, will generate a surface which is evidently the correct blank for the wheel upon the above supposition.

The meridian outline of this surface is the curve TVF, in Fig. 239, which, it is to be noted, *intersects* the pitch hyperboloid at V. But OM, of Figs. 238 and 239, the generatrix of that hyperboloid, is also a line of the front tooth-surface. This is now cut off at V' by the transverse plane through V, and it is clear that between this plane and the gorge the teeth cannot possibly engage with each other. Practically, then, they should terminate still farther from the gorge circle.

391. Suppose now, however, that the pitch is so much increased that the section of which $B'R'$ is the front may be carried up to the full height and made pointed at R'. The reverse involute through R' will cut the circle through B' at a point N; the angle OCG must then be made equal to that measured by the arc $B'N$, and a straight line from this new position of G, through N, of which NQ is a portion, will contain the lowest acting points of the backs under the new condition, as did GD under the old one. Below R' the state of things will be analogous to that which previously existed below P'; and we may proceed to construct as before a curve corresponding to $JLCP'$, and to ascertain the limit beyond which the teeth cannot be continued in the direction of the gorge.

But above R' we shall have, in the first place, a right line, $R'P'$, which is, of course, a portion of $P'O'$ of Fig. 238, the generatrix of the exterior hyperboloid, TWF, of Fig. 239, which is now the form of the blank. In the next place the tooth above R' will not be pointed, but will have a sensible thickness at the top. Thus, the arc $A'Q$ being made equal in circular measure to $B'N$ and QT similar and opposite to $A'P'$, the tooth outline on the upper plane is bounded at the top by the arc $P'T$. Other planes perpendicular to the axis may be passed at any convenient intervals, and the arc limiting the section of the tooth at the top being found in like manner for each, the line $R'T$ is determined, which is the outer edge of the back of the tooth, and lies, of course, upon the hyperboloid generated by $R'P'$. It is not a straight line, but in most, if not all, practical cases, its curvature within the limits employed will be very slight.

392. Nature and Action of the Back of the Tooth.—Evidently, when we thus assume any plane as the lower base of the frustum to be used for a wheel, we need not make the section of the tooth by that plane pointed ; for instance, we might have made $B'M$ greater, so that the lower, as well as the upper, section would have been blunted. This, however, would not in the least affect the mode of proceeding, nor the nature of the surface which forms the back of the tooth. This surface can not be generated as was that forming the front, as will be seen from the consideration that the tangents to the involutes at D, E, and F, in Fig. 240, are not parallel. Therefore it is not of single curvature ; it is not of double curvature, for it contains the right line GD ; it is, then, a warped surface, and generated by sliding the right line upon any three of the involutes as directrices.

Nor does it work in right-line contact with the engaging tooth of the other wheel. Supposing the latter for the moment to be of the same nature, it will be seen, by reference to Fig. 236, that the only point of contact between the involutes of any two transverse sections of the fundamental cones which have a common tangent (which may be called conjugate sections), is found by revolving either of those circles about that line until it lies in the plane of the other, and then drawing the line of centres, which will cut the tangent at the point in question.

Were the wheels made up of laminæ, each acting upon its conjugate only in this manner, we should have a series of such contacts, the first occurring in the planes nearest the gorge, the last in those most remote ; the teeth touching each other in a single point, which would travel endlong, in a line parallel and near to the common element of the pitch hyperboloids ; very much as in the case of twisted spur or bevel wheels.

393. Peculiar Action of the Backs of the Teeth.—This, however, is not the case, as the supposed laminæ are not capable of such independent action. Let a conjugate pair at any instant touch each other as above ; then at the next instant, the consecutive point of one section will come into contact with the engaging tooth at a point not situated in the conjugate plane but in a plane beyond. But since the above described contacts between conjugate planes occur at successive instants, we have in effect that successive phases' of the action are simultaneously represented, the result being that the two teeth at any given instant touch each other along a curved line of limited length ; each point of which during the action travels along the tooth from end to end in the manner previously explained.

394. Construction of the Back of the Conjugate Tooth.—Under these circumstances, if the back of the tooth of one wheel be made up of the reverse involutes, as at first supposed, it is not demonstrable that the back of the engaging tooth should be, but its proper form may be determined as follows : pass any plane perpendicular to the axis of the second wheel, and upon it draw the outline of the section cut by it from the assumed tooth of the first. Rotate this plane through a small angle, and turn the first wheel through the corresponding angle, determined by the given velocity ratio. The outline of the new section of the given tooth is now to be traced upon the plane, and the process repeated until the section of the conjugate tooth, which, of course, is the envelope of the various curves thus traced upon the perpendicular plane, is mapped out with the desired degree of accuracy. A new transverse plane is now to be passed, and the same series of operations again performed ; and this once more repeated, gives us finally three directrices, upon which a right line being made to slide will generate the back of the required conjugate tooth.

395. Construction of the Front of the Conjugate Tooth.—The front also of this conjugate tooth, as stated in **(385)**, will be different from that of the first wheel, whose generation was illustrated in Fig. 238. Referring to that diagram, it is seen that in every position of the generatrix, the tangents to all the involutes are parallel to OM ; and they are all parallel to the gorge plane. Therefore OM is the trace upon the gorge plane of the plane tangent to the tooth-surface at any phase of the action ; the latter plane turning upon OM as upon a hinge, during the rotation of the wheel. And in each of its positions it will cut a right line from any transverse plane of the second wheel, which, meantime rotating about its own axis, will thus have traced upon it a series of lines, to which the front outline of the section of the conjugate tooth by that plane must be tangent.

Thus, in Fig. 241 $O'C'$ is the vertical axis of the first wheel, corresponding to C of Fig. 238, $O'M'$ corresponds to OM of the same
diagram, $O'D'$ is the axis of the second wheel, and the right-hand
portion of this figure is a projection upon a plane perpendicular to
the inclined axis; in which PAB is the path of a marking-point which
generates an involute pertaining to the first wheel, this path corresponding to PMA' of Fig. 238. The arrow indicating the direction of
the rotation, P is already the lowest possible point of the acting face
of the tooth of the second wheel, and PM is the position of the plane
tangent to the tooth of the other at this phase. When P has moved
to A this plane will have the position AM, and the radius DP will be
found at DE, the arc PE being equal to PA. So when P reaches B,
the plane will appear as BM, and DF will be the new position of DP,
the arc PF being equal to PB. Thus the outline of the tooth being

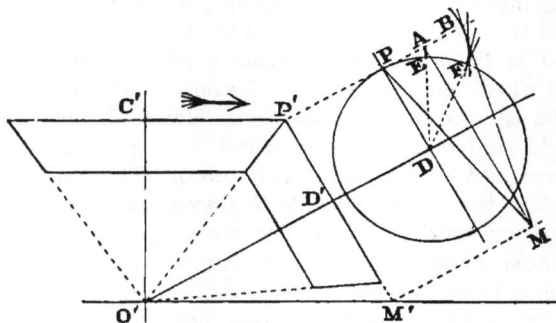

FIG. 241.

tangent to the trace of the plane in each of its successive positions,
we are enabled to map out its form as shown at FB.

Evidently this curve is not a true involute, which could be produced
only on the supposition that the tangent plane should always be parallel to AM; but in most practical cases the angle PMB will be small,
and the deviation of the tooth-outlines from the involute form will be
comparatively slight.

396. Different Actions on Opposite Sides of the Gorge.—A singular fact
in relation to the teeth thus constructed is that on opposite sides of the
gorge plane the fronts and backs are transposed, as will be seen by reference to Fig. 238; so that if a double pair of such wheels be employed,
turning in a given direction, the work of one pair will be done by the
single curved surfaces of the teeth, that of the other pair by the
warped surfaces, the distribution being reversed if the directions of
the rotations be changed.

Due attention should be paid to this point if a single pair only is to be used, and unless the arrangement of other parts of the machine prevents, the wheels should be placed on that side of the gorge plane which will secure the advantage of having the acting surfaces of the teeth touch each other throughout their length.

397. Practically, then, this construction enables us to make a pair of skew wheels which in one direction of the driver's rotation will work in right-line contact. The location of the upper bases of the hyperboloidal frusta being arbitrary, they should be placed as far as

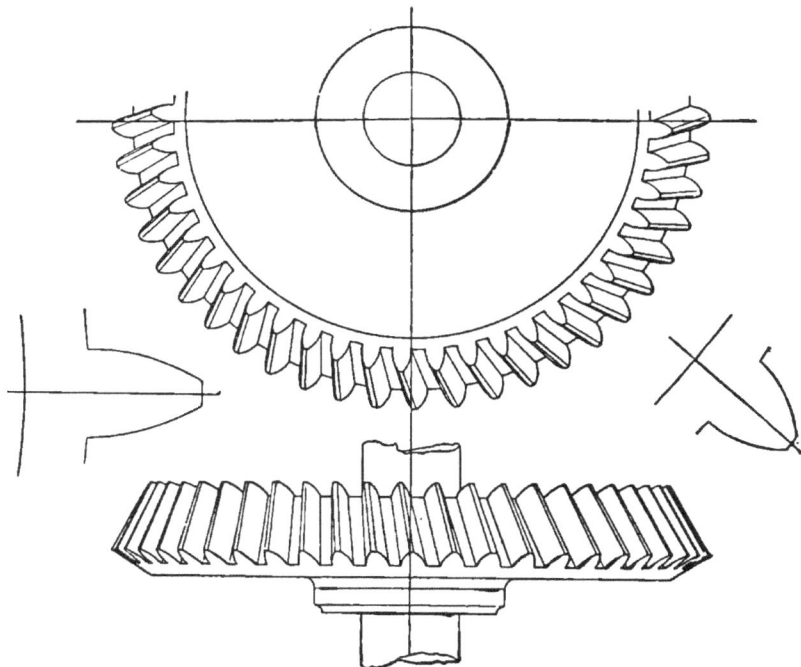

FIG. 242.

possible from the gorge planes, in order to reduce the transverse obliquity to the lowest limit.

The blanks should, of course, be terminated by cones normal to the pitch surfaces, as previously explained ; the traces of the teeth upon these cones are most readily found by passing a series of transverse planes, and drawing in each the outlines of the tooth ; which will intersect the circles cut from the cones in points of the required curves.

The appearance of the complete wheel as thus laid out is shown in Fig. 242, which represents the larger of a pair having respectively 48 and 32 teeth, the angle between the projections of the axes being 60.° On the right is shown the development of the tooth-outline on the outer normal cone of the larger wheel, the corresponding development for the smaller one being given on the left. These outlines, it will be observed, are by no means symmetrical, owing to the transverse obliquity of the teeth, although in one wheel the sections by planes perpendicular to the axis may be symmetrical with respect to the central radii.

398. It will be apparent on reflection that the analogy between conical and hyperboloidal wheels, in order to be perfect, requires that the teeth of the latter *should* vanish at the gorge, as do those of the former at the vertex. It is true that models have been made, as, for instance, those by Schroeder of Darmstadt, in which teeth of sensible magnitude are given to wheels whose mid-planes coincide with the gorge circles of hyperboloids, whose elements are represented by wires passing through the teeth ; it is also true that these wheels work, transmitting the rotation with perfect constancy of velocity ratio.

But it does not follow that these are true skew teeth, nor that they work in right-line contact at all on either side. These wheels, as made, are thin ; and, as will be seen presently, screw teeth formed upon pitch cylinders tangent to these hyperboloids at the gorge circles will curve so little in the small portion used, that it would be difficult by mere inspection to detect the curvature or to ascertain whether contact existed in more than one point. And in fact such teeth are very often, if not always, actually made by means of a milling cutter travelling in the direction of the tangent to the helix at the mid-plane of the wheel, without rotating the blank during the operation.

But these wheels cannot be made of any considerable length in the direction of the axis, and the suspicion of their identity with those in the models above mentioned has yet to be removed by the production of a pair in which straight-line teeth extend past the gorge circles from end to end of long hyperboloids.

399. Skew-bevel wheels are not often met with in practice. The usual expedient, when two axes lie in different planes, is to introduce a counter-shaft, whose axis intersects both the others, and to use two pairs of bevel wheels. And when they are at a great distance from each other this may be unavoidable ; but if they be not, there can be no question that the loss of power due to the imperfect rolling of the

pitch surfaces, and the transverse obliquity of the teeth, with a pair
of properly constructed skew wheels would in many cases be less than
that incurred when the arrangement above mentioned is adopted ; to
say nothing of the superiority of the single pair in respect to neatness,
lightness, and compactness.

400. Twisted Skew Wheels.—If we suppose a pair of tangent hyperboloids to be made up of a series of transverse laminæ, it is clear that
by twisting them uniformly, as in
Fig. 243, the rectilinear elements
will be changed into hyperboloidal
helices, while the surfaces still touch
each other along a right line as
before. And had teeth been added
previously to the twisting, then, as
was the case with the spur and the
bevel wheels, these teeth would continue to act with the proper velocity
ratio. .

This at first glance appears a useless
addition to an already ample degree
of complexity, but upon closer examination it will be seen that such twisted
teeth can actually be made more easily
than straight ones. The hyperboloidal helix can be traced by supposing a marking point to travel uniformly along an element, while the
surface turns uniformly upon its axis.
And if the teeth begin and end their
contact upon that line, as in Dr.
Hooke's spiral wheels, this will be the

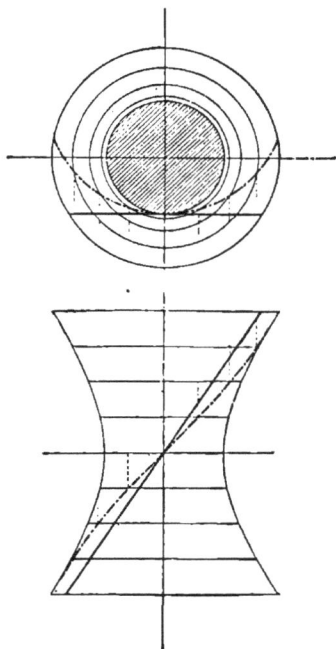

FIG. 243.

only line of their surfaces whose form is of essential importance.

Practically, therefore, it is requisite merely to arrange proper mechanism for simultaneously moving a milling cutter along the line of an
element of the pitch surface, and rotating the blank upon its axis,
both motions to be uniform.*

The same mode of proceeding holds good when, as in Fig. 101, the
hyperboloids retain the limiting forms of a cone and a plane, for if
we imagine the disc to be made up of as many concentric rings as
there are laminæ in the cone, each one will be driven round by the
twisting of the latter, so that the original line of tangency on the

---

* Scientific American Supplement, No. 178.

plane will be distorted into a curve of a spiral form, which may also be traced as above described by the uniform motion of a marking point along the right line of contact, while the disc rotates uniformly about its axis.

The teeth thus formed, it is evident, will have but a single point of contact, which will travel along the common element of the pitch surfaces, just as in the case of other twisted wheels; but this is sufficient to make the action continuous, if the amount of twist in the length of the tooth be a little greater than the pitch; and the action will be peculiarly smooth, since the amount of sliding friction will be no greater than that between the pitch surfaces.

The Teeth of Screw Wheels.

401. The most common example of Screw Gearing is the arrangement familiarly known as the Endless Screw, or Worm and Wheel. In this case the axes are situated in planes which are perpendicular to each other, and the nature of the action will be readily seen by inspection of Fig. 244. *C* being the centre of a pitch circle, and *TT* the pitch

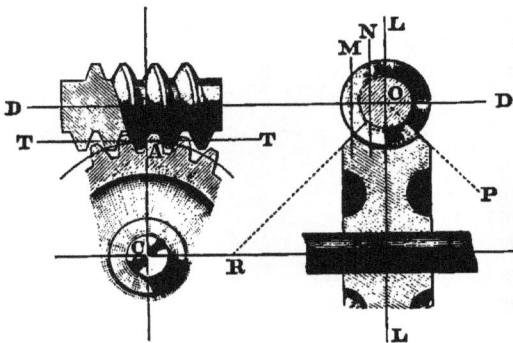

Fig. 244.

line of a rack, let teeth be constructed of any of the forms proper for spur gearing. In the plane of the paper draw any line *DD* parallel to *TT*, and taking it as an axis, let the outline of the rack be made the meridian section of a screw whose pitch is equal to that of the rack teeth. This screw is still a rack, and if moved endlong will turn the wheel; but if instead of this the screw itself be turned, the effect will be precisely the same. For, supposing the wheel to be very thin, its tooth is confined between the threads of the screw, all of whose meridian sections are alike, but each successive one is in advance of the preceding, so that when the screw has made one revolution, the wheel-tooth must have been driven through an angle measured by the pitch arc. In short, the screw is a rack which advances by rotation; and this is the fundamental principle of all screw gearing, with the exception of one combination, which will be described hereafter.

402. Distinctive Peculiarities of the Action.—The line TT, by revolving about the axis DD, generates the pitch cylinder of the screw, which is tangent to that of the wheel at the point A.

Three characteristic features distinguish the action from that of twisted gearing, viz. :

1. The velocity ratio is independent of the relative diameters of the pitch cylinders, and depends wholly upon the screw pitch.

2. The directional relation depends upon the direction of the twist ; the screw, turning in a given direction, will drive the wheel one way if right-handed, the other way if left-handed.

3. The rotation of the wheel is caused solely by the end thrust of the screw.

403. Wheels with Similar Transverse Sections.—In giving sensible thickness to the wheel, we may proceed as follows : the elements of the two pitch cylinders which pass through A determine the common tangent plane represented by MN, in Fig. 245. The screw helix through A will develope upon this plane into a right line, which, when the plane is wrapped upon the pitch cylinder of the wheel, will become another helix lying on that surface ; these two helices will be either both right-handed or both left-handed. Through each point in the outline of the wheel-tooth, already laid out, draw a helix of the same pitch ; we shall thus have a wheel precisely like one of the twisted pair shown in Fig. 109, all the transverse sections being similar.

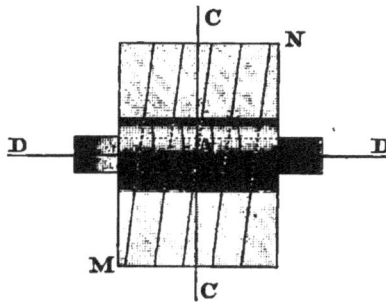

Fig. 245.

The thickness of this wheel is to be determined only by considerations relating to the pressure to be transmitted and the strength of the material; it has no bearing upon the kinematic action, since at any instant each tooth touches the engaging screw-thread only in a single point in the original transverse plane through the axis of the screw. Therefore, this form of wheel is open to the objection that the wear will be comparatively rapid, being confined to one line upon each thread and each tooth.

404. Close-fitting Tangent Screws.—A worm wheel to which this objection does not apply can be practically made in this manner : an exact copy of the screw in steel is notched and hardened so as to become a cutting tool, which is used to finish the teeth, usually roughly cut upon the blank with an ordinary milling cutter. The cutting screw is often made to drive the worm wheel during this operation ; but a better plan is to have the wheel-blank driven at the proper speed by independent means.

When this method is adopted it is necessary, after taking one cut, to press the axes nearer together, and then take a lighter finishing cut. Therefore an involute wheel-tooth, working with a straight-sided sloping rack tooth, is to be preferred, because this change in the position of the axis does not affect the velocity ratio.

In this manner a perfect worm wheel is practically made with great facility. The accurate delineation of it is more difficult and tedious, but it can be made by a process illustrated on the right in Fig. 244, where is shown a section by a plane through the axis of the wheel perpendicular to that of the worm. *LL* is the mid-plane of the wheel, in which the teeth were first laid out, as above explained.

405. If now we pass any plane as *N* parallel to *LL*, it will cut from the screw a curved line ; this being taken as a rack tooth, the form of the wheel tooth which will work with it may be determined by the process of Fig. 162, and this will be the outline of the wheel tooth in that plane. Another parallel plane at *M* will give a different section of the screw, from which we derive, as before, the conjugate form of the wheel tooth, and this may be repeated as many times as is deemed necessary.

The blank for the wheel is usually of the form shown ; the line *OR* describes a cone, from which the parallel planes, *M* and *N*, cut circles, and the intersections of these circles with the outlines of the wheel-tooth in these planes will be points in the visible contour of the tooth.

406. Superior Action of the Close-fitting Screw.—Not only are all meridian sections of the screw alike, but all sections by planes parallel to and equidistant from its axis are alike. The whole screw being a rack which advances by rotation, it is clear that at each instant there will be points of contact not only in the plane *LL,* but in other consecutive planes as *N, M,* etc. These points constitute a line of contact which, though not a true helix, will evidently be a line of double curvature of kindred nature, and during the rotation it will travel along the wheel-tooth from the point toward the root. Between each tooth and its thread, then, we have contact along a line, and the wear is distributed over a surface.

407. Practical Proportions.—Abstractly considered, both the diameter of the worm and the number of teeth in the wheel are optional. But it is found that in practice the results are not satisfactory if the wheel has less than from twenty-five to thirty teeth ; with a straight-sided rack and involute wheel-tooth, the obliquity being 15°, thirty-six teeth

will give a total angle of action greater than twice the pitch, with the
arc of recess one and a half times as great as that of approach, which
is all that could be desired. This is, of course, for a single-threaded
worm ; in regard to its diameter, a simple and a good practical rule
is to make the radius of the blank about twice the pitch. If the wheel
be a simple twisted one with all its transverse sections alike, its blank
will be a cylinder, whose thickness should be from two and a half to
three times the pitch ; if it be cut by the screw, and of the form
shown in Fig. 244, the angle ROP may be made from 60° to 90°, the
thickness being as just given.

408. Physical Embodiment of Sang's Theory.—The employment of
the screw to cut its own wheel at once suggests the formation, in the
same manner, of guide templates suitable for use in connection with
the pantagraphic cutter engine of Pratt & Whitney. In cutting such
a template the screw practically and automatically executes the process
of finding the form of a tooth conjugate to that of a given rack, as
illustrated in Fig. 162.

Now if, as in Fig. 244, we use a straight-sided sloping rack tooth,
the result will be the formation of a series of involute templates ; if
cycloidal arcs be substituted for straight lines, as in Fig. 120, we shall
have a set of templates for epicycloidal teeth with a constant describing circle. But the outline of the screw-thread may be made of any
other reasonable form ; and according to Sang's Theory (**283**), if it be
bounded by any four similar and equal curves in alternate reversion,
the series of templates produced will be interchangeable, and thus
by means of the pantagraphic engine, the cutters for wheels upon any
desired basis or system may be readily and accurately duplicated.

409. Multiple-threaded Screw Wheels.—Thus far we have supposed
the screw to be single-threaded, with a pitch equal to that of the fundamental rack tooth.

Now the helical pitch may be doubled,
as shown in Fig. 246 ; this will double
the angular velocity of the wheel, but
if no other change be made, the alternate
teeth only will come into action. This
difficulty is obviated by making the screw
double-threaded, as in Fig. 247 ; which
at the same time reduces by one half the
pressure upon each tooth.

In like manner we may make the helical pitch three, four, or any whole num-
ber of times as great as the tooth-pitch, increasing the number of

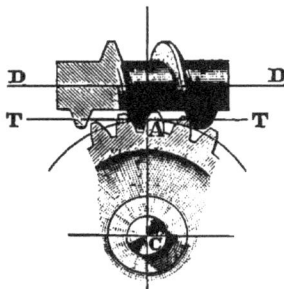

Fig. 246.

threads accordingly, and taking care to make the diameter great enough to avoid excessive obliquity of action.

And in this way we may give to the screw as many threads as there are teeth upon the wheel, or even more ; the combination then having but slight resemblance to the single-threaded endless screw, as will be seen by referring to Fig. 110, which represents a pair of screw wheels properly so called.

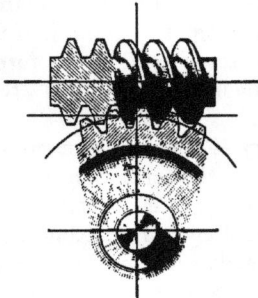

Fig. 247.

When the numbers of the threads and teeth are equal, Prof. Willis states that the two wheels may be made exactly alike ; * this we imagine to be a mere slip of the pen, since that eminent writer was the first to point out the true construction, which requires the section of the screw thread to be a *rack* tooth, although it need not be disputed that the difference would not be conspicuous, nor that if they were exactly alike they would engage and transmit rotation, but with a slight fluctuation in the velocity ratio.

410. Screw and Rack.—There is no limit to the increase in the number of teeth upon the worm wheel, and if it be made infinite, the wheel becomes a rack, which if cut by the screw itself, will be identical with a portion of an ordinary nut.

Now the exterior surface of the screw and the interior surface of the nut are precisely the same ; and this affords an illustration of the extreme case in deriving the conjugate to a given rack tooth, mentioned in (**282**), for if we split the nut and screw longitudinally through the axis, the meridian sections will correspond to the two conjugate racks shown in Fig. 163, being exactly converse to each other.

Oblique Screw Gearing.

411. The axis of the screw thus far has been supposed to lie in the plane of rotation of the wheel. This, however, is not essential, for even if it cross that plane obliquely, rotation can still be transmitted with a constant velocity ratio, by the end thrust of the screw.

And in this new relative position of the axes, as before, the screw may have two, three, or any number of threads, and thus we pass from a simple endless screw to the disguised forms of oblique screw wheels.

---

* Principles of Mechanism, p. 163.

The fundamental principle remains unchanged, however ; the screw is still a rack which advances by rotation, and the first step in the construction is to determine the form of its conjugate tooth with reference to the new conditions.

412. The Oblique Rack and Wheel.—That this is the case, may be perhaps most readily seen by first considering the things which may be accomplished by the rack and wheel alone.

Ordinarily, as is well known, the rack travels in the plane of rotation of the wheel. But this, again, is not a matter of necessity ; without the slightest change in the forms of the teeth, it may be made to travel obliquely across that plane, the velocity ratio remaining absolutely constant, although its value will be changed.

This is illustrated by Fig. 248 ; if we suppose a rack to be made by cutting teeth across a broad rectangular plate, *MN*, indicated by the dotted lines, it will at once be seen that it cannot only move from right to left, causing the wheel to turn as usual, but is also free to slide in the direction of the axis, and that it may receive both motions at once.

If now a strip be cut diagonally from this broad rack, and made to travel by guide rollers, as shown, or by any other means, in the direction of that diagonal, the effect is precisely the

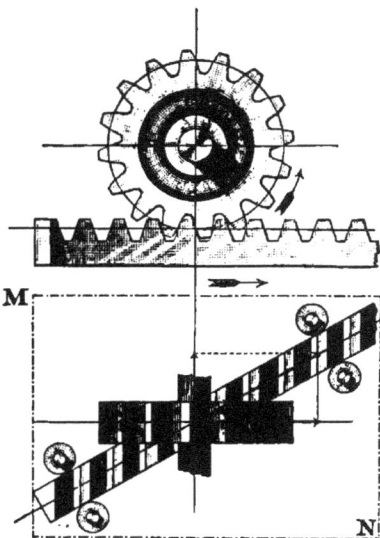

Fig. 248.

same. The action between the teeth is unchanged ; but assigning a definite linear velocity to the rack in the new direction, that motion may be resolved into two components, one lying in the plane of rotation, the other perpendicular to it. The latter does not affect the wheel, but the former does, and causes it to rotate ; and the linear velocity of the pitch circumference is equal in magnitude to this effective component.

It will readily be seen that another rack may be cut upon the back of this one, with teeth perpendicular to its sides, which may engage with another wheel in the ordinary manner ; and thus we have a new means of transmitting a limited rotation with a constant velocity ratio between two axes in different planes, by the use of common spur gearing only.

413. The resolution above mentioned is represented in Fig. 248; the obliquity of the rack's travel, when assigned, gives the direction of the resultant, and if the component in the plane of rotation be made equal to the pitch arc, the magnitude of the resultant determines the distance through which the rack will advance while the wheel turns through the pitch angle.

Let us assume this distance as the helical pitch in constructing from the rack an oblique single-threaded worm, as shown in Fig. 249,

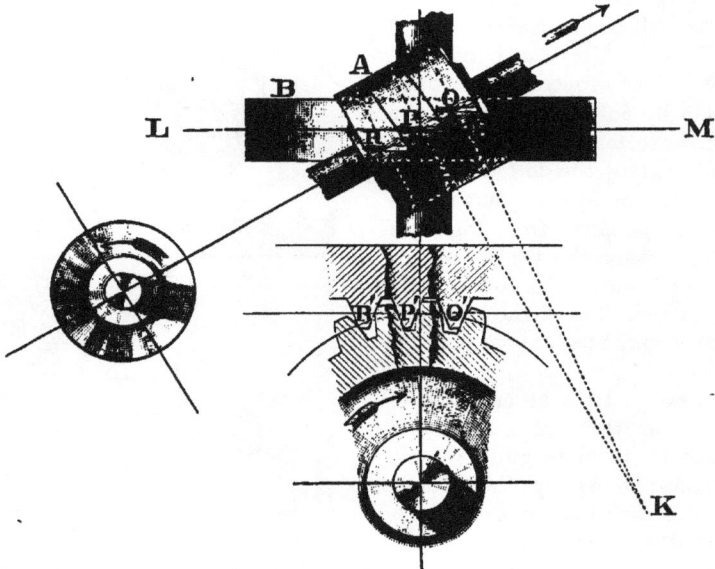

FIG. 249.

A being its pitch cylinder, *B* that of the wheel, *P* their point of tangency. Let *P* be also the present point of contact between a thread and a tooth, as shown at *P'*, below; when cut by the plane *LM*, *normal to the axis of the wheel*, let the section of the thread have the form of the rack tooth in the preceding figure, and that of the wheel-tooth be conjugate to it as before.

As the screw rotates there will always be a section of its thread similar to this, similarly situated with regard to its axis. This will

travel along with uniform speed, as indicated by the straight arrow, advancing in one rotation to the new position, O.

Now in order that the velocity ratio may remain strictly constant, this travelling section of the thread must be always acting against a tooth outline of the same form. Consequently, *in every transverse section of the wheel, the teeth will be bounded by similar curves,* although, as will subsequently appear, they will not necessarily be of uniform height.

414. We have now to consider the twist of the wheel itself, which depends upon the same principles as in the common arrangement. Thus, suppose the thread and the tooth in contact at P, to be gradually reduced in size; they will ultimately become two helical lines, lying one upon each pitch cylinder, still tangent to each other, and therefore developing upon the common tangent plane into the same straight line. Thus, if the pitch of the screw-helix be given, that of the other is found as in Fig. 245, and is the same for all the helices of the teeth of the wheel, which, if cut from a cylindrical blank, will be a simple twisted one, as in Fig. 110, with all its transverse sections alike.

415. Careful consideration of the action will show that the surfaces generated as above explained are precisely such as the screw would cut for itself under the assumed conditions, and thus confirm the previous statement (**401**) that the proper forms of the tooth in this, as well as in the common arrangement of screw gearing with axes in planes mutually perpendicular, are to be determined by the principles which apply to the case of a wheel working with a *rack*, and not with another wheel of any finite radius whatsoever; notwithstanding the fact that the latter construction is the one given by Prof. Rankine.* We would not be understood to assert that two oblique screw-wheels will absolutely refuse either to engage or to work with each other because otherwise fashioned, but that the method of construction here set forth is the only one by which perfect theoretical precision can be attained.

416. Peculiar Features of the Action.—The teeth of the wheel in Fig. 249 are sections by successive transverse planes through R, P, O; their conjugates, R', P', O', being cut from the screw by the same planes.

We now observe that one rotation of the worm will drive the wheel through more than the original pitch angle, although the helical pitch is equal to the diagonal pitch of the rack in the preceding

---

* Machinery and Mill-Work, p. 163.

figure. In that case, the rack-tooth was always acting against a surface whose elements were right lines perpendicular to the plane of
rotation. But the worm acts against helices which cross the plane of
rotation obliquely, and with different degrees of obliquity, for they all
have the same pitch, but lie at different distances from the axis.
Since, however, the velocity ratio is constant, we may confine our
attention to the helices upon the pitch cylinders, and study their
action as represented in the development of these surfaces upon the
common tangent plane.

Thus, in Fig. 249, having drawn PK, perpendicular to PO and
equal to the circumference of the pitch cylinder A, OK is the developed helix through O, and PE is the length of the new pitch arc for
the wheel. If then a complete wheel is to be made, the proportions
must be such that PE is an aliquot part of the pitch circumference.

417. And all this accords precisely with the fundamental principles
relating to the composition and resolution of motions. In Fig. 250
let DD and CC be the elements of the pitch cylinders of the worm
and the wheel respectively, inter-
secting at P ; PO the pitch of the
worm, PK perpendicular to PO,
and equal to its pitch circumference, and OK the developed helix,
all as in the preceding figure ; then
PM parallel to OK is the developed
helix through P, which point we
will assume as before to be the present point of contact between a
tooth and a thread. By the turning of the worm, the point P
virtually advances uniformly in the
direction PD, going in one revolution to the new position O, while
meantime the coincident point P
of the wheel must move in the direction PE. Regarding PM as a

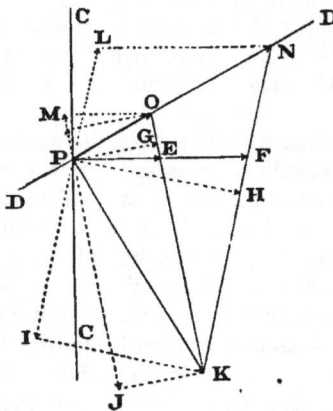

Fig. 250.

helix of the wheel, and supposing the screw to be pushed endlong in
the manner of a *rack*, we observe that PO may be resolved into the
components PM, PE. Of these, the first is simply the sliding component and ineffective ; but PE is the one which must represent the rotation of the wheel, due to the supposed rectilinear motion of the screw.

When, however, we suppose the worm to have a motion of rotation
only, in the direction shown by the arrow in Fig. 249, let PK repre-

sent the linear motion of its driving point P, which acts against the helix PM of the wheel. The normal and tangential components of PK are respectively PG perpendicular to PM, and PJ coincident with it. The resultant motion of the point P of the wheel must be in the direction PF, perpendicular to CC, and of such magnitude, PE, as to have the same normal component, PG.

The same method might have been used in relation to the rectilinear motion PO of the screw when used as a rack, and with concordant results, since PG is the normal component of that motion also.

418. In all the foregoing the diameter of the pitch cylinder of the worm was assumed at pleasure, which it may be, subject to the condition above pointed out, that PE must be an aliquot part of the pitch circumference of the wheel.

It is perfectly possible, then, that a double as well as a single thread might be made upon a pitch cylinder of given size. Supposing this to be desired, and also that the subdivision of the given wheel should remain unchanged, it is then evident that in Fig. 250 we must double PE, the pitch arc, and not PO, the helical pitch of the screw, which will now become PN, determined by producing KF to cut DD in N. We thus determine new helices, KN, PL, with reference to which the motion PK is to be resolved as before, PH being the normal or effective, PI the tangential or sliding component.

419. The diagram, Fig. 250, is drawn without regard to practical proportions, the conditions being selected with a view to illustrating one particular in which oblique screw gearing may, in some cases, differ from the ordinary arrangement. We have seen that when the axes lie in planes perpendicular to each other, the helices on the worm and the wheel, whatever the number of threads and teeth, are either both right-handed or both left-handed (**403**).

Now in Fig. 250, both KO and KN will form right-handed helices upon the pitch cylinder of the screw ; but on wrapping the tangent plane down upon the wheel, KN will become a right-handed, KO a left-handed, helix. Evidently there is an intermediate position in which the developed worm-helix will be parallel to CC, and will, therefore become a *rectilinear* element of the wheel's pitch cylinder. In that case the screw will work with a common spur-wheel ; and the proportions which must obtain in order to secure this result are shown in Fig. 251. If the pitch arc, PE, and the obliquity, CPD, are both assigned, the pitch and circumference of the screw are determined by drawing through E a perpendicular to PE, cutting DD at O, and PK perpendicular to DD at K. If the pitch arc, PE, and the circumference PK are assigned, we first describe an arc about P with radius

PE; draw KE tangent to it, and also PD perpendicular to PK. Then KE produced cuts PD in O, determining the pitch PO and the obliquity OPE.

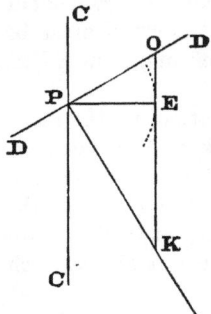

FIG. 251.

420. Close-fitting Oblique Tangent-screw.—The oblique worm may be used to cut its own wheel, just as in the common arrangement; and the wheel itself, instead of being cut from a plain cylindrical blank, may, for the sake of a neater appearance, conform somewhat to the curvature of the screw, and be terminated by conical frusta instead of transverse planes. The form of the blank is fixed by very simple considerations; regarding the screw as a thread wrapped around a cylindrical core, we perceive that the tops of the wheel-teeth must just go clear of that core during the rotation. Now imagine the whole screw to revolve about the axis of the wheel; then the element of the core which lies nearest to that axis will generate a hyperboloid, and if the blank for the wheel be turned to this exact size, the tops of the teeth will just touch the core. It should therefore be made a little smaller.

It will be seen that the screw still travels obliquely across the plane of rotation, so that the transverse sections of the teeth are all of the same form, contact existing between each thread and its tooth at a single point only; whence the only advantage gained by the adoption of this form of blank lies in the fact that the teeth become higher as they recede from the mid-plane, and, therefore, continue longer in action.

421. Oblique Screw and Rack.—Further, it will be seen that the diameter of the oblique worm-wheel may be increased at pleasure until it eventually loses its curvature, and assumes the form of a rack. Under these circumstances Prof. Rankine's instructions,* if we interpret them correctly, are to the effect that the normal section of the worm-thread should be that of the tooth of a wheel working with a rack, the tooth-outline of the latter to be adopted as the normal section of the rack to work with this worm; and he gives specific directions for finding the radius of that wheel in any given case.

Without disputing that the thread and tooth thus formed will work, we would remark that it is not the only nor yet the best manner of forming them. The tooth-surfaces of the rack must be made up of parallel rectilinear elements, and since in the case of a rack and wheel the teeth have but a single point of contact, the

---

* Machinery and Mill-Work, p. 290.

tooth-surface of the rack will, in Prof. Rankine's construction, touch each other in that point only. Which would hold true at the limit when the obliquity vanishes ; whereas it is perfectly patent that the rack may then be a part of a nut, and its whole surface in contact with that of the screw-thread, whatever the form of the latter. This superficial contact is not attainable when the rack travels obliquely, it is true ; but *line-contact* between the acting surfaces may be secured with a screw-thread·of any reasonable form, as will appear from the following considerations.

422. Taking a common triangular-threaded screw in illustration, we have, in Fig. 252, a meridian section on the left, and an outside view on the right, the axis being parallel to the paper. The apparent contour of the completed screw is the trace of its projecting cylinder, whose elements are perpendicular to the paper, and tangent to the surface. If then a rack be made, as shown in section below, its teeth bounded

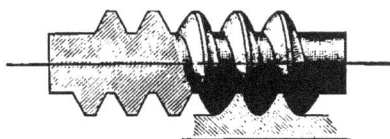

FIG. 252.

by the same outlines and their elements also perpendicular to the paper, these teeth will always touch the screw along the line of its visible contour, whether the rack be made to travel endlong by turning the screw, or to slide transversely, or both. And if these actions occur simultaneously, we have the oblique motion desired.

Line-contact is thus obtained, and the normal section of the rack-tooth is determined by merely drawing the screw, in its simplest position. But this is not the only form which will effect the same result ; for in whatever direction we chose to look at the screw, if the space between the threads be visible at all, the line of apparent contour will give the normal section of a tooth for a rack which will work in line-contact with it, either longitudinally or obliquely.

423. Among so many forms, the selection of the best may be safely left to the screw itself, which will assert its preference if given the opportunity. This may be afforded by allowing it to cut its own rack, the latter being moved by independent means in the required direction, and at the same speed as that which it is eventually to receive from the screw.

It is clear that the amount of metal removed will be the least possible ; whence the deduction that the normal section of the rack-tooth will be the visible contour of the space between the screw-threads when viewed from the direction which will give the greatest apparent breadth to that space.

What that precise direction is, may depend somewhat upon the meridian section of the screw, but probably will not vary appreciably from that of the tangent to the helix of mean obliquity.

It will be seen by reference to Fig. 251, that if the pitch, diameter, and inclination of the screw to the line of travel be properly proportioned, the rack may be made of the usual form, that is, with the teeth cut transversely across its face. But whatever the arrangement, proportions, or form, the kinematic action consists of sliding contact pure and simple; there is not, as has been sometimes erroneously stated, the slightest admixture of rolling contact, or of anything even distantly resembling it.

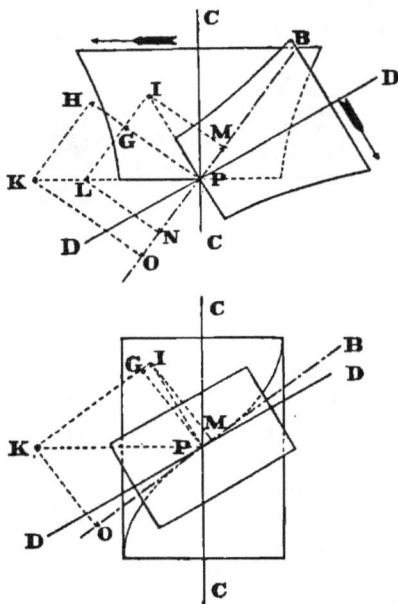

424. To Secure the Least Amount of Sliding.—Although the velocity ratio depends upon the pitch and not upon the size of the screw, yet it is evident that for a given pair of axes, and a given velocity ratio, there must be some definite ratio between the diameters of the pitch cylinders, which will involve less sliding than any other.

What this proportion is, may be thus deduced. If upon the given axes we construct a pair of rolling hyperboloids with the assigned velocity ratio, these surfaces will work in contact with no sliding other than that along the common element. This element passes through the common point of the gorge circles, and lies in a plane tangent at that point not only to both hyperboloids, but to their inscribed tangent cylinders of which the gorge circles are the bases.

Figs. 253 and 254.

Now, if these be taken as the pitch cylinders of the screw wheels, and the common element as the developed helix, the amount of sliding will be the same as that between the two hyperboloids when rotating with the given velocity ratio.

And this amount is then a minimum. In Fig. 253, let PI represent the motion of the point P of the inclined surface; if the vertical one be driven by it, we find in the usual manner the motion of the coincident point to be PL, and that the sliding is represented by NM on the line of contact. PI remaining unchanged, let the vertical hyperboloid be compelled to revolve, by independent means, faster than the other should drive it ; and let PK be the new velocity of its point P. The sliding component along PB now becomes PO, which is greater than PN ; and there will be also a sliding perpendicular to the common element, represented by GH. A similar result would follow, were PI to be increased while PL remained the same ; whence it appears that the sliding is least when the velocity ratio is that for which the hyperboloids were constructed ; and this holds true of the tangent cylinders, the obliquity of the helices being determined as above.

425. The sliding action represented by GH cannot actually occur when teeth are used ; but the linear motions PI and PK can still be retained in the case of screw gearing, as shown in Fig. 254, by making the developed helix PB parallel to KI; all the sliding must necessarily be in this direction, and its amount OM is, of course, greater than before. This proceeding, it will be observed, is merely the converse of that explained in connection with Fig. 250, the results being exactly concordant.

If the axes lie in planes perpendicular to each other, the hyperboloids have two common elements, either of which may be taken as the developed helix, according to the directional relation desired ; both wheels will be right-handed in one case, and left-handed in the other.

426. Resemblance to Skew Wheels.—The appearance of screw gearing thus constructed is quite unlike that of the familiar worm and wheel. In that combination the wheel cannot move the screw, nor is it usually desirable that it should, since in most cases the latter is required not only to turn the wheel, but to hold it in any given position. But in the arrangement now under consideration either wheel may be used as the driver ; it is quite apparent that a single-threaded worm made in the manner above described could not work unless it were of great length and its axis nearly parallel to that of the wheel, but that this construction is adapted only for the teeth of *screw wheelwork* properly so called ; and equally evident that for this it is the best.

The mid-planes of these wheels are the gorge circles of the tangent hyperboloids ; the pitch helices are tangent to the elements of those surfaces ; and from the considerations presented in **(420)**, it will be seen that both blanks may be made of hyperboloidal outline. From all which, there results a resemblance between these and skew wheels sufficiently close to be misleading.

This resemblance, however, is superficial ; the teeth are made up of cylindrical helices, touch each other in a single point only, can be extended only to a short distance either way without ceasing to engage, and their acting sections are respectively those of a spur wheel and a rack.

And the one whose section corresponds to that of the rack-tooth may be made of any length, and if desired may be used as a rack, driving the other wheel not by rotation, but by moving longitudinally, whatever the inclination of the axes or the number of threads.

Hindley's Screw, or Hour-glass Worm.

427. A Worm may be so shaped as to conform to the curvature of the wheel, assuming a figure somewhat like that of an hour-glass ; in Fig. 255, the complete worm is represented on the left, and on the right is given a section through its axis by a plane perpendicular to that of the wheel. The surface of the thread is of a complicated and peculiar nature, but practically it is very easily made, thus : a tool, with a cutting point of the contour of the wheel's tooth, is so clamped to a disc that its upper surface lies in the meridian plane of the worm, and both the disc and the worm blank are driven by intermediate gearing at their proper relative velocities.

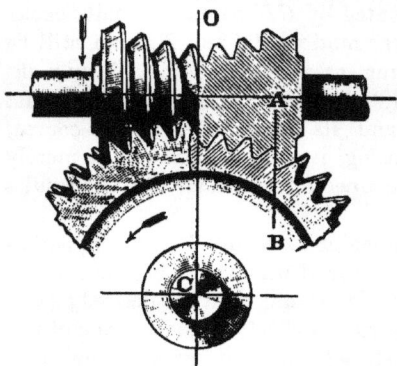

Fig. 255.

In this manner the screw was made by Hindley, who first introduced it. For some reason it has never come into general use, although in smoothness and steadiness of action it would appear to be superior to the common form of tangent-screw.

The worm itself may now be formed into a cutter, and made to finish its own wheel in the usual manner ; which is the method adopted by Messrs. Clem & Morse of Philadelphia, Penn., who have recently constructed a very elegant engine specially designed for cutting this description of screw gearing.

428. The outline of the pitch surface of this worm is an arc of the pitch circle of the wheel. Upon this suface the helix EFP, in Fig. 256, is traced by a point which moves about the centre C through the arc EP, while the worm makes one revolution, both motions being uniform in velocity. The longitudinal advance, UV, is not equal to EP ; it is not uniform, nor is it the same for successive convolutions.

The projection of this helix upon a plane, OC, perpendicular to DD, as shown at the right, is of a spiral form, resembling that of the hyperboloidal helix in Fig. 243 ; to which this curve is very similar. Other points in the outline of the tooth, either within or without the pitch circle, also describe curves of the same nature, lying upon surfaces whose meridian outlines are circular arcs ; these curves are easily drawn, and their envelope is the visible contour of the worm.

The form of the threads is, abstractly, arbitrary ; their meridian sections are

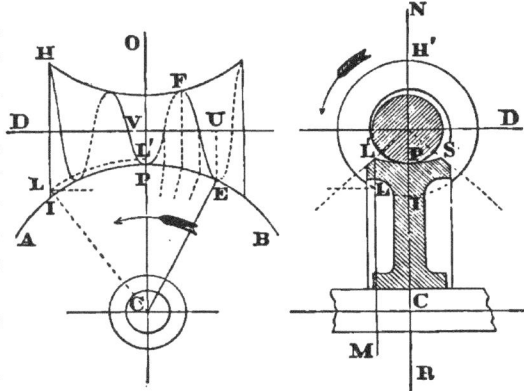

FIG. 256.

exactly converse to those of the wheel-teeth, the spaces between which, as seen in Fig. 255, they fit and fill entirely, throughout the action.   Thus there is no relative motion in the manner of a rack or otherwise, but the worm is locked into the wheel, and can move only by revolving about one axis or the other.   A good and simple practical form for the teeth of the wheel is that given in Fig. 255, the sides being straight sloping lines, equally inclined to the radius of symmetry ; the amount of inclination is fixed by the consideration that, in the extreme position of the action, the outer side should lie in a line, AB, perpendicular to DD, in order that the worm may be readily disengaged from the wheel.

429. The Action Confined to the Mid-plane of the Wheel.—The advantage of the hour-glass worm lies in the fact just stated, that the whole side of every thread, *in the meridian plane*, is always in contact with the adjacent tooth.

It has been asserted that the teeth of the wheel, when cut by the worm, touch the threads of the latter at all points ;* or, in other words, that the whole surfaces are in contact.

Either one of two considerations is sufficient to show that this is

---

* "Mechanics" for January 14, 1882.

impossible.   Let any side plane be passed, parallel to the axis of the
worm and perpendicular to that of the wheel ; then, in the first place,
the sections of the successive threads will not be similar ; and, in the
second place, the section of the pitch surface of the worm will not be
a circle, whereas the teeth of the wheel travel in circular paths.
Therefore, such superficial contact is attainable only when the radii
of these paths become infinite, in which event the pitch surface of
the worm becomes a cylinder, the worm itself a common screw, and
the wheel a portion of a nut ; but, in general, the action is confined
to the central plane of the wheel.

430. The Pitch Surface of the Wheel may to some extent conform to
the curvature of the worm, as shown in section in the right-hand
part of Fig. 256. Any side plane, LM, parallel to the mid-plane, NR,
cuts the base, III, of the worm's pitch surface in a point L, which
when revolved about the axis of the wheel into the gorge plane OC,
takes the position L', thus giving a point in the contour $L'P'S$.

By a similar process, reducing the radius VP of the gorge of the
worm, and increasing the radius CP, according to the height of the
tooth, we may determine the outline of the blank.   This conforma-
tion of the wheel does not, it must be noted, secure any additional
contact, or in any way affect the action, but merely gives a neater
finish than if the blank were made a plain cylinder.

431. The Teeth of the Wheel are automatically shaped by the cutting
worm with perfect ease ; but the accurate delineation of the form
thus determined is a rather complicated and tedious matter.   Still, if
necessary, it can be accomplished in the following manner.

When the outline of the blank has been found, as above, it will be
seen that any transverse plane will cut from it a circle, which will
limit the height of the teeth.   The same plane will cut from the
threads of the screw a series of sections of varying form, each lying
a little in advance of the meridian section of the same thread.   The
sections of the wheel-teeth by the same plane must lie in the spaces
between those of the screw-threads ; and since each revolution of the
worm turns the wheel one tooth ahead, it follows that each tooth-
section must take in succession all the different positions thus determined.   We may, then, proceed thus : lay a piece of tracing paper
over the drawing, and secure it by a pin fixed at the centre of the
wheel.   Upon this, trace the sections of two consecutive threads, and
also the circular arc bounding the top of the tooth in this plane.
Then rotate the paper about the central pin through the pitch angle,
again trace the sections of the two adjacent threads, and so on until
all have been traced ; the clear space within the lines thus drawn

will be the required section of the tooth by the given plane. Absolute certainty of definition will require the screw to be placed in several positions, cut in each by the same plane, and the above process to be repeated for each, using the same tracing throughout ; because the sections of these peculiar screw-threads in the different phases of rotation are dissimilar. And finally, this whole operation must also be repeated with several different planes, the number depending upon the degree of accuracy aimed at : all of which we leave the reader to execute at leisure.

432. Multiple-threaded Hour-glass Worm.—Though we are not aware that this form of worm has ever been made with two or more threads, there seems to be no abstract objection to increasing the number. Also it is clearly possible, by making the screw pitch sufficiently large, to use the wheel as the driver : in which case the teeth might preferably be made in the form of turned pins, set into the periphery of a cylinder as in face gearing. Evidently a greater number of these pins, or teeth, would be simultaneously engaged than in the common form of screw wheels, which would certainly tend to increase the steadiness of the motion ; but the whole action is confined, as before, to the meridian plane of the worm perpendicular to the axis of the wheel.

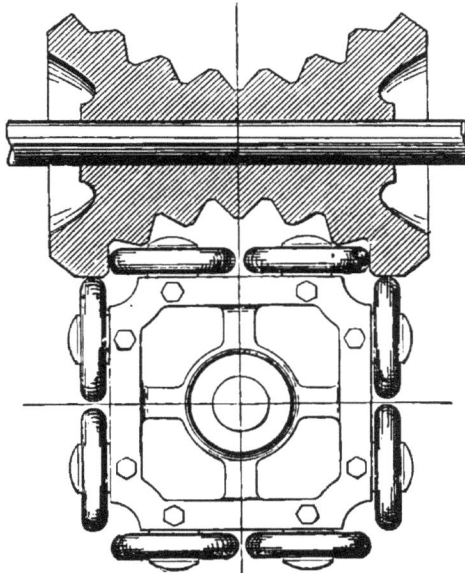

Fig. 257.

433. Rollers Substituted for Teeth.—With the purpose of reducing the sliding friction as much as possible, the singular device illustrated in Fig. 257 has been proposed.

The meridian section of the screw-thread is bounded by right lines at right angles to each other, and the worm is made long enough to embrace more than one quarter of the circumference of a wheel, the number of whose teeth must be a multiple of four ; for this wheel is substituted a frame carrying rollers, arranged in the manner shown in the figure.

The object of diminishing the friction is certainly accomplished; but, on the other hand, only one thread of the worm is in action at once, during the greater part of the time. And that one has at no instant more than a single driving point; for, since the warped surface of the screw cannot be placed in right-line contact with a surface of single curvature, these rollers cannot be made cylindrical, as they have sometimes been represented; but their contours must be slightly convex curves tangent to the meridian sections of the threads.

Taking these drawbacks into consideration, it would appear that this arrangement is, from a practical point of view, more curious than useful; although it might serve a purpose in very light-running mechanism intended rather for the modification of motion than for the transmission of power.

The Teeth of Face-Wheels.

434. Let two Wheels, exactly alike, whose teeth are cylindrical pins fixed in the faces of circular discs, be so placed that each axis lies in a plane perpendicular to the other, at a distance from it equal to the diameter of the pins, as in Fig. 258.

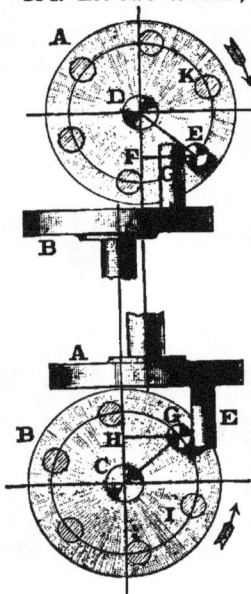

Under these circumstances, the angle FDE will always be equal to the angle HCG, as long as the pin E of the driver A is in contact with the pin G of the follower B : the velocity ratio is, therefore, perfectly constant.

The length of the pin E must be such that the next pin I of the other wheel shall not catch upon its end in going into gear. And it will also be noted that although the pin I, at the instant when the next pin K of the driver begins to act upon it, may also touch the pin E upon the back, it cannot continue to do so. It is not possible, therefore, even theoretically to secure entire freedom from backlash.

FIG. 258.

The maximum length of one pin having been ascertained, it is of course the same for all. And it is next to be observed that if the number be increased this length must be diminished; also, that there will in every case be a limit beyond which the number cannot be increased without at the same time diminishing the diameter, and in consequence the distance between the axes.

Ultimately, then, the axes will intersect at right angles, and the pins will become consecutive points in the circumferences of two equal circles rolling together like the bases of the pitch cones of a pair of mitre-wheels. In other words, there are no pitch *surfaces,* these degenerating into *lines,* and the elementary teeth into points.

435. Equal Wheels with Axes at Right Angles.—The preceding is the simplest form of face gearing, but it cannot be used if the axes intersect, as is often necessary. When they do, however, the teeth of one wheel may still be made cylindrical : those of the other being solids of revolution, whose outlines may be determined as follows.

In Fig. 259, let the two wheels be of equal size, their axes perpendicular to each other. Let E be a pin of no sensible diameter, fixed in the wheel A, and F a similar one fixed in the wheel B, the distance between the two being arbitrary. Let the wheels turn through equal angles as indicated by the arrows ; the pins will come into the new positions G, H, and I, K, and in the meantime their common perpendicular will continually change both in position and in magnitude.

But it is always easily determined ; and it is evident that if we now make F the axis of a surface of revolution, the radius of each transverse section being this common perpendicular, we shall have the form of a pin or tooth for B, which will work with the pin E of no sensible diameter, the velocity ratio being constant throughout.

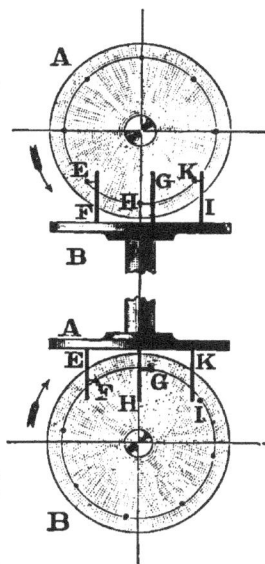

Fig. 259.

This process is more fully illustrated in Fig. 260, in which portions of the pitch circles only are shown. The action is readily traced : the curve EI will be generated while the pins E and F traverse the equal arcs EP, FO, and the curve LG, while they move through the arcs PG, OR, which are also equal to each other ; the ordinates x, z, being found by placing the pins in the intermediate positions shown, any required number may be determined in a similar manner.

436. Upon assuming a sensible diameter for the pin of the upper wheel, we proceed to derive the outline of the working tooth of the other, precisely as was done in the case of pin gearing, as shown on

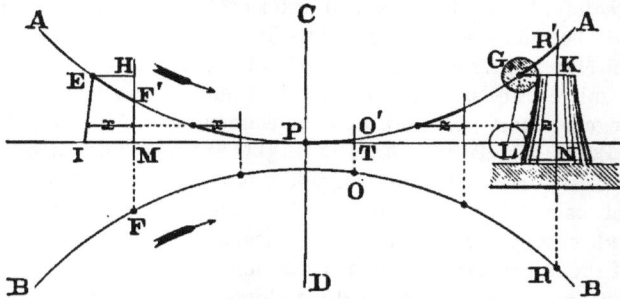

Fig. 260.

the right in the figure ; drawing a curve at a constant normal distance from LG, equal to the radius of the pin.

Since on the equal circles AA, BB, we have the arcs $EP = FO = F'O'$, and $PG = OR = O'R'$, it follows that the arcs EF', PO', GR', etc., are also equal to each other. Whence, EP being equal to PG, we find that EH is greater than GK; similarly x is greater than z, and, in general, all the ordinates of EI are greater than the corresponding ordinates of GL, the difference diminishing as we descend, until finally IM is equal to LN.

Consequently, the working tooth for the lower wheel must be de-

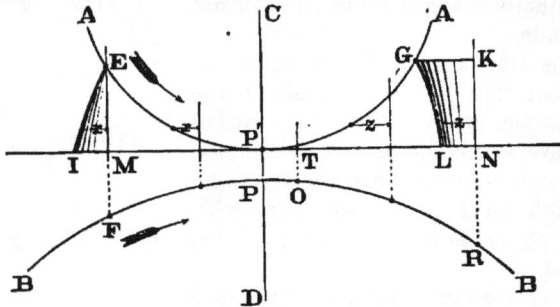

Fig. 261.

rived from the curve GL and not from EI ; and in order to secure receding instead of approaching action, the cylindrical pins must be given to the *driver*, and not to the follower as in pin gearing.

437. Unequal Wheels with Axes at Right Angles.—In Fig. 261, let the cylindrical pins be given to the smaller wheel, of which $AP'A$ is the pitch circle. Assuming any point E upon this circumference, draw through it a parallel to CD, cutting the larger pitch circle BPB in F. Then the arc EP' is greater than the arc FP; therefore, if we make FO equal to EP', a parallel to CD through O will cut the tangent at P' in some point T on the right of P'; and IM will be the lowest ordinate of the curve EI, which, as in the preceding figure, will be generated while a pin at E traverses the arc EP' and one at F moves to O. Also $P'T$ will be equal to LN, the lowest, and evidently the least, ordinate of the curve LG generated while the pins traverse the equal arcs OR and $P'G$. The whole process, including the determination of the intermediate ordinates, is the same as before; but since the least of the two elementary surfaces must necessarily be employed, we perceive that the working tooth in this case must be derived from EI instead of LG, and also that the cylindrical pins must be given to the follower, as in pin gearing, and not to the driver, in order that the action may be receding.

438. As the diameter of the driver becomes greater its curvature becomes less, and at the limit will disappear; the combination then becoming identical with that of a rack driving a pin-wheel. The curves EI and GL will in that case evidently be similar and equal, each being the cycloid of which AA is the generating circle.

There is, in fact, a close analogy between pin-gearing and face-gearing; for if in the latter we suppose the axes to be parallel, the teeth of one wheel will project radially from the curved surface of a cylinder, and their contours will be precisely like those of a spur-wheel working with the round staves of a trundle, or lantern pinion.

439. But the cylindrical pins may be given to the larger of the two unequal wheels. The case is then very nearly the converse of the preceding one: the rotations being in the same direction, the pointed tooth, as seen in Fig. 262, appears on the opposite side of the plane of the axes, and if the action is to be receding,

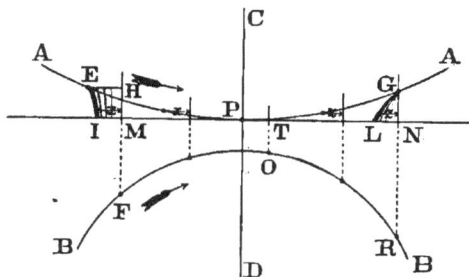

Fig. 262.

the larger wheel, with the cylindrical pins, must drive.

The greater the difference between the diameters of the pitch circles, the shorter will be the teeth of the follower for a given arc of action. And, in consequence of this, the size of the driver cannot be indefinitely increased, since upon reaching the limit, its pins would move in a plane parallel to the face of the follower, and the tooth of the latter would reduce to a series of concentric circles lying in the same plane.

440. Unequal Wheels with Axes in Different Planes.—Precisely the same method of construction is applicable also when each axis lies in a plane perpendicular to the other, although they do not meet, and whether the wheels be equal or unequal in size. Thus in Fig. 263,

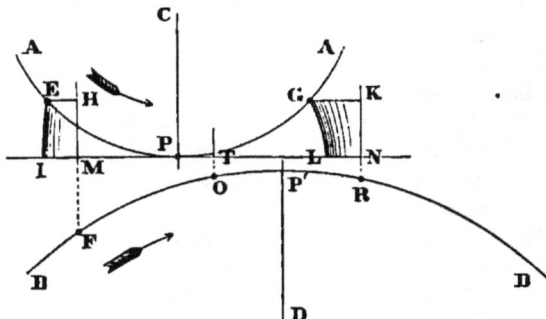

Fig. 263.

supposing the cylindrical pins to be assigned to the smaller wheel, the axis E of one of them is placed arbitrarily with reference to the axis F of a tooth of the other, whose contour is derived from a curve EI, determined as in Fig. 260. It is clear that under some conditions this elementary tooth may be pointed, as it was in Fig. 261; and in Fig. 264 the conditions are so selected that this is the case; but, as before, the cylindrical pins being given to the large wheel, this pointed tooth makes its appearance on the other side of the axis, the direction of the rotation being unchanged.

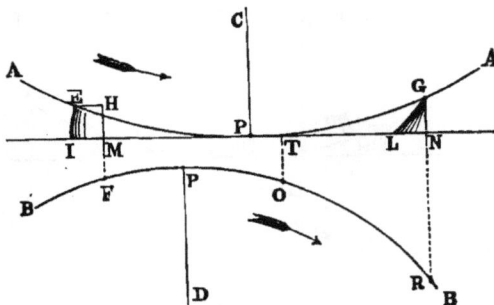

Fig. 264.

Hence we perceive that in order to secure receding action, the larger wheel must drive, no matter to which one the cylindrical pins are assigned.

Had the wheels been made of the same size, but with the least distance between the axes greater than the diameter of the cylindrical

pin, the same process would have enabled us to determine the form
of the working tooth for the other wheel : in short, the construction
is general for the relative positions of the axes above mentioned, and
Fig. 258 represents only a special case.

441. Miscellaneous Arrangements of Face Gearing.—It is not essential
that either the pins or the teeth, because they are turned in the lathe,
should be inserted into plane surfaces ; and we append a few examples in which they are otherwise arranged.

Let the axes in Fig. 265 intersect at any angle ; let the velocity
ratio be given, and suppose the
cylindrical pins to be assigned to
the wheel with the vertical axis.

Draw AB, dividing the angle
DAC according to the velocity
ratio as in bevel gearing. Through
any point E of AB draw a vertical line as the axis of a pin ; and
also a horizontal line cutting the
inclined axis in C ; then EC by

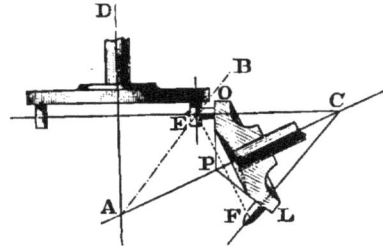

FIG. 265.

revolving around AC will generate the cone ECF. The teeth of the
inclined wheel are to be solids of revolution, whose axes will evidently
be elements of this cone; and they may be fixed in the surface of
another cone OPL, normal to ECF.

A pin of the vertical wheel is shown at E in contact with such a
tooth, whose contour may be determined as follows : First, suppose
the cylindrical pin to have no sensible diameter ; then if the vertical
wheel turn through a given angle, the other will also turn through
an angle which is known, because the circumferential velocities of
the circles which roll in contact must be equal.

Consequently, the relative positions of the axes of the given pin
and the required tooth may be determined for any phase of the
action, and their common perpendicular at that instant may be ·
found.

Having repeated this process as many times as may be deemed necessary, these common perpendiculars will obviously be the radii of the
transverse sections of the required tooth for a pin of no sensible diameter. From these the meridian outline can be constructed, and the
contour of the working tooth is derived from this in the usual manner
by assigning any radius at pleasure to the cylindrical pin.

442. A modification of this arrangement is shown in Fig. 266, the pins
being fixed radially in the periphery of the cylinder with the horizontal
axis ; but the process of constructing the tooth is the same as before.

19

From the employment of the conical surfaces, which afford the most convenient and natural means of supporting the teeth of one wheel of a pair, these combinations have been erroneously supposed to contain the germ from which bevel gearing was developed.

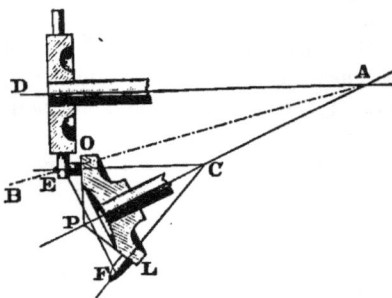

Fig. 266.

But a moment's study of these figures will show that the principle of rolling cones, upon which bevel gear wholly depends, is not here involved in any way whatever; the fundamental idea appears to have been that of causing the teeth or pins to present themselves to each other, at the instant of passing the plane of the axes, in the same relative position as though the axes were parallel.

But, as shown in Fig. 267, the same methods and processes are applicable also to the construction of what may be called bevel face-gearing, the axes of the pins coinciding with the elements of one pitch cone, and projecting normally from the surface of the other.

Face gearing is for general purposes practically obsolete; hence it has not been deemed necessary to discuss in relation to it the questions of limiting lengths or numbers of teeth, angles of action, and the like. For models and light machinery it is sometimes employed, on ac-

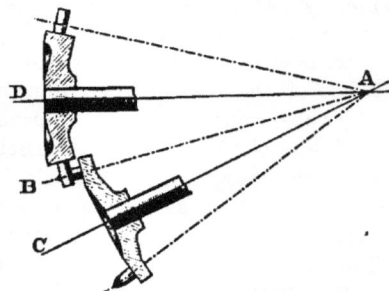

Fig. 267.

count of the facility of forming the teeth in the lathe; a special construction would in any event be almost necessary for each individual case, and this can readily be made by the aid of the preceding explanations.

443. Screw and Face Gearing Combined.—An example has already been given, in Fig. 257, of the use of rollers for reducing the friction which attends the action of the worm and wheel. In that case the screw was of the hour-glass form; but the same expedient has been employed in connection with. the common variety, as for instance in Van der Mark's hoisting apparatus; the axes of the rollers projecting radially from the periphery of a cylinder, and their outlines being the same as those of the spur-teeth, conjugate to the rack-teeth cut from the screw by a meridian plane. Since the teeth of the wheel, whether they be rollers, capable of rotating, or not, are solids of revolution, this arrangement combines the peculiarities of both screw and face gearing.

In Fig. 268 is represented another combination of essentially the same nature. Upon the upper face of the disc B is formed a spirally coiled rib, constituting what may be properly called a plane or face screw.

The section of this rib by a plane normal to the axis of the wheel C should, as in the figure, have the form of a rack-tooth, the conjugate to which is the meridian section of the teeth of C. Thus, the distance between the corresponding edges of the adjacent coils being equal to the pitch, the action is precisely equivalent to that of a rack and spur-wheel.

444. This arrangement has been erroneously represented as having the rib coiled in the form of an Archimedean spiral, and the axis of B lying in the same plane with the axes of the teeth of C.

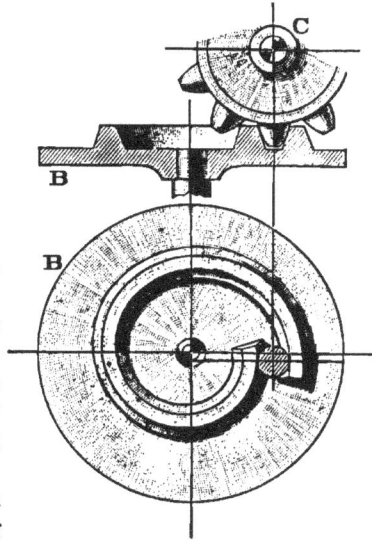

Fig. 268.

But the radiant of that curve does not cut the coils normally ; and in order to secure perfect kinematic action, the spirals, as shown in the figure, should be involutes of a circle whose circumference is equal to the tooth-pitch, and the axes of the teeth of C must lie in a plane parallel to the axis of B at a distance from it equal to the radius of that circle.

This is, of course, upon the supposition that, as in the case selected for illustration, the spiral expands at the rate of one tooth-pitch in each convolution, thus making the action equivalent to that of a common single-threaded worm. We may, however, double or treble the rate of expansion, increasing the angular velocity of the face-wheel in the same proportion, and introducing a corresponding number of intermediate ribs or coils ; if this be done, the diameter of the base circle of the involutes must be also doubled or trebled, as the case may be, and the position of the plane of action changed accordingly.

445. Spherical Screw and Wheel.—A curious, and we believe a novel, modification of the preceding arrangement is shown in Fig. 269, the rib being coiled upon the concave surface of a hollow sphere, forming what may be called a spherical screw; the construction is as follows.

The spherical involute OFP is generated by a point O in the plane ACB, while the plane rolls around the cone ACK to which it is tangent. Since the element of contact is the instantaneous axis, the tangent to the curve at any point is perpendicular to the rolling

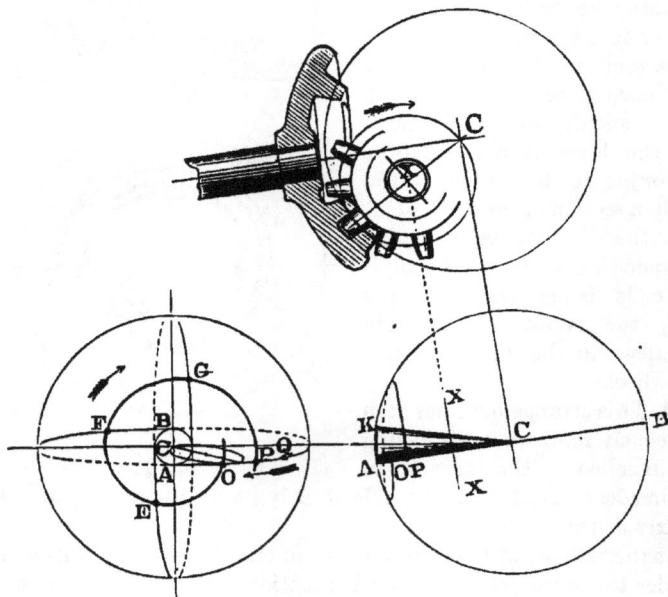

Fig. 269.

plane; which latter cuts from the sphere a great circle tangent to the base of the cone. Therefore, the points G, F, E, etc., of the curve, will in revolving about the horizontal axis of the cone, come normally into the plane ACB, at points in the circumference of the great circle cut by it from the sphere.

Or otherwise: let the cone ACK and a disc ACB move in rolling contact about fixed axes; then while the former makes one revolution to the right, the latter will turn to the left through an arc PO equal to the circumference of the cone's base AK, and if the sphere revolve with the cone, the point P will trace upon its surface the same curve PFO in a reverse order.

If then a groove of this form be cut in the concave surface of a portion of a spherical shell, and O, P, Q, etc., represent equidistant pins projecting from the edge of the disc ACB, it is clear that a rotation of the shell about the horizontal axis will impart to the disc a rotation about an axis perpendicular to its plane, with a constant velocity ratio.

446. Let us now take the circumference of the disc as the pitch circle of a wheel, and lay out upon it the form of a tooth suitable for driving in its own plane a pinion in inside gear, as shown in the upper diagram in the figure. It will then be apparent from what precedes that this tooth outline will be the normal section of the thread of the spherical screw, whose surface it will sweep up if the disc roll around the cone as before ; and that this screw will correctly drive the internal pinion if its teeth be made solids of revolution as shown in the cut. The axis of the screw being horizontal, the axis, XX, of the pinion, is not vertical, but perpendicular to the plane ACB.

It is hardly necessary to point out, finally, that the pinion may be placed in outside gear, and the screw cut upon the convex surface of the sphere. In short, the construction is a general one, the face-screw and wheel being but the special case in which the radius of the sphere becomes infinite and its surface degenerates into a plane.

APPENDIX.

1. **In Graphic Operations** there is frequent occasion to rectify circular arcs, and to lay off arcs of given linear values, either upon circles of given radii or subtending given angles.

In such cases the following constructions, which we borrow from Prof. Rankine, will be found extremely useful and convenient ; the results being obtained much more expeditiously than by calculation, and with a degree of accuracy amply sufficient for all ordinary purposes.

I. *To lay off on a right line a distance approximately equal in length to a given circular arc.*

Let AB, Fig. 1, be the given arc, and AH its tangent at A. Draw the chord BA, and produce it; bisect AB in D, and set off AE equal

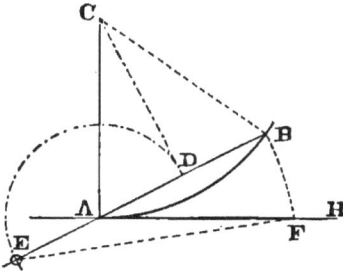
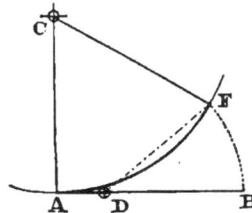

FIG. 1. FIG. 2.

to AD. About E as a centre, with EB as radius, describe an arc cutting AH in F : then $AF =$ arc AB, very nearly.

II. *To lay off on a given circle an arc approximately equal in length to a given right line.*

In Fig. 2, let AB, the given line, be tangent at A to the given circle. Set off $AD = \frac{1}{4} AB$, and about D as a centre, with radius $DB = \frac{3}{4} AB$, describe an arc cutting the given circle in F. Then arc $AF = AB$, very nearly.

III. *To find the radius of a circular arc which shall subtend a given angle, and be approximately equal in length to a given right line.*

Let AB, Fig. 3, be the given right line. Draw AG perpendicular to AB; and also AH, making the angle BAH equal to half the given angle. Set off $AD = \frac{1}{4} AB$, and about centre D, with radius $DB = \frac{3}{4} AB$, describe an arc cutting AH in F. Bisect AF by the perpendicular EC, which will cut AG in C, the centre of the required arc AF.

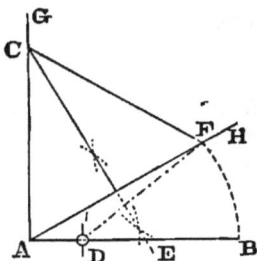

2. Amount of the Error in the above Processes.—It is stated by Prof. Rankine that in the application of either of these rules the straight line is a little less than the arc.

Fig. 3.

The magnitude of the error is given as about $\frac{1}{800}$ part of an arc of 60°; but it varies as the fourth power of the subtended angle, and may be reduced to any desired extent by subdivision. Thus, for an arc of 30°, the error will be $\frac{1}{800} \times \frac{1}{16} = \frac{1}{14400}$; and for one of 20°, only $\frac{1}{800} \times \frac{1}{81} = \frac{1}{72800}$.

Practically, therefore, if the given or required arc subtends an angle of over 60°, subdivision should be resorted to in applying either process.

The first two rules are applicable also in many cases to other curves than circles, provided that the change of curvature in the part to be dealt with be small and gradual.

GRAPHIC CONSTRUCTIONS RELATING TO TANGENT LINES.

3. The Determination of a Tangent to any curve with absolute precision, must of course depend upon a previous investigation of its mathematical properties.

But a sufficient degree of accuracy for most practical purposes can be attained by purely graphic methods, very simple, and quite independent of the special nature of the curve.

The most direct and expeditious of these consists in finding by trial the centre and radius of a circular arc which shall sensibly coincide with the given curve in the immediate vicinity of the point of tangency. This is more particularly eligible when the change of curvature in that part is small and gradual, as in Fig. 4.

If it be desired to draw the tangent at a given point P, we find by trial the centre C, as above, and the required line will be perpendicular to CP.

If it be required to draw a tangent in a given direction, or through a given point O not upon the curve, the line is drawn mechanically with the aid of a ruler, which can be done as accurately as a line can be drawn through two given points. Then in order to locate the point of tangency, we find the centre C as before, and draw the perpendicular CP, cutting the tangent in the required point.

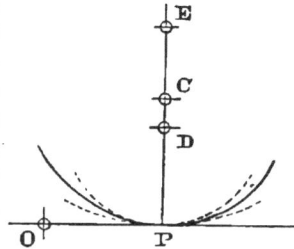

FIG. 4.

4. This tentative method is not very reliable in respect to the exact *location* of the centre of curvature, especially if the curve be flat; in which case a considerable variation in the radius may be made without sensibly affecting the curvature of the circular arc. But the *direction* of the radius of curvature is thus ascertained with considerable precision, as may readily be verified by seeking for two other centres, D and E, of circular arcs which shall lie respectively within and without the given curve, and deviate from it equally on opposite sides of the given or apparent point of tangency. If the manipulation be made with care, these three centres will be found to lie very nearly, as they should, in one straight line. And this fact may be practically utilized in dealing with a very flat curve; in which case, should the centre C fall at an inconvenient distance, the centre D of the inner circle may be used instead with equal confidence.

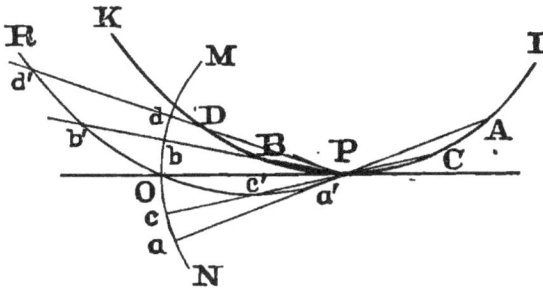

FIG. 5.

5. The following constructions may sometimes be preferred, although they are less direct and more laborious.

In Fig. 5, describe with any convenient radius a circular arc MN about P, the point upon the curve KL at which the tangent is to be drawn. From various points, as A, B, etc., taken at pleasure upon KL, draw through P right lines cutting MN in a, b, etc. From these intersections set off, on these lines, distances equal to the corresponding chords of the curve measured from P, and in the same direction ; thus, for instance, aa' is laid off to the right, and equal to PA ; while bb', equal to PB, is laid off to the left, and so on. Through the points d', b', etc., thus located, draw the auxiliary curve ROP, cutting MN in O ; then OP is the required tangent.

Should the intersection at O be too acute, the distances aa', bb', etc., may be made twice as great as the chords, thus changing the direction of the auxiliary curve.

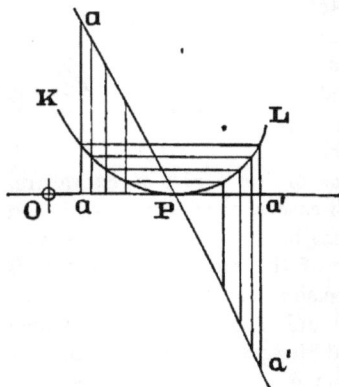

6. A similar principle is involved in the process for finding the point of contact when the tangent is given, as in Fig. 6. Draw any number of chords parallel to the tangent, and, through their opposite extremities, draw in opposite directions a series of parallel lines, which may or may not be perpendicular to the tangent. Upon each ordinate set off from the tangent a distance equal to the corresponding chord or some multiple thereof, and through the points thus determined draw the auxiliary curve, which will cut the given one at the required point of tangency.

Fig. 6.

NORMALS.

7. **For the General Problem** of drawing a normal to a given curve from a given point without, no solution, by graphic or any other means, has yet been discovered.

When the given point lies upon the curve, the mathematical properties of the latter may or may not be such as to enable us to draw the normal by a direct and independent construction. But, in general, the graphic operation in this case depends upon the previous determination of the tangent, by one or other of the methods above described.

THE ELLIPSE.

8. *First Method.*—About the centre C, Fig. 7, describe two circles whose diameters are respectively equal to the major and minor axes, AB and MN.

Draw any radius CE at pleasure, cutting the inner circle in D and the outer one in E. Draw through D a parallel to one axis, and through E a parallel to the other : the intersection O of these lines will be a point upon the ellipse.

This is the most accurate of all methods of constructing this curve by points, all the intersections, D, E, O, being right angles.

To draw the tangent at a given point.—By reversing the above process, we find that the given point P would have been determined by the radius CGH. At G and H, draw tangents to the two circles, cutting the axes produced, in L and R respectively : then LR is the required tangent at P.

Conversely, to find the point of contact : if the tangent be given, pro-

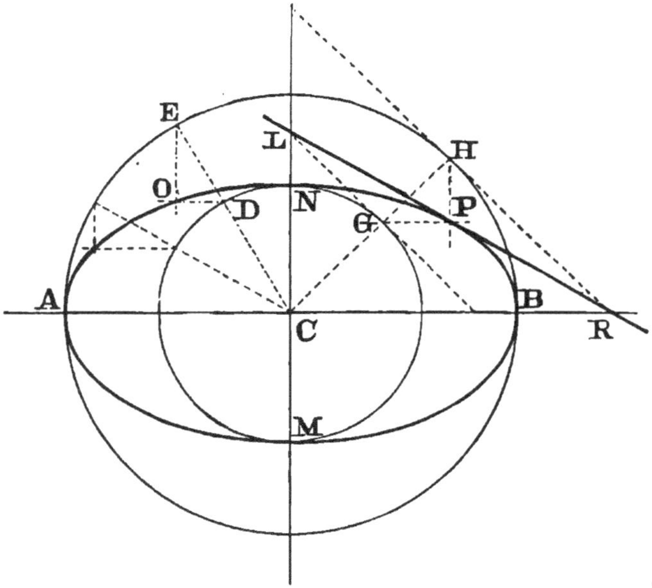

FIG. 7.

duce it to cut MN produced in L, and AB produced, in R. From L draw a tangent to the inner circle, from R a tangent to the outer one, and from C draw a perpendicular to these lines, thus determining G and H; draw through G a parallel to AB, and through H a parallel to MN; these will intersect in P, the point sought.

9. *Second Method.*—Let C, Fig. 8, be the centre, AB the major axis, DE the minor axis. About D with radius AC describe an arc cutting AB in the foci F, F'. About F as a centre, describe an indefinite arc with any radius FH, greater than AF and less than BF. About F' describe another arc, with a radius $F'G = AB - FH$;

this arc will cut the one first drawn in O, O', two points of the required curve.

This method is not eligible for the ordinary purposes of the

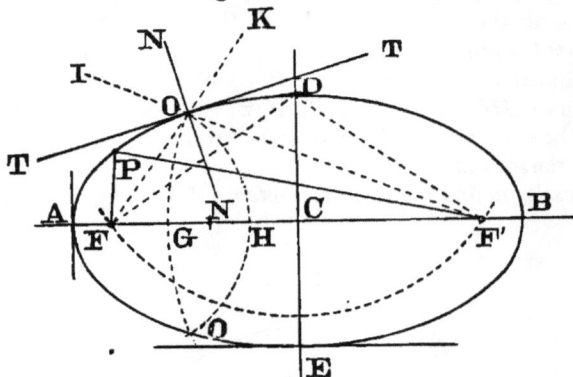

Fig. 8.

draughtsman, since it involves much more labor than the preceding one, and the intersections of the arcs are in many cases too acute to be reliable. But it is of interest as depending upon the property of the curve that the sum of the focal distances is the same for every point upon it, and equal to the major axis; thus,

$$PF + PF' = OF + OF' = AF + AF' = BF + BF' = AB.$$

10. *Third Method.*—Upon this property also depends the operation of drawing the "gardener's ellipse" by the aid of a string and two pins.

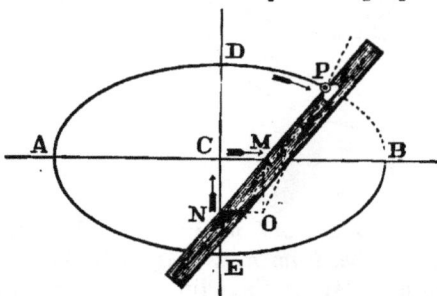

Fig. 9.

Let two fine pins be fixed in the drawing board at F and F'; around these pass a loop of waxed sewing silk, of which the total length is $AB + FF'$: if this loop be kept constantly taut by a pencil P, the latter in moving will trace the curve.

To draw a tangent at any point, as O: produce FO and $F'O$, and bisect the exterior angles FOI, $F'OK$, by the line TT. To draw the normal at the same point, bisect the angles IOK, FOF', by the line NN. Obviously, the axes cut the curve normally at their extremities A, B, D, E.

11. *Fourth Method.*—Fig. 9 illustrates the principle of a common elliptographic trammel. The three points P, M, N, are in one right line, the distance PM being equal to CD, and PN equal to CA. The point M being then kept always upon the line of the major axis, and the point N upon the line of the minor axis, the point P will at all times lie upon the ellipse.

This method is extremely convenient when no great precision is required, the three points being selected upon the graduated edge of a scale, or marked upon the edge of a smoothly cut strip of paper.

To draw the normal at P : at M draw a perpendicular to AB, at N a perpendicular to DE; these perpendiculars intersect at O, and OP is the normal required. For MO and NO are the planes normal to the paths of the two moving points, and O is consequently the instantaneous axis of the whole line NP.

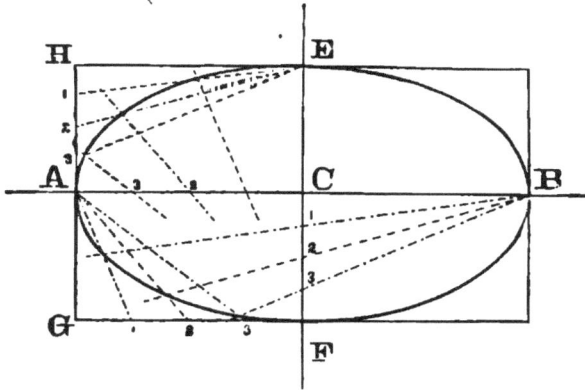

Fig. 10.

12. *Fifth Method.*—To inscribe an ellipse in a given rectangle, Fig. 10. Join the middle points of the opposite sides by the right lines AB, EF; these will be the axes, and intersect in the centre C. Divide the semi-minor axis CF, and the half side GF of the rectangle, into the same number of equal or proportional parts; through the points of subdivision on CF, draw right lines from B, and produce them to intersect the lines drawn from A to the corresponding points on GF; these intersections will lie upon the ellipse.

Or, divide the semi-major axis AC and the half side AH in like proportion, and proceed in a similar manner, the two series of intersecting lines converging in the extremities E and F of the minor axis.

13. The same process is applicable, when it is required to inscribe the ellipse in any given parallelogram, as shown in Fig. 11. But in

this case *AB*, *EF*, will not be the axes.   They are, however, conjugate to each other, for each is parallel to the tangents at the extremities of the other : and since the parallelogram can always be constructed if *AB* and *EF* are given, we have thus a simple and ready .
method of constructing the ellipse upon any pair of conjugate diameters.

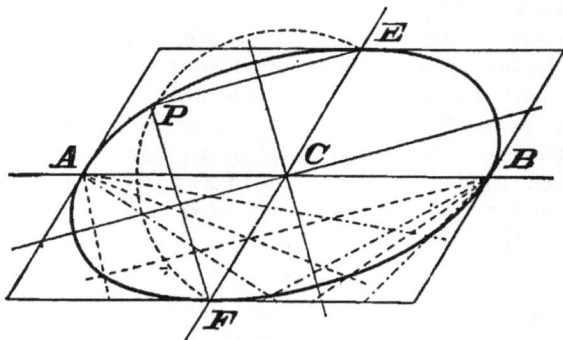

FIG. 11.

In order to determine the directions of the axes, describe about the
centre, *C*, a circle upon either of the given conjugate diameters, as
EF: its circumference cuts the ellipse in *P*, and the supplementary
chords, *PE*, *PF*, are parallel to the axes.

14. *Sixth Method.*—To construct the ellipse by means of ordinates
of the circle. In Fig. 12, let it be required to draw the ellipse of
which *AO* is the semi-major and *AE* the semi-minor axis. Describe a
circle with radius *CR = AE*, and divide *CP* and *OA* into any number
of proportional parts.   At each point of subdivision, erect a perpendicular to *OA*, equal to the ordinate of the circle at the corresponding

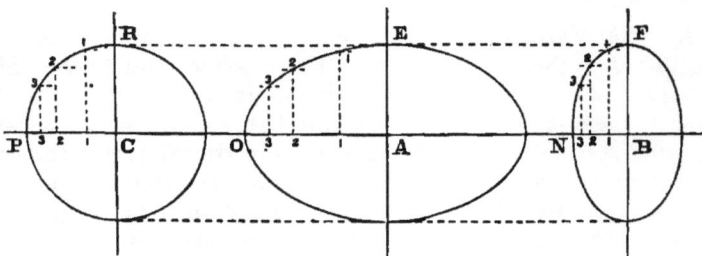

FIG. 12.

point of *CP*; the curve *OE* thus determined is the required ellipse.
Or *CR* may be made equal to the semi-major axis *BF*; then if the
radius *CP* and the semi-minor axis *BN* be similarly subdivided, the

ordinates of the circle will be equal to the corresponding ordinates of the ellipse *NF*.

15. *In a given ellipse, to find the conjugate to a given diameter.* Let *PO*, Fig. 13, be the given diameter. Draw any chord *EF* parallel to *PO*, and bisect it, then the required conjugate diameter *MN* passes through the point of bisection, and *TT*, the tangent at *P*, is parallel to *MN*.

Otherwise: draw the chord *EF* parallel to *PO*, and also the diameter *EG*: then *MN* and *TT* are parallel to the supplementary chord *FG*.

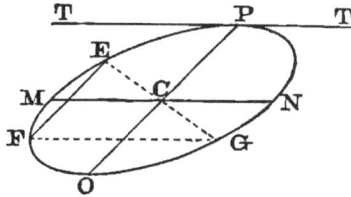

16. *To draw the tangents to an ellipse from a point without.* Let *AA*, Fig. 14, be the major axis, *C* and *O* the foci, *P* the given point. About *P* describe an arc through the nearer focus *C*; intersect this at *M* and *N* by another arc whose centre is *O*, with a radius *ON* equal to *AA*. Draw *OM* and *ON*, cutting the ellipse in *G* and *H*; then *PG* and *PH* will be the required tangents. For,

$$OH + HN = AA = OH + HC; \quad \therefore \ CN = HN;$$

also, $PC = PN$, therefore, *PH* is perpendicular to *CN*, and bisects the angle *CHN*.

Similarly, *PG* is perpendicular to *CM*, and bisects the angle *CGM*.

17. In Fig. 14, draw *PC*, *PO*; then the angles *CPG*, *OPH*, will be equal. For *MN* is a common chord of the circular arcs whose centres are *O* and *P*, therefore, *PO* is perpendicular to *MN*; also, *PH* is perpendicular to *CN*; whence the angles *OPH*, *CNM*, are equal.

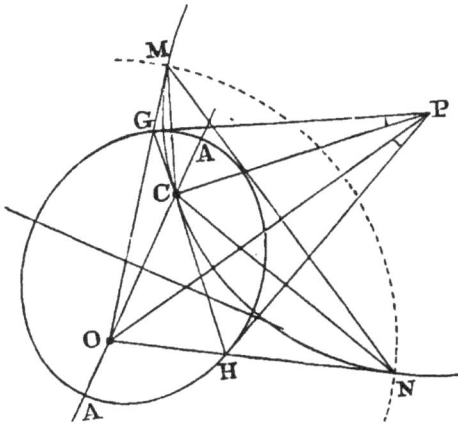

But *CNM*, in the circumference, is one half of *CPM*, at the centre, which again is equal to *CPG*; therefore, $OPH = CPG$.

18. In Chapter XI. (348), the argument in connection with Fig. 213 depends upon the following demonstration.

Let the foci C, E, of the ellipse whose major axis is RR, Fig. 15, be so situated with respect to the foci D, F, of the equal and similar ellipse whose major axis is BB, that the right lines CD, EF, which intersect at P, shall be equal to each other; then the common tangents to the two ellipses, HK and GL, will also intersect in P.

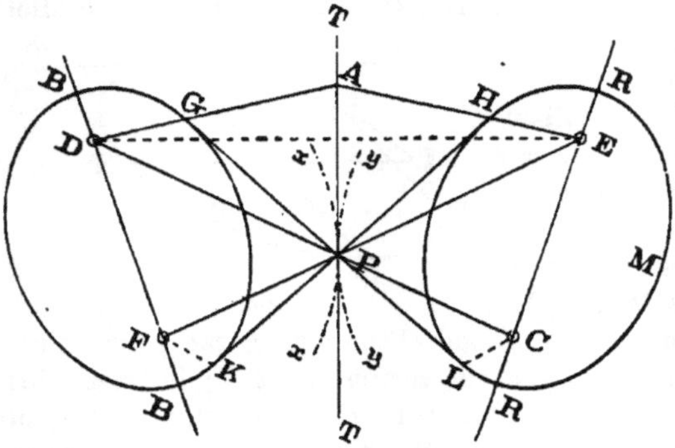

FIG. 15.

For, first, joining DE, the triangles DEF, DEC, have two sides of one respectively equal to two sides of the other, and the third side is common, whence the angle at F is equal to the angle at C; and in the triangles DPF, CPE, the angles at P are equal, whence the angles PDF, PEC, are also equal: and since $EC = DF$, we have also $DP = PE$, and $FP = PC$.

Next, draw TT bisecting the angles DPE and FPC; by the preceding construction (16) draw PG tangent to one ellipse, and find the point of tangency G; draw DG, and produce it to cut TT in A; draw AE cutting the other ellipse in H, and join HP.

Then the triangles DPA, EPA, have the angles at P equal, the side $PD =$ side PE, and the side PA common; the triangles are therefore similar, and we have

$$ADP = AEP;$$
but $\overline{PDF = PEC,}$
$\therefore \quad \overline{ADF = AEC}:$

consequently, since the ellipses are similar, $HE = GD$, and the triangles PHE, PGD, having the two sides and the included angle of one equal to the two sides and the included angle of the other, are

FIG. 16.

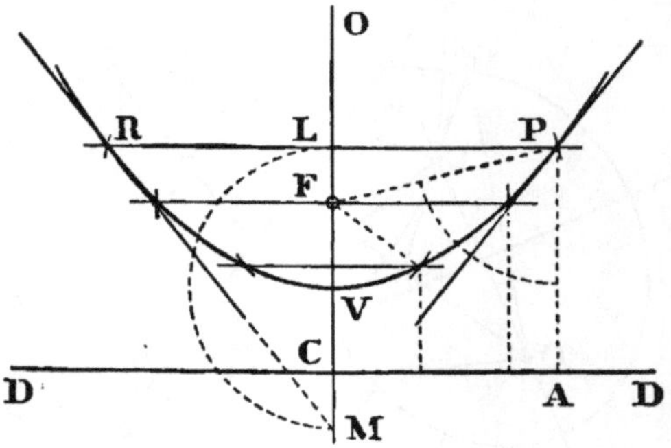

similar and equal; whence PH is equal to PG and tangent to the ellipse at H, and also the angle GPD is equal to the angle HPE.

Now draw PK tangent to the ellipse at K; then (17) the angle FPK is equal to the angle GPD, and, therefore, to the angle HPE. But FE is a right line by hypothesis, therefore, HPK is also a right line and tangent to both ellipses. In like manner it may be shown that PL, tangent to the ellipse on the right, is a prolongation of PG.

It will also be observed that if xx be an arc of an ellipse whose foci are D, F, the major axis being equal to $DP + PF$, and yy an arc of a similar and equal one whose foci are C, E, then TT will be tangent at P to both those ellipses.

THE PARABOLA.

19. The Parabola, Fig. 16, is a curve every point of which is equally

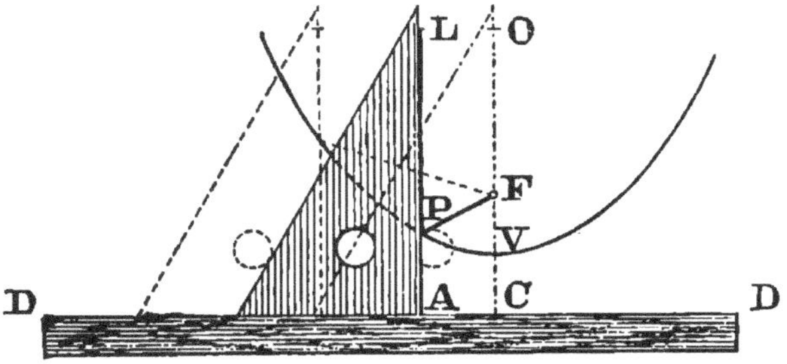

Fig. 17.

distant from a given point F, called the focus, and a given right line DD, called the directrix. Hence the axis OFC is perpendicular to DD, and the vertex V lies at the middle point of FC. Draw any parallel to DD, as RLP; then the points in which this line is cut by an arc described about F, with a radius equal to LC, will lie upon the curve.

To draw a tangent at any point, as P: Draw FP, and also PA perpendicular to DD: then the required tangent bisects the angle FPA.

Otherwise: let R be the given point. Draw RL perpendicular to the axis, and on the axis make $VM = LV$; then MR is the required tangent.

Second Method.—The parabola may be traced mechanically, as shown in Fig. 17. Let DD be a ruler fixed to the drawing-board. Let a fine thread whose length is equal to LA be fixed at one end to the point L on the vertical side of the right-angled triangle, and at the

20

other end to the drawing-board at F, the focus of the required curve. Then by sliding the triangle along the ruler, keeping the thread taut by a pencil P which always touches the side AL of the triangle, the

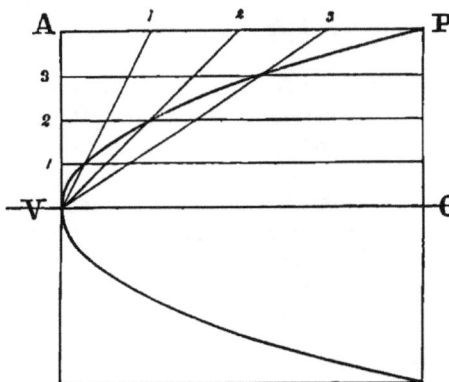

FIG. 18.

motion of the pencil will be so controlled as to trace a parabola, since PF is always equal to PA.

20. *Third Method.—* In Fig. 18, let V be the vertex, VO the axis, and P a point in the required curve.

Draw PO, VA, perpendicular to the axis and equal to each other; divide VA in any manner, and through the points of division draw parallels to VO: divide AP in like proportion, and join the points of division with the vertex V. The intersections of the lines thus drawn through corresponding points upon VA and AP will lie upon the parabola VP.

21. *Fourth Method.—*In Fig. 19, let V be the vertex, VO the axis,

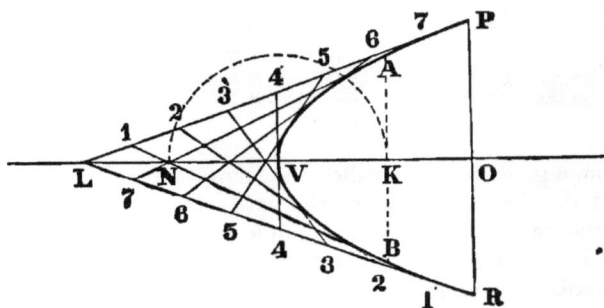

FIG. 19.

and P a point through which the curve is to pass.

Draw PR perpendicular to VO, and make $RO = PO$. On the axis set off $VL = VO$; draw PL and RL, divide them into the same number of equal parts, number the points of division in opposite directions, and join the points correspondingly numbered, as

1, 1, 2, 2, etc.: the lines thus drawn will be tangents to the required curve.

To find the point of tangency on any one of these lines, for instance 1, 1. This line cuts the axis at N; set off on the axis, $VK = VN$, and draw KB perpendicular to VO, cutting 1, 1, in the required point B.

THE HYPERBOLA.

22. The Hyperbola is a plane curve generated by the motion of a point subject to the condition that the *difference* of its distances from two fixed points called the foci shall always be equal to a given line, whose length must be less than the distance between the foci.

In Fig. 20, set off $CF = CF'$, and let F, F', be the foci: set off also $CA = CB$, and let AB be the given constant difference. It is evident that

$$FB - BF' = AF' - AF = AB;$$

therefore A and B satisfy the conditions and are points on the curve. With any radius FO greater than FB, describe an arc about F as a centre; then about F' describe, with a radius $F'O = FO - AB$, another arc, which will cut the one first drawn in two points O, O', of the required hyperbola. Since with the same radii, arcs may be described about the other foci, it follows that the curve is composed of two equal and opposite branches; and since FO may be of any length, these branches are infinite. The point C is called the centre, and the line AB the

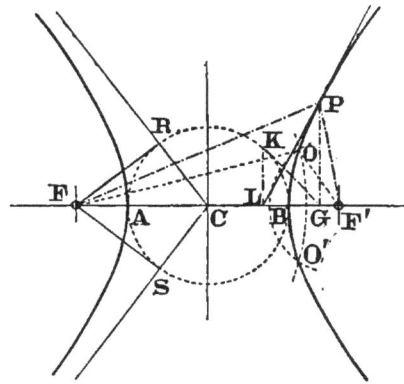

FIG. 20.

major axis, to which the minor axis is perpendicular.

To draw a tangent at any point of the hyperbola, as P: draw PF, PF', and bisect the angle between them.

Otherwise thus: describe a circle upon AB as a diameter, let fall PG perpendicular upon the major axis, and from its foot G draw GK tangent to that circle. Find the point of tangency K, and draw KL perpendicular to AB and cutting it in L; then LP is the required tangent.

By the converse operation we may find the point of contact if the tangent be given.

To find the asymptotes. From either focus, as F, draw tangents to the circle described upon AB : find the points of tangency R and S, then CR and CS are the asymptotes.

23. *Second Method.* In Fig. 21, let C be the centre, AB the major axis, and O a point through which the hyperbola is to pass.

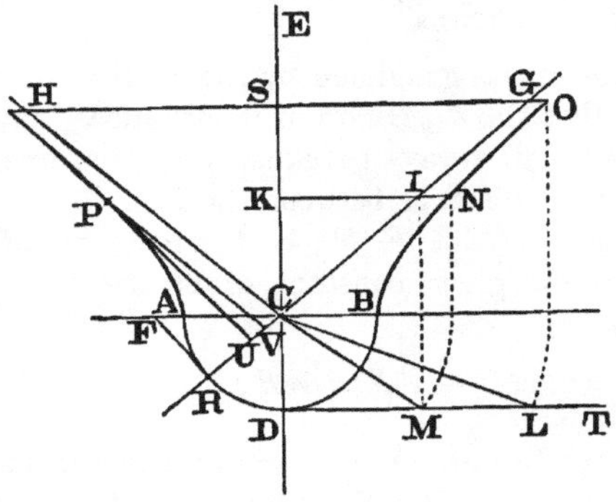

Describe upon AB the semicircle ADB, and draw DCE perpendicular to AB, also DT tangent to the semicircle. Draw OS perpendicular to DE, and with centre C and radius $CL = OS$, describe an arc cutting DT in L; on SO set off $SG = DL$, then CG is an asymptote to the curve.

Fig. 21.

Draw any line KI parallel to AB, and produce it ; set off DM on the tangent, equal to KI, then on KI produced set off $KN = CM$, and N will be a point on the curve.

To find the focus. Produce GC to cut the semi-circumference in R, at which point draw a perpendicular to GR ; which will cut the major axis in the focus F.

To draw a tangent at any point, P. Draw PV parallel to the asymptote HC, cutting the other asymptote GR in V. On GR set off $VU = CV$, and draw PU, which will be the tangent required.

24. *Third Method.* Given, the asymptotes and one point in the curve. The construction depends upon the property that if any line be drawn cutting both asymptotes, the parts intercepted between each of those lines and the curve are equal.

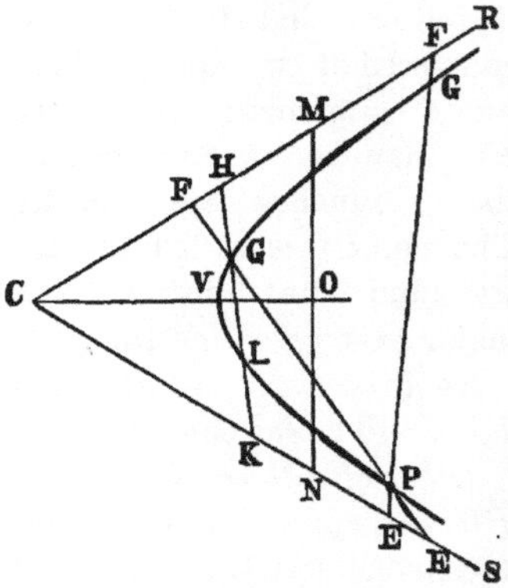

In Fig. 22, let CR, CS, be the asymptotes, P the given point. Through P draw any line at pleasure, as EF, and make $FG = EP$; then G will lie upon the curve. Any

Fig. 22.

point thus found may then be treated in like manner, thus : drawing *HGK*, make $KL = HG$, then *L* is a point upon the hyperbola ; and so on.

To find the vertex. Set off upon the asymptotes any equal distances *CM, CN* ; draw *MN*, and bisect it by a perpendicular *CO*, which will be in the direction of the major axis, and cut the curve at the vertex *V*.

25. *Fourth Method.* Given, in Fig. 23, the major axis *AB,* and *P* a point in the required curve.

Draw *PO, BE*, perpendicular to *AB* and equal to each other, and join *PE*. Divide *PO* and *PE* into the same number of equal parts, numbering the points of division from *P* upon each line. From *A* draw lines to the points upon *PO*, and from *B*, lines to the points upon *PE* ; the

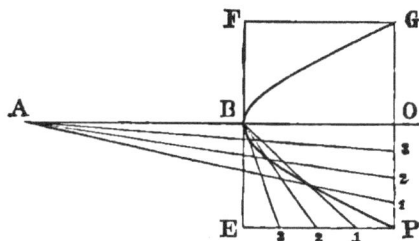

Fig. 23.

intersections of the lines thus drawn to corresponding points of division, as for instance, *A*1, *B*1, will lie upon the hyperbola required.

THE SPIRAL OF ARCHIMEDES.

26. **A Spiral** is a plane curve traced by a marking point which moves along a right line, while at the same time the right line revolves about one of its points as a fixed centre.

This fixed centre is called the *pole,* and a right line drawn from it to any point of the curve is called a *radiant,* or *radius vector.* Supposing the angular velocity of the revolution to be uniform, the linear motion of the marking point along the radius vector may be governed by any law at pleasure, and thus an infinite variety of spirals may be produced.

In Fig. 24 both motions are uniform, and the resulting curve is the well-known spiral of Archimedes, also called the *equable spiral,* because the rate of expansion is constant, so that the distance between any two consecutive coils, measured on a radiant, is the same.

If this distance or rate of expansion be given : with it as radius, describe a circle about the pole as centre, and divide its circumference into any number of equal parts by radial lines ; divide the given distance into the same number of equal parts, and set out from the pole *P,* upon consecutive radii, as *I, II, III, IV,* distances equal to one, two, three, etc., of these subdivisions: the spiral is then drawn through the points thus determined.

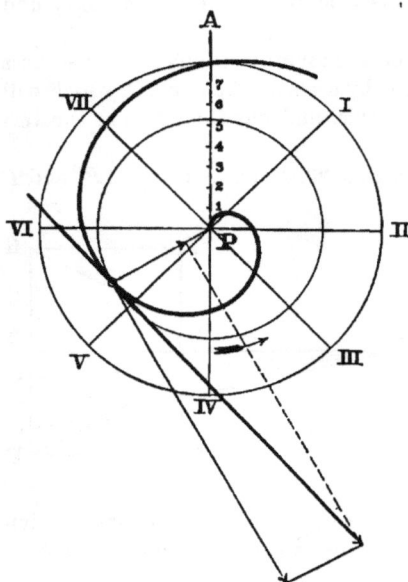

FIG. 24.

If any two radiants be given: divide the included angle into any number of equal parts, and the difference between the given radiants into the same number; then set off on the first dividing radial line, a distance equal to the least radiant plus one of the subdivisions of the difference, and so on, each successive radiant being greater than the preceding, by one of these subdivisions.

To draw a tangent at any point. The tangent to this spiral can readily be drawn by geometrical construction; for the motion of the tracing point is the resultant of two known components. The direction of the circular motion is that of the tangent to the circle through the given point, of which the pole is the centre, and its magnitude may be represented by the circumference of that circle. While making one revolution, the tracing point also travels along the radiant through a given distance, which is the other component; and the resultant, representing both in magnitude and direction the actual motion of the point at the instant, is the required tangent to the spiral path. It is not necessary to set off the entire circumference in making this construction; should this be inconveniently large, we may use one-half, one-third, or any other fraction, reducing the radial component in the same proportion.

27. In Fig. 24 PA is the zero line from which the circular divisions are reckoned in constructing the curve in the first manner above explained. And as the length of this radiant is zero, the circumference of the circle described with it as radius is also zero, but the other, or radial component in the determination of the tangent is constant; the zero line, therefore, is tangent to the curve at the pole. This is called the *axis* of the spiral; and it will be noted that

1. The length of the radiant varies *directly* as the angle of rotation from this fixed axis.
2. The lengths of successive radiants which include equal angles, having a constant difference, form a series in arithmetical progression.
3. This spiral consists of two infinite branches, curving in opposite directions, and symmetrically placed with reference to a line passing through the pole, perpendicular to the axis.

THE RECIPROCAL SPIRAL.

28. This is the exact converse of the Archimedean spiral, the lengths of the radiants varying *inversely* as the angle of rotation from a fixed axis.

In Fig. 25 describe about the pole B a circle with any convenient radius, and beginning at the axis BA, divide it into any number of

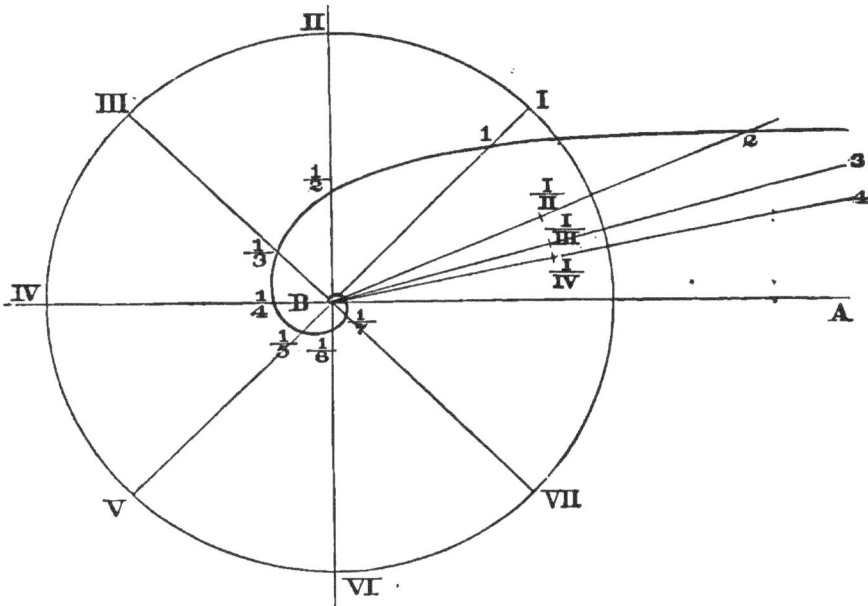

Fig. 25.

qual parts at the points *I, II, III,* etc. On the radii through these points set off in their order distances measuring respectively 1, $\frac{1}{2}$, $\frac{1}{3}$, $\frac{1}{4}$, etc., by any scale of equal parts. Draw also lines from B, making with BA angles equal to $\frac{1}{4}$, $\frac{1}{3}$, $\frac{1}{2}$, etc., of the angle ABI; and set off on them by the same scale distances measuring 4, 3, 2, etc.; the spiral is then drawn through the points thus determined.

This curve, evidently, makes an infinite number of convolutions about the pole, which it continually approaches, but never reaches.

THE LOGARITHMIC SPIRAL.

29. This curve presents to the Archimedean spiral the contrast that the successive radiants which include equal angles form a series in *geometrical* progression : each being greater or less than the preceding in a certain constant ratio, instead of by a constant distance.

Hence the radiant which bisects the angle between two others is a mean proportional between them ; thus in Fig. 26, if the angles APK, KPH, HPC, be equal, we have

$$AP : PK :: PK : PH.$$
$$PK : PH :: PH : PC, \text{ and so on.}$$

Had the radiants AP, PH, and their included angle, then, been given, the intermediate point K would have been found by bisecting

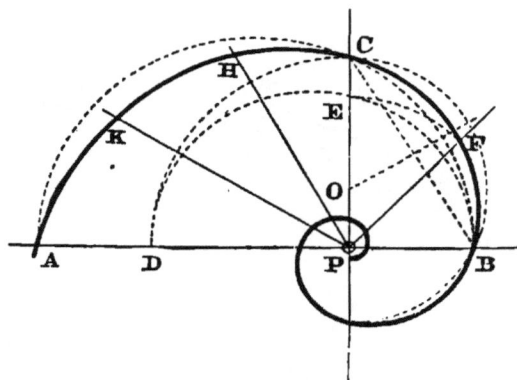

Fig. 26.

the given angle, and setting off PK, a mean proportional between the two given radiants.

If the given radiants lie in the same straight line, as AP, PB, the construction is the same— this angle of 180° is bisected by CP perpendicular to AB, and describing on AB a semicircle, cutting this perpendicular in C, we have $CP^2 = AP . PB$, as required, therefore C is a point upon the spiral. Toward A set off $PD = PC$, and describe a semicircle on DB as a diameter, cutting CP in E. Bisect the angle CPB, and on the bisector set off $PF = PE$, then F is also a point on the curve.

In like manner the spiral may be extended as far as desired : thus, drawing the chord CB, bisect it by a perpendicular cutting CP in O, and about O describe a semicircle passing through C and B ; this semicircle will cut CP produced in G, which will lie upon the spiral, since by this construction $PB^2 = CP. \ CG.$

30. In view of the practical application of curves of a spiral form in the construction of cams for communicating definite motion to one piece by the rotation of another, it is of interest to note that by setting up the successive radiants as equidistant ordinates, any spiral may be transformed into a curve capable of transmitting motion with corresponding changes in velocity, while the driver moves in a right line. And conversely, any curve may be transformed into a spiral possessing analogous mechanical properties, by setting out its equidistant ordinates as the radiants, taking care that the successive ones include equal angles.

THE HELIX.

31. The Helix is a curve traced upon the surface of a cylinder of revolution by a point which moves uniformly around the axis, and at the same time travels uniformly in a direction parallel to the axis; the rates of the two motions being entirely independent of each other.

Thus in Fig. 27, let the relative motions be such that the marking

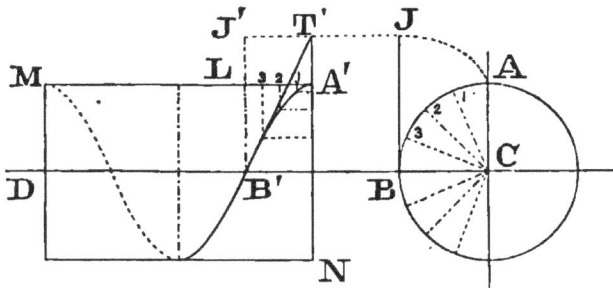

Fig. 27.

point shall traverse the distance $A'M$ while going once around the cylinder; then in going half way around, it will travel half as far; in going one quarter around, or through the arc AB, it will travel one quarter as far, or through a distance equal to AL, which will, evidently, bring it to the position B'. Intermediate points are readily found by subdividing the arc AB and the distance AL into the same number of equal parts at 1, 2, 3, etc., and projecting the former points of subdivision to lines drawn through the latter, perpendicular to the axis, as shown.

If the cylinder be cut along the line $A'M$, and unrolled into a plane, it will develope into a rectangular sheet whose length is equal to the circumference, and the helix will develope into the diagonal of this rectangle.

To draw the tangent at any point, as B'. The curve pierces the base of the cylinder at A', corresponding to A in the end view. Project B' to B, and perpendicular to the radius BC, set off BJ equal to the arc BA. Project J to T' upon NA' produced, then $B'T'$ is the required tangent.

By projecting J to J' upon a perpendicular to the axis through the given point B', it will be readily perceived that the same result would have been reached by compounding the two motions of the point, the resultant being the tangent.

A curve analogous to the helix may also be traced upon the surface of a cone, or of a hyperboloid, by a point moving uniformly along an element, while the surface at the same time rotates uniformly about its axis. (For illustration of the conical helix, see Fig. 226; the hyperboloidal helix is represented in Fig. 243.)

DRAWING OF ROLLED CURVES.

32. In Fig. 28 AA is a curved ruler fixed to the drawing-board, and BB is a free one rolling along it. Let a pencil be fixed to and carried

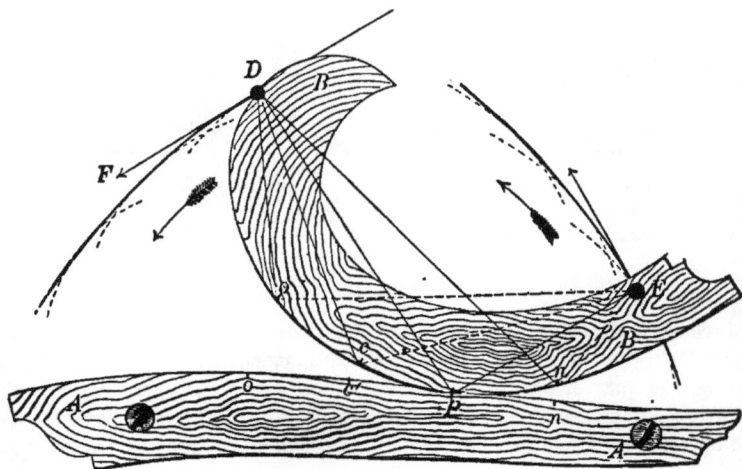

Fig. 28.

by the latter, either in the contact edge, as at D, or at any distance from it, as at E. Since P, the present point of contact, is the instantaneous axis, the motion of D is in the direction DF, perpendicular to DP, the contact radius. DF is, therefore, tangent to the path of D, traced as the ruler BB rolls ; but it is also tangent to the circular arc whose centre is D and radius PD, consequently the path of D is also tangent to that arc.

Let the arcs Pc, Po, of BB, be equal to the arcs Pc', Po', of AA, then cD will be contact radius when c reaches c', and oD when o reaches o'. If, then, we describe with these radii circular arcs about c' and o', the curve traced by D will be tangent to those arcs ; and that traced by E will be tangent to arcs about the same centres with aE and oE as radii.

Curves thus described by points carried by one line which rolls upon another are called *rolled curves, roulettes,* or *epitrochoids ;* and the drawing of a series of tangent arcs as above explained is the readiest and most reliable known method of laying them out.

The line which carries the tracing point is called the *generatrix* or *describing line,* and the one in contact with which it rolls is called the *directrix* or *base line ;* either of these may be straight, or both may be curved.

THE CYCLOID.

33. This curve is traced by a point in the circumference of a circle which rolls upon its tangent.

In Fig. 29 find Aa', the length of a convenient fraction Aa of the circumference ; step this off upon the tangent the required number of

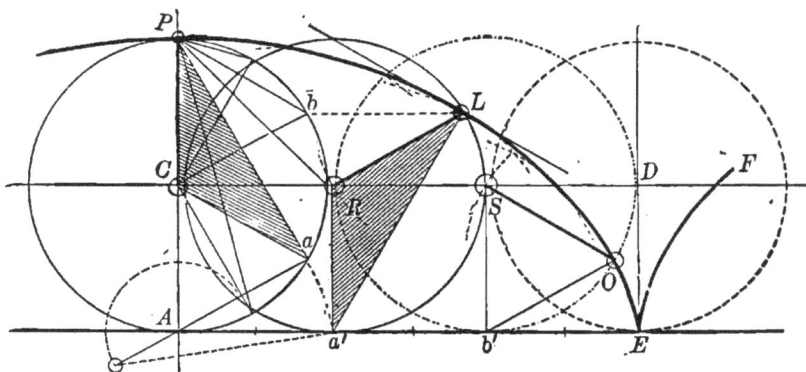

Fig. 29.

times, making AE equal to the semicircumference. Divide each into the same number of equal parts, draw chords from P to the points of division on the semicircle, with which as radii, strike arcs about the corresponding points on AE ; the cycloid is tangent to all these arcs.

To find points on the curve : When aC becomes contact radius, it has the position $a'R$, perpendicular to AE. The angle aCP remains unchanged; therefore, make $a'RL$ equal to it; then RL is the *generating radius* in its new position, and L is a point on the cycloid. Also, $a'L$, the *instantaneous radius,* is normal, and a perpendicular to it is tangent to the curve at L.

Conversely : Let O be any point on the curve; about this as a centre describe an arc with radius equal to CP. This arc cuts CD, the path of the centre of the rolling circle, in S; then OS is the generating radius ; Sb', perpendicular to AE, is the contact radius, and $b'O$ is normal to the cycloid.

THE EPICYCLOID.

34. The describing circle, in Fig. 30, rolls on the *outside* of another whose centre is G.

Draw the common tangent at A, set off upon it the length of Aa

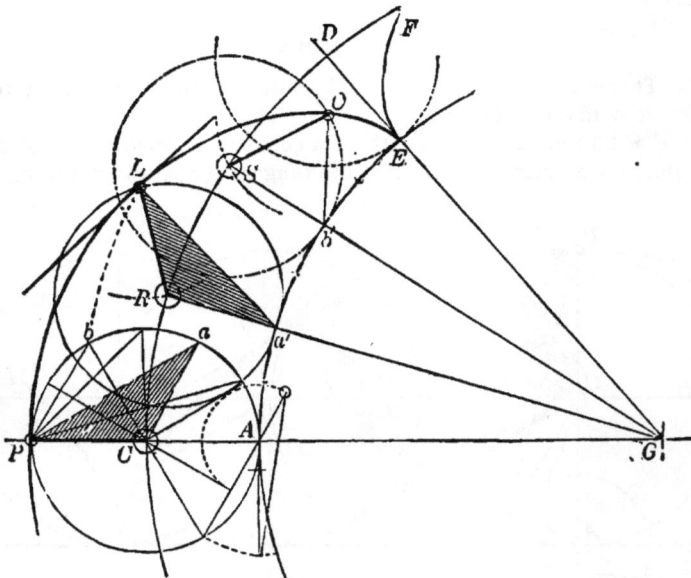

Fig. 30.

(any convenient fraction of semicircumference AP) and find the arc of the base circle equal to that length. Step this off as above, making $AE =$ semicircumference AP. The curve is drawn by tangent arcs in the same manner as the cycloid.

The path of the centre of the describing circle is in this case another circle whose centre is G; and the contact radii $a'R$, $b'S$, are prolongations of the radii Ga', Gb', of the base circle; which slightly modifies the processes of finding the point of the curve corresponding to a given point of contact and the converse.

THE HYPOCYCLOID.

35. Traced, as shown in Fig. 31, by a point in the circumference of a circle rolling *inside* another.

The construction is in all respects the same as in the case of the

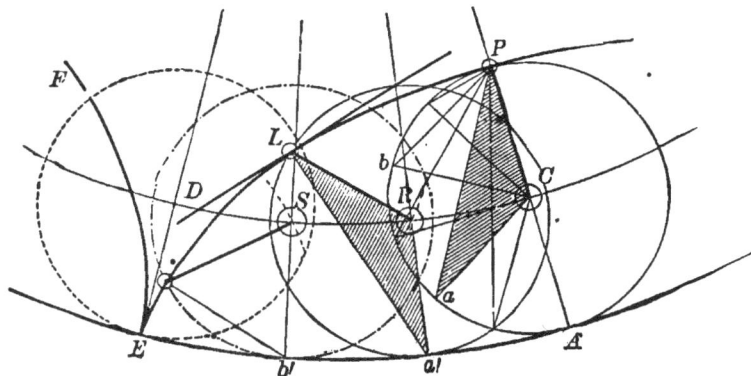

Fig. 31.

epicycloid, and the diagrams being lettered to correspond throughout, no further explanation is needed.

In all three of these curves, if the rolling continue beyond E, a new branch springs up, which is of course perfectly symmetrical with EL.

It is to be particularly noted that these branches are *tangent* to ED, and therefore to each other at E.

THE INVOLUTE OF THE CIRCLE.

36. This is in a manner the converse of the cycloid, being generated by the rolling of a tangent right line upon a circle. Thus in Fig. 32, the ruler, carrying in the line of its edge the pencil P, while rolling around the cylinder describes the curve in question.

It may also be regarded as generated by unwinding an inextensible fine thread from a cylinder: the thread being always taut and always tangent to the circle, its length is equal to that of the arc from which it was unwound; thus, beginning at O, make the tangents AP, BE, DF, respectively equal to the arcs OA, OAB, OBD, and so on; then the curve passes through the extremities P, E, F, G, etc., of these tangents.

The tangent OG, then, will be equal to the circumference of the circle, and if the unwinding be continued, the result will be the formation of a spiral, the distance between the successive convolutions,

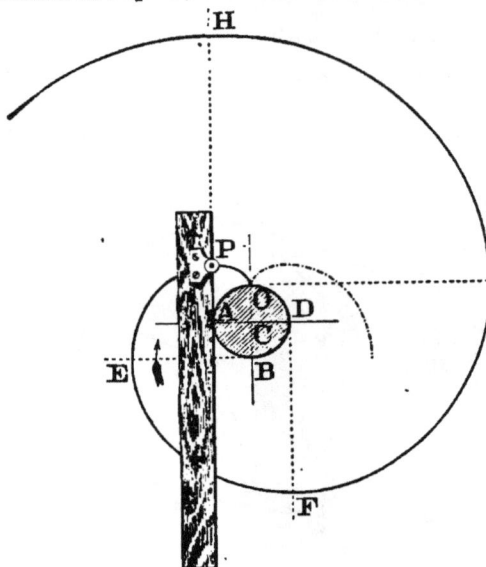

FIG. 32.

measured on the tangent to the base circle, as for instance PH, being constant and equal to the circumference.

Considering it as traced by the ruler as in the figure, it will be seen that as the point of contact A is the instantaneous axis, the edge AP is normal to the curve. This being true for all positions of the ruler, we have the simple construction that the normal at any point of the curve is tangent to the base circle.

If the rolling of the ruler continue in the direction of the arrow, it is evident that after P reaches O a new branch will be formed as shown by the dotted line; the two branches being tangent to each other, and to the radius CO at its extremity.

THE EPITROCHOID.

37. Although the term epitrochoidal is used in a general sense, including all rolled curves, yet custom sanctions also a special sense, and the curve traced by the rolling of one circle upon another, when the marking point is not situated upon the circumference, is the one ordinarily meant when " *The Epitrochoid* " simply is mentioned with no qualifying word in connection with it.

In Fig. 33, if the circle whose centre is C, roll upon the circle whose centre is D, carrying the marking point P situated without the circumference, it describes the looped curve PLE, called the *curtate epitrochoid.* If the tracing point be situated at *V*, within the circumference, the resulting waved curve *VWX* is called the *prolate* epitrochoid.

Since the rolling circle measures itself off upon the base circle as in the preceding cases, the position of the generating radius can always be found as in the construction of the epicycloid, and its length being constant, points on the curve are readily found ; and the instantaneous radius being always normal to the epitrochoid, the tangent at any

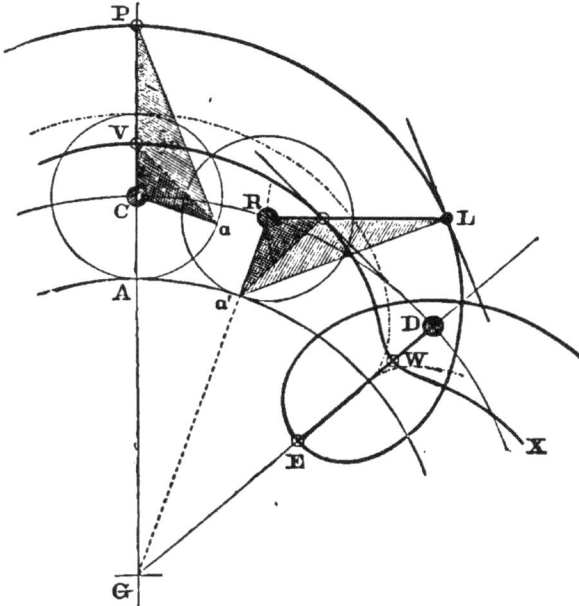

FIG. 33.

point may be drawn with the same facility. For example, let it be required to draw the tangent at *L* ; with radius equal to *CP* describe an arc cutting the path of the centre in *R*, draw *RG*, cutting the base circle in *a'* : then *a'R* is contact radius, *RL* is generating radius, *a'L* is the normal, and a perpendicular to it is the required tangent.

DOUBLE GENERATION OF THE EPICYCLOID.

38. By way of distinction, the curve traced as in Fig. 30, by the rolling of one circle upon another in external contact, is called an *external* epicycloid. But if the contact be internal, the curve traced by the rolling of the larger upon the smaller is called an *internal* epicycloid ; and in Fig. 34, the circle whose centre is B, rolling upon the fixed circle whose centre is D, and carrying the marking point F, thus describes this curve, of which FL is a portion.

Now let the same point F be carried by the circle whose centre is C, whose diameter FA is equal to the difference between the diameters FG and AG of the other circles : it will then trace the same path FL.

First, let the three centres, C, B, D, the two points of contact A and G, and the tracing point F, lie in one straight line FG. Then through A draw EAH in any direction at pleasure, and draw FI par-

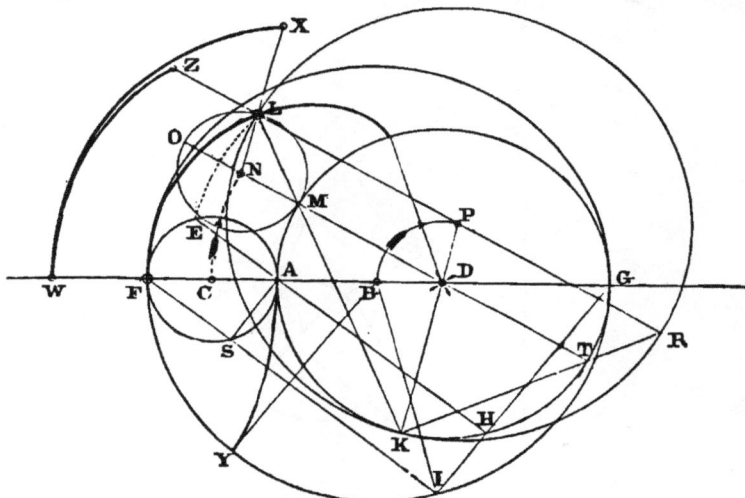

Fig. 34.

allel to it. Then the three chords EA, AH, FI, subtending equal angles in the three circles, are proportional to the radii ; therefore,

$$EA + AH = FS + SI.$$

Also, the triangles AHG, FIG, are similar.

Set off the arc $GHK =$ arc GI. Draw the chord $KM = HA$, and prolong it to L, making $ML = AE$. Draw the diameter MDT, and a parallel to it through L, cutting KT produced in the point R. Then the triangles MKT, LKR, are respectively similar and equal to the triangles AHG, FIG.

Now KD bisects MT in D, and when produced will bisect LR in P ; therefore, the circle round LKR will be tangent at K to the circle round MKT and AHG. Consequently, if the circle whose centre is

B rolls upon the circle whose centre is D until I reaches K, its centre will then be at P, and KL will be the new position of the chord IF; the point F meantime tracing the internal epicycloid FL.

39. Produce TM to O, making $MN = NO = AC$. Then the circle whose centre is N and radius NM, will pass through L, and the arcs OL, FE, will be equal ; because $ML = AE$, and the angles CAE, NML, are equal.

The point F will, therefore, reach L, if it first describe the arc FE about centre C, and then the arc EL about centre D. And it will be perceived that this is equivalent to the rolling of the circle whose centre is C, upon the circle whose centre is D, if it be proved that the arcs AM, OL, are equal.

In order to do this, we have, *first*,

$$BD = DP = AC = NL, \text{ by hypothesis,}$$

and

$$LP = ND, \qquad\qquad \text{by construction ;}$$

$$\therefore \text{ the angles } ONL, MDP, \text{ are equal.}$$

Draw BI; then since arc $KHG = $ arc GI, we have, *second*,

$$KDG : IBG :: BG : DG ;$$

but

$$IBG = 2\,(IFG) = 2\,(CAE) = 2\,(NML) = ONL = MDP,$$

also

$$KDG = ADP.$$

Whence

$$ADP : MDP :: BG : DG ;$$

or

$$\therefore ADP - MDP : MDP : BG - DG : DG,$$

$$ADM : MDP :: BD : DG,$$

$$=\qquad ADM : ONL :: ON : AD$$

$$\therefore \text{ arc } AM = \text{ arc } OL = \text{ arc } FE.$$

Q. E. D.

40. Although the epicycloids thus traced by the rolling of the two circles upon the same base circle are identical, it is not to be assumed that the epitrochoids generated by marking points not in the circumferences of the describing circles will be the same.

On the contrary, they will be quite different, as shown in the diagram. If the generating radius BF be extended to W, the latter point will trace the *internal epitrochoid* WZ, during the generation

21

of the internal epicycloid *FL*, the final position of the generating radius being *PLZ*. On the other hand, prolonging the radius *CF* to the same point *W*, we perceive that during the generation of the external epicycloid *FL* by the rolling of the smaller circle, the *external epitrochoid WX* will be described, and *NLX* will be the final position of the generating radius *CW*.

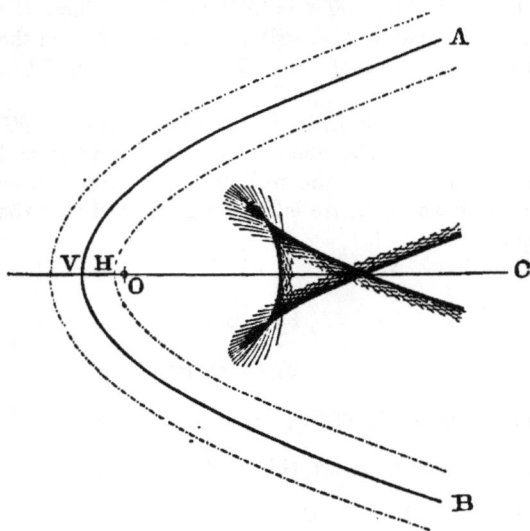

Fig. 35.

41. Every internal epicycloid, then, is also an external one ; and it may be remarked that the latter mode of generation is usually more convenient in practical execution.

Similarly, every hypocycloid is capable of two generations. Thus, if in Fig. 34 we take *FIG* as the fixed base circle, the hypocycloid *AY* will be traced by the point *A*, whether it be carried by the semi-circumference *FSA* of the smaller circle, which is equal to *FY*, or by the semicircumference *GHA* of the larger, which is equal to *GIY*.

PARALLEL CURVES.

42. **Parallel Curves** are those whose normal distance from each other is everywhere the same. If one curve and the length of the normal be given, the other is readily mapped out by merely describing any number of circular arcs with their centres upon the first curve, and a radius equal to the normal : the envelope of these arcs is the parallel curve.

At first thought it is natural to suppose that two curves thus related will be similar in form, like two concentric circles. And this will really be the case, if the derived curve be exterior to the first. But if it lie within, that is, upon the concave side of the fundamental curve, quite curious and unexpected results may arise, of which Figs. 35 and 36 are sufficiently remarkable illustrations.

In the former, the original curve is the parabola *A VB*, of which *VC* is the axis and *O* the focus.

The two dotted curves resemble it in form, but though both are symmetrical, neither is a true parabola. The normal distance *VH* is less than *VO* ; but the result of assuming a greater one is the formation of the figure of which the construction is shown, and its resemblance to the original curve is at least not striking.

In Fig. 36 the ellipse of which the axes are *AB*, *DE*, is the fundamental curve ; and as before, when the normal distance is small, the parallel curves are somewhat similar to it, although neither is a true ellipse. But again, upon increasing the normal, the derived curve loses all resemblance to the original, and developes the four-cusped

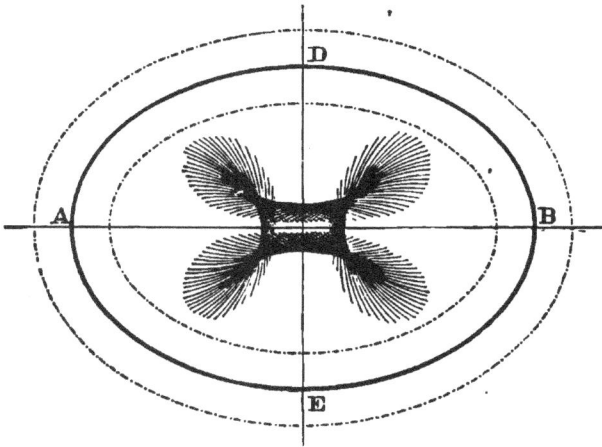

Fig. 36.

figure shown within, symmetrical about the centre, and also with respect to the axes of the ellipse.

THE LIMACON AND THE PARALLEL TO THE EPICYCLOID.

43. Mention has previously been made of these curves, whose peculiarities, which were shown to have direct practical bearing upon the true theory of pin gearing, merit for that reason further examination.

In Fig. 37 the epicycloid *PBE* is traced by the rolling of the circle from *C* to *D*, and the extremity *I* of the normal *PI* meantime traces the parallel curve *IOHF*, consisting of two branches.

It is worthy of note that this curve will always exhibit these two branches, however small the normal distance chosen may be ; which will be readily seen if we suppose it to be traced in the opposite direc-

tion by the rolling of the circle from D to C, in which case the initial
motion of the circle having the direction of the arrow at E, and the
point E being the centre of rotation, it is apparent that the initial
motion of the extremity of the normal, be that line long or short, will
have the direction of the arrow at F; so that under no circumstances
will there be an interior curve similar to the original, as was the case
with the ellipse and the parabola.

Beginning at F, then, this curve descends to some point H, and
then begins to rise, the normal taking successively the positions EF,
WU, SH. The fact that there will be a *cusp* at the lowest point, is

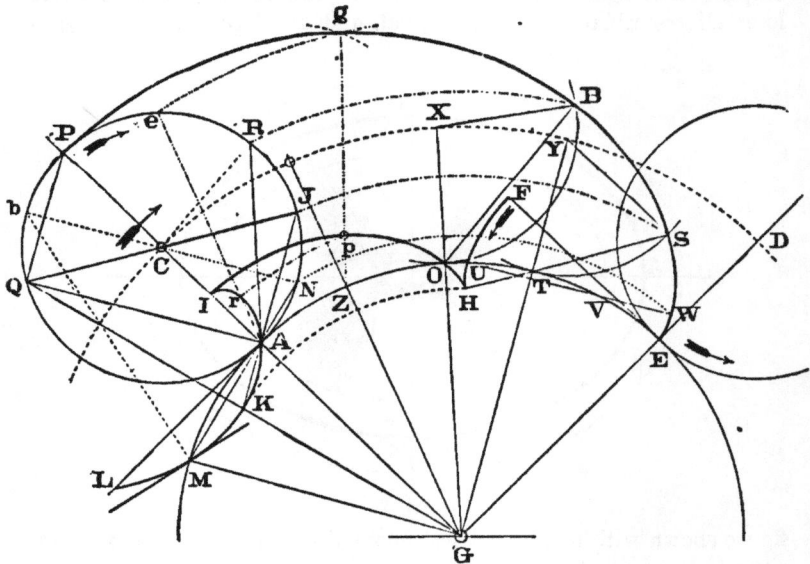

Fig. 37.

shown by the consideration that were there either a wave or a loop,
there would also be a *tangent* in some direction nearly coincident with
HS, the impossibility of which is perfectly obvious.

44. Now again beginning at P: the rolling of the circle is a compound motion, consisting of a rotation around the travelling centre C,
and a revolution around the fixed centre G. We may then suppose
these two motions to take place separately, in succession.

Thus if the generating circle be first turned through the angle
PCR, and then be made to revolve about G through an angle measured by an arc AO of the base circle equal to PR, the tracing point

P will have reached the position B on the epicycloid, and OB, the normal to that curve, will be the new position of the chord AR.

We may reverse this process; if, for instance, we select any point S on the epicycloid, and describe a circle about G, cutting the generatrix at J, then AJ is the chord which, by the preceding operation, will become the normal ST, and the arc PJ will be equal to the arc AT.

45. A series of chords in the generating circle, drawn through the point A, then, are the lines which will eventually become normals to the epicycloid. If now we set off upon each of these chords, from the circumference toward A, a distance equal to PI, as er, JK, etc., the points thus located will determine a new curve $IAKL$, called the **Limacon**.

And just as points upon the epicycloid are derived as above from the upper extremities of these chords, so points upon the parallel curve are derived from their lower extremities. Or in other words, as points of the former are derived from points upon the generating circle, so points on the latter are derived from points upon the limacon. For instance, the point g of the epicycloid would be determined by rotating P about C to e, then revolving the whole generatrix about G, the angle being measured by the arc AZ equal to Pe: during this revolution the point r of the limacon goes through the same angle, giving the position p of a point upon the parallel curve, which lies upon gZ, the final position of eA.

By the aid of the limacon, then, we may determine the direction of a normal at any point of the parallel curve : let p, for instance, be the given point; we first describe an area bout G through p, cutting the limacon at r, then draw Ar and produce it to cut the generating circle at e, and finally draw through e an arc cutting the epicycloid in g; then pg is the required normal.

46. To Draw a Normal to a Given Epicycloid from a Given Point Without.—It may be pointed out that the foregoing indicates a method, circuitous, it is true, and probably more curious than useful, of graphically solving the problem just enunciated.

Supposing the epicycloid given as in the figure, and any point p assigned, the normal *distance* of this point from the curve, even if not given, may practically be ascertained with great accuracy by drawing a circle about p as a centre, tangent to the curve ; for the eye is capable of appreciating the *fact* of tangency with extreme nicety, if the lines be fine, although wholly unable to locate the *point* of contact. The radius of this tangent circle then is used in the construction of the limacon, by means of which, as above, the normal is drawn. It may be added, that should the intersection of the epicycloid at g, by the arc through e, in this construction, be too acute, a better determination can be made as follows : about p, with radius rA, describe an arc cutting the base circle in Z, or set off the arc AZ equal to Pe ; GZ produced will then be the corresponding contact radius, thus locating the position of the centre of the generatrix ; an arc about this centre with radius CP will then cut the arc eg less acutely than that arc cuts the epicycloid ; and if gZ be found to pass, as it should, though the given point p, the determination may be accepted as at least accurate enough for all practical purposes, if any such there be.

47. By the aid of the limacon, however, we can determine with absolute precision the location of the point of cuspidation H, and of the points O and U at which the parallel curve crosses the base circle, as well as the directions of the normals at these points.

In order to do this we will first examine the limacon itself more particularly. If we suppose CJ to be a crank turning about C as a fixed centre, and JK to be a rod jointed to it at J, and capable of sliding endwise through a socket pivoted so as to rotate freely about the fixed centre A ; then if JK be equal to PI, it is clear that a pencil fixed at the end K of this rod will mechanically trace the limacon during the rotation of the crank.

Now, the motions of the points J and A being always known, the instantaneous axis of the rod can be found for any given position, and we are thus enabled to draw the tangent and normal to this curve at any point with geometrical accuracy. The motion of J being perpendicular to CJ, the plane normal to the path of this point is CJ itself ; the motion of the point A of the rod JK at the instant is in the direction JA, therefore, AQ perpendicular to JK is the plane normal to its path. But since both A and J lie in the circumference of the circle whose centre is C, the intersection Q of these normal planes, that is, the instantaneous axis, will also lie in that circumference throughout the action, and will at any given instant be diametrically opposite to the position of the crank at that instant.

48. *To draw the tangent to the limacon at any point.* In illustration of the above, let it be required to draw the tangent to the given limacon at the point M. Draw through A the chord MA and produce it to cut the generating circle in N, and draw the diameter Nb ; then b is the instantaneous axis of NM, bM is normal, and a perpendicular to it is tangent to the curve at M as required.

49. Now when, as in this figure, the normal distance PI is less than PA, it will be perceived that the limacon must cross the base circle of the epicycloid at A ; and since the final chord AL lies outside of that base circle, the limacon must also cross its circumference again at some point M between A and L. And further, there must be some point K, intermediate between A and M, nearer to G than any other point of the curve; at this point, therefore, the limacon must be tangent to a circle whose centre is G.

Obviously, the point O of the parallel curve is derived from the point A of the limacon; and making AR equal to IP, and the arc AO equal to the arc PR, we may find the direction of the normal OB either by describing about G an arc through R cutting the epicycloid in B, or by producing GO to cut CD, the path of the centre, in X, describing the generating circle in that position, and making the chord $OB = AR = PI$. Also, the intersection U is derived from the point M. We here observe that the chords MA, AN, of the two·circles which are tangent at A, lie in one right line; therefore, the triangles ACN, AGM, are similar, and we have the known magnitude MN divided at A into segments directly proportional to the given radii AC, AG. The arc PN being thus determined, we lay off the arc AV equal to it, then with centre V and radius equal to AM describe an arc cutting the base circle in U, at which point UVW is the normal.

50. Finally, the point of cuspidation H is derived from the lowest point K of the limacon. And this point must be so situated that GK produced shall pass through the instantaneous axis Q of the rod KAJ when the limacon is mechanically traced as above described.

But Q is then diametrically opposite to J, and P is diametrically opposite to A; therefore, PQ is parallel to AJ, and consequently to AK, which is a prolongation of JA; and, moreover, PQ is equal to AJ.

Then from similar triangles, AKG, PQG,

$$\frac{AK}{PQ} = \frac{AG}{PG}, \text{ whence } \frac{AK}{AJ} = \frac{AG}{PG}.$$

The magnitude of AJ being thus determined, the arc AT is made equal to the arc PJ, whence the position of the contact radius TY and the normal TS to the epicycloid are found as before, and STH being made equal to JAK, the point in question is located with geometrical precision.

FINIS.

INDEX.

Table of Cutters for Teeth of Gear Wheels,

MADE BY

THE PRATT & WHITNEY COMPANY,

HARTFORD, CONN., U. S. A.

All Gears of the same pitch cut with our Cutters are perfectly interchangeable.

Diameter of Cutters.		Diametral Pitch.	Price of Cutters.	Size of Hole in Cutters.		SET OF 24 CUTTERS. For each pitch coarser than 10.		
5	inches.	1½	$25 00	1¼	inches.	No. 1	cuts	12 T
4¼	"	2	20 00	"	"	No. 2	"	13
4	"	2½	18 00	"	"	No. 3	"	14
3¾	"	3	15 00	"	"	No. 4	"	15
3½	"	3½	12 00	1	"	No. 5	"	16
3¼	"	4	9 00	"	"	No. 6	"	17
3⅛	"	5	7 00	"	"	No. 7	"	18
3	"	6	6 00	"	"	No. 8	"	19
2⅞	"	7	5 00	"	"	No. 9	"	20
2¾	"	8	4 50	⅞	"	No. 10	"	21 to 22
2⅝	"	9	4 00	"	"	No. 11	"	23 " 24
2½	"	10	3 50	"	"	No. 12	"	25 " 26
2⅜	"	12	3 50	"	"	No. 13	"	27 " 29
2¼	"	14	3 50	"	"	No. 14	"	30 " 33
2⅛	"	16	3 00	"	"	No. 15	"	34 " 37
2	"	18	3 00	"	"	No. 16	"	38 " 42
1⅞	"	20	3 00	"	"	No. 17	"	43 " 49
1 13/16	"	22	3 00	"	"	No. 18	"	50 " 59
1¾	"	24	3 00	"	"	No. 19	"	60 " 75
1¾	"	26	3 00	"	"	No. 20	"	76 " 99
1¾	"	28	3 00	"	"	No. 21	"	100 " 149
1¾	"	30	3 00	"	"	No. 22	"	150 " 299
1¾	"	32	3 00	"	"	No. 23	"	300 Rack.
						No. 24	"	Rack.

The cutters are made for diametral pitches. By diametral pitch is meant the number of teeth per inch in the diameter of the gear at pitch line. Two pitches should always be added to this diameter in preparing a gear for cutting. For example: a gear of 80 teeth, 8 to the inch, diametral pitch, would be 10 inches on pitch circle, but the gear should be turned 10 2/8 (or ¼). The teeth should always be cut two pitches deep beside clearance.

The cutters are made for a clearance of 1/16 of the depth of the tooth; example: 8 to the inch has a clearance of 1/64; therefore the tooth should be cut two pitches (¼) and 1/64 deep. The gears must be set to run with this clearance to give the best results.

In ordering bevel gear cutters, give the diameter of gear at outside pitch line, and number of teeth, also the width of face. For the present all cutters are made to order

www.ingramcontent.com/pod-product-compliance
Lightning Source LLC
Chambersburg PA
CBHW021451210326
41599CB00012B/1026